HVAC
Systems
Evaluation

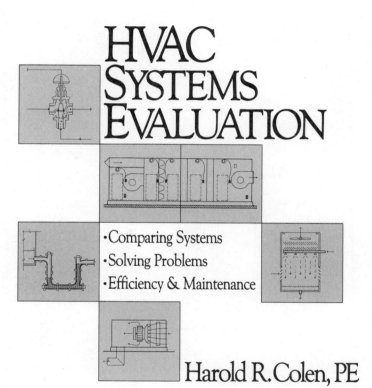

- Comparing Systems
- Solving Problems
- Efficiency & Maintenance

Harold R. Colen, PE

HVAC
SYSTEMS
EVALUATION

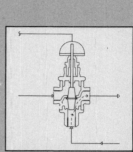

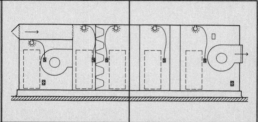

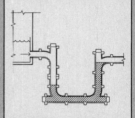

- Comparing Systems
- Solving Problems
- Efficiency & Maintenance

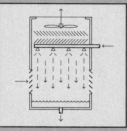

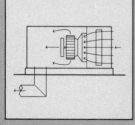

Harold R. Colen, PE
Illustrated by Carl W. Linde

R.S. MEANS COMPANY, INC.
CONSTRUCTION CONSULTANTS & PUBLISHERS
100 Construction Plaza
P.O. Box 800
Kingston, MA 02364-0800
(617) 585-7880

© 1990

In keeping with the general policy of R.S. Means Company, Inc., its authors, editors, and engineers apply diligence and judgment in locating and using reliable sources for the information published. However, no guarantee or warranty can be given, and all responsibility and liability for loss or damage are hereby disclaimed by the authors, editors, engineers, and publisher of this publication with respect to the accuracy, correctness, value and sufficiency of the data, methods and other information contained herein as applied for any particular purpose or use.

The editors for this book were Mary P. Greene and John J. Moylan. Typesetting was supervised by Joan Marshman. The book and jacket were designed by Norman R. Forgit. Illustrations by Carl Linde.

Printed in the United States of America

10 9 8 7 6 5 4 3 2

Library of Congress Cataloging in Publication Data

ISBN 0-87629-182-5

Dedication

To Claire Levy, who helped me pursue academics; Arlene Lesser of New York University, who encouraged me to publish my material; and most of all, to Thelma Colen, who is the wind beneath my wings.

TABLE OF CONTENTS

FOREWORD

This book has been created for HVAC designers and contractors, facilities managers and their staffs, design/build contractors, architects, sheet metal contractors, and environmental control contractors. It provides a comprehensive HVAC systems evaluation for cost effective selection and installation, and for optimum operation, maintenance, and retrofit of in-place systems. The book also serves as an aid to electrical designers and contractors for coordinating design disciplines and ensuring that the electrical requirements for a project's HVAC systems are not overlooked.

Part I, "Systems Comparison," includes 17 chapters which compare installation, operation, and maintenance cost considerations for basic HVAC systems. Each chapter not only points out the advantages and disadvantages of alternative HVAC systems, but also makes installation recommendations for trouble-free operation.

Part II, "Avoiding and Resolving Operational Problems," addresses each HVAC system from another point of view. These chapters focus on common problems that occur in each system (and within the various system components), and offer strategies to prevent and correct such occurrences. Included are maintenance and inspection recommendations.

Part III, "Electrical Coordination," covers electrical requirements commonly associated with HVAC equipment. This information is important not only to the HVAC contractor (who is responsible for the complete system), but also to the electrical contractor (who must perform the actual electrical installation). These items are frequently overlooked at the design stage, thereby causing conflicts, delays, and claims. This section describes and illustrates what is required for each of several common electrical components.

HVAC Systems Evaluation contains a wealth of reference information, in the form of more than 400 tables and illustrations, in a comprehensive Glossary of HVAC terms, and a list of abbreviations and symbols. These references will be useful not only in conjunction with this reference, but also in reading and preparing plans and specifications, and in selecting and maintaining HVAC systems.

Part One:
SYSTEMS COMPARISON

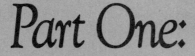

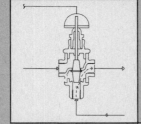

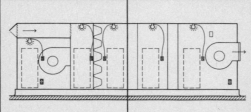

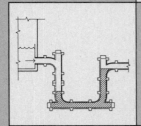

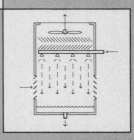

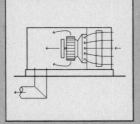

Part One, "Systems Comparison," includes 17 chapters which compare installation, operation, and maintenance cost considerations for basic HVAC systems. Each chapter not only points out the advantages and disadvantages of alternative HVAC systems, but also makes recommendations for avoiding future problems.

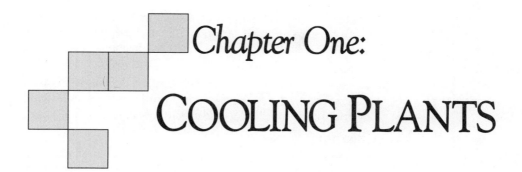

Chapter One:

COOLING PLANTS

When selecting a cooling plant, the first step is to analyze the advantages, disadvantages, installed cost, operating cost, and the user's ability to operate and maintain each considered system. Using this criteria, we will now compare the types of cooling plants in the following order.

- Compression (or Refrigeration) - Type Chillers
- Absorption-Type Chillers
- Self-Contained (or Packaged) Refrigeration Units
- Water or Air Source Heat Pumps

Compression (or Refrigeration) Chillers

There are four main types of compression, or refrigeration, chillers. Each uses a different type of compressor:

- Reciprocating
- Helical Rotary (Screw)
- Scroll
- Centrifugal

Reciprocating Chillers

Reciprocating compressors contain pistons that are driven by a connecting rod from a crankshaft. If the compressor is a hermetic-type, the compressor motor is contained within the same sealed enclosure as the refrigeration compressor. The motor inefficiency produces heat that would normally be rejected to the surrounding air with a non-hermetic type compressor. Since the refrigerant is warmed by the motor inefficiency, less cooling is generated per kw (kilowatt) than with an open type compressor.

Open compressors are 10-15% more efficient than hermetic machines. This is because the compressor motor is exposed to the open air and coupled to the compressor shaft, which extends through a seal to the crankcase. The heat produced by the motor's inefficiency is rejected to the atmosphere, rather than to the refrigerant. A mechanical seal is required to prevent refrigerant from escaping the "open" compressor at the rotating shaft. Modern seals are quite durable and require almost no maintenance. Smaller sized (5-25 hp) reciprocating hermetic compressors produce 12 btu/hour per watt.

Installation Costs

The installed cost of a 20-40 ton air-cooled reciprocating chiller is approximately 5% more than for a water-cooled reciprocating chiller, not including a cooling tower. However, when the cost of installing cooling towers, pumps, and condenser water piping is added to the water-cooled reciprocating chiller, it becomes 15% more expensive than air-cooled 20-30 ton systems, and slightly more than water-cooled 100 ton systems. See Figure 1.1 for a comparison of system installed costs.

Air-cooled chillers are completely packaged with air-cooled condensers, whereas the packaged water-cooled chiller requires field installation of the cooling tower, condenser water pump, and piping.

Operating Costs

The energy cost for water-cooled units at an overall electric cost, including demand, of $0.10 per kwh (kilowatt-hour) is $.09 per ton-hour. At the maximum design outdoor air temperature of an air-cooled condenser, the energy cost is $0.15 per ton-hour. When many of the operating hours of any air-cooled unit are at ambient temperatures less than design, seasonal energy costs of air-cooled units can be less than water-cooled units. See Figure 1.2 for a comparison of system energy use.

Installed Costs of Various Chilled Water Plants Including Cooling Tower, Pump, and Piping Installed Cost ($ per Ton)

Tons	Reciprocating Air-Cooled Chiller	Reciprocating Water-Cooled Chiller Only No Tower, Pump, or Piping	Reciprocating Chiller with Tower, Pump, and Piping
20	790	730	890
30	640	610	760
65	570	440	660
75	560	430	550
100	530	410	530
150	540	400	520

Tons	Reciprocating	Helical Rotary	Centrifugal	Single-Stage Absorption	Two-Stage Indirect-Fired Absorption	Two-Stage Direct-Fired Absorption
100	530	500	610	1060	1660	1260
150	520	490	560	800	—	—
200	500	480	510	810	1010	860
300	—	500	440	660	810	800
400	—	450	420	660	810	760
600	—	400	380	600	730	730
1000	—	—	320	540	610	670

Note: 1990 budget costs.

Figure 1.1

Air-cooled units operate at higher condensing pressures than water-cooled units, at peak ambient temperatures. Unit efficiency increases as the condensing pressure is lowered. Colder water or air temperatures entering the condenser will lower the condensing pressure and kw/ton, as Figure 1.3 shows.

Advantages:

Low installed cost
Figure 1.1 shows that the installed cost of reciprocating chillers is 5% less than helical rotary, and 15% less than centrifugal chillers, in sizes up to 200 tons.

Ton-Hour Energy Cost ($) Electricity cost $0.10/kwh, Oil $0.50/gal, Gas $0.50/therm (100,000 btu)					
Reciprocating	Helical Rotary	Centrifugal	Single-Stage Absorption	Two-Stage Indirect-Fired Absorption	Two-Stage Direct-Fired Absorption
$0.086	$0.07	$0.06	$0.103	$0.053	$0.045

Example of chilled water (CHW) system energy cost calculation.

200 ton cooling load, 10 hours/day (h/d), 1000 full load ton-hours (flth) per year, 20 days/month (d/m), electric energy $0.075/kwh, electric demand $30/kw, oil $0.80/gal

Electric drive centrifugal chiller

Electrical demand load:		
200 tons x .65 kw/ton		130 kw
Chilled water pump primary	0.022 kw/ton	
secondary	0.240 kw/ton	
Total CHW pumps	0.262 kw/ton x 200 tons = 52.4	
Condenser water pump	0.040 kw/ton x 200 tons = 8.0	
Cooling tower fan	0.086 kw/ton x 200 tons = 17.2	
Supply and return evaporator fans	0.400 kw/ton x 200 tons = 80.0	
Total electrical demand		287.6 kw
Operating energy cost		
Chiller energy 130 kw x 1000 flth x $0.075/kwh	=	$15,000
Demand 287.6 kw x $30/kw x 6 months (mo)	=	51,768
CHW pumps 52.4 kw x 10 h/d x 20 d/mo x 6 mo x $0.75	=	4,716
Cond. water pump 8.0 kw		
Cooling tower fan 17.2		
25.2 kw x 1000 flth x $0.075/kwh	=	1,890
Evaporator fans 80 kw x 10 h/d x 20 d/mo x 6 mo x $0.075/kwh	=	7,200
Total cooling season electrical energy cost	=	$80,574
$80,574/200 tons x 1,000 flt = $0.40 per ton-hour		

Figure 1.2

5

Many manufacturers to choose from
There are more manufacturers of reciprocating chillers than any other type of chiller.

More trained mechanics
All refrigeration service schools use reciprocating compressors in their training programs. More reciprocating compressors are in service than any other refrigeration machine.

Availability of parts
More reciprocating compressors are installed than any other refrigeration machine. Manufacturers and service wholesalers stock the common replacement parts.

Can be air-cooled
Packaged air-cooled chillers offer many advantages over water-cooled units, such as lower energy costs in moderate climates, lower maintenance costs, ease of seasonal changeover from heat to cooling, and no water treatment or water use. (Refer to Chapter 2 for a more detailed comparison of air- and water-cooled condensers.)

No field-installed refrigerant piping
Packaged chillers have all refrigerant piping installed and tested at the factory. This eliminates the following problems associated with field-installed refrigerant piping:

    a. If dry nitrogen is not passed through the piping during the soldering process, scale and other contaminants are produced. These contaminants will collect at orifices and in the compressor, causing operational problems, including possible compressor burnout.

    b. The refrigeration piping system must be purged of all moisture by evacuating the piping system to 250 microns. Moisture within the refrigeration system contributes to compressor failure.

    c. Improper soldering, or the use of soft solders, such as 50-50, leads to premature pipe joint failure.

Figure 1.3

Lower Entering Air and Water Temperatures Reduce Condensing Pressures and KW/Ton Power Input	
Water Temperature Entering Condenser	**KW/Ton**
85°F	0.88
80°F	0.82
75°F	0.77
Air Temperature Entering Condenser	**KW/Ton**
95°F	1.04
85°F	0.92

d. Precise refrigerant pipe sizing and oil trap installation are mandatory to permit oil return to the compressor.

Disadvantages:

Somewhat shorter expected life than centrifugal or helical rotary chillers
The service life of reciprocating chillers is 20-25 years, whereas centrifugal chillers usually operate for 25-30 years.

More maintenance than centrifugal, scroll, or helical rotary chillers
Reciprocating chillers have many more moving parts than centrifugal, helical rotary, or scroll compressors. Although there are more parts to replace, major repairs are not as costly as with centrifugal chillers. Installation of phase failure relays for reciprocating chillers greatly reduces the need for major repairs.

Helical Rotary Chillers

The helical rotary or screw-type compressor uses positive displacement. The compressor consists of a main cylindrical rotor working with a pair of star wheels, or gate rotors. (See Figure 1.4.) The compressor is driven through the main rotor shaft, and the star wheels follow by meshing with the rotor. Refrigerant vapor is drawn into the rotor space. As rotation continues, the flute volume decreases and compression occurs. Since compression takes place simultaneously on each side of the rotor, there is no gas pressure load on rotor bearings.

No power (other than that needed to compensate for small friction losses) is required to be transferred across the meshing points to the star wheel. As a result, wearing of components is negligible.

Advantages:

Lower installed cost
The installation costs are approximately 10% less than for reciprocating chillers in sizes over 100 tons.

Lower energy cost than reciprocating chillers
.70 kw/t vs. .87 kw/t for a 100 ton unit

Full load efficiency equal to single-stage centrifugal chillers

Part load efficiency 10-15% better than single-stage centrifugal chillers
50% capacity, 35% power

Compact units
Chillers up to 150 tons will fit through a 36" door.

No damage to the compressor from compressing liquid refrigerant or oil
Reciprocating compressors, on the other hand, will be severely damaged by compressing liquids.

Excellent for air-cooled applications

Surge not a factor
Since it is a positive displacement compressor, it will not encounter surge.

A relatively quiet unit
Inherently smooth operation of the helical rotary chiller design results in a relatively quiet unit.

Disadvantages:

Fewer manufacturers
There are fewer choices and a smaller range of availability when purchasing helical rotary chillers (vs. reciprocating chillers).

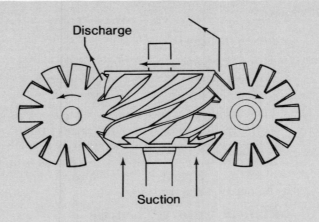

Discharge

Suction

Schematic of a Single-Screw Compressor

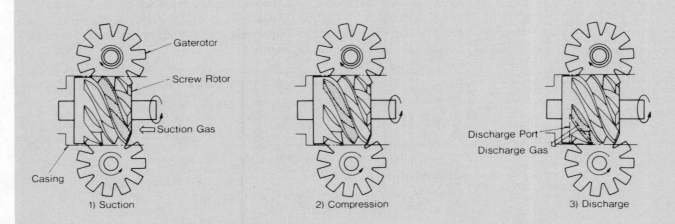

Gaterotor

Screw Rotor

Suction Gas

Casing

1) Suction

2) Compression

Discharge Port

Discharge Gas

3) Discharge

Sequence of Operation of a Single-Screw Compressor

Figure 1.4

Greater installed cost
Compared to smaller-sized reciprocating chillers, helical rotary chillers are more expensive.

Less efficient than two-stage centrifugal chillers
.70 kw/t versus .60 kw/t for 200 tons

Scroll Compressors

The scroll compressor was first developed 100 years ago. It operates using an orbiting scroll that is intermeshed with a fixed scroll. Suction gas enters from outside the scrolls. As the moving scroll rotates, compression chambers are formed, as the volume is reduced. The compressed refrigerant is discharged from a port in the center of the fixed scroll disc. No suction or discharge valves are required.

Compliant designed compressor scroll surfaces touch lightly. Non-compliant compressor scrolls do not touch, but maintain a minimal clearance. The compliant design is more efficient and reliable, because the small tolerances required for non-compliant compressors do not have to be maintained.

Scroll compressors are now made in up to 70-ton capacities, for both air- and water-cooled applications. See Figure 1.5 for an illustration of a scroll compressor.

Advantages:

Fewer parts
The scroll compressor contains one-third the number of parts of the same capacity reciprocating compressor. Fewer parts make servicing simpler, and extend service life.

No valves required
Scroll compressors are also more tolerant of liquid refrigerant than are reciprocating compressors. While liquid refrigerant, oil, or dirt taken into a reciprocating compressor will damage it, the scroll compressor can handle non-compressible liquids and dirt.

Smooth, quiet operation
This is the result of low torque variation through the overlapping compression cycle. The reciprocating compressor has one or more pistons moving back and forth.

Potential for high and variable speed applications
This is possible because of the scroll compressor's rotary motion, large ports, and absence of valves.

Five to ten percent higher efficiency than reciprocating compressors
This is possible because there are no valve losses, and fewer parts.

Good part load performance
At 40% of capacity, only 27% of the full load power is used. At 50% of capacity, a reciprocating unit uses 50% of full load power.

Good performance
Although scroll compressors for air-conditioning use are relatively new, unit performance has been good.

Disadvantages:

Few manufacturers
At the present time, only a few major manufacturers produce the scroll compressor.

Limited number of available sizes
Units are now made in capacities up to 70 tons.

Sequence of Operation for the Scroll Compressor
Reprinted by Permission from 1988 ASHRAE Handbook – Equipment

Figure 1.5

Centrifugal Chillers

Centrifugal compressors are part of the family of turbomachines, which includes fans, propellers, and turbines. Because its flow is continuous, the centrifugal compressor has a greater volumetric capacity, size-for-size, than positive displacement compressors. Although centrifugal compressors operate at higher rotating speeds, there is little vibration and wear because of the steady motion and the absence of contacting parts. See Figure 1.6 for an illustration.

Centrifugal compressors can be constructed as either open or hermetic machines. In open machines, the motor is coupled externally to the compressor. Electric motors, diesel engines, steam turbines, or any source of power can be used to drive the compressor. Open compressors must have a seal to prevent refrigerant leakage at the compressor drive shaft. Hermetic unit motors are contained within the same housing as the compressor and therefore do not require a shaft seal. The motor is cooled by the surrounding refrigerant.

Centrifugal compressors can be directly driven by the motor, or by gear trains. Direct-driven machines are usually 3,600 rpm, two- or three-stage compressors. Gear-driven compressors operate at speeds of 6,000 to 20,000 rpm, with many more moving parts than direct-drive compressors. Gear-driven compressors are reliable and are more economical because a few basic compressor sizes can deliver varying capacities.

If refrigerants that operate above atmospheric pressure are being used, a purge pump is not required to remove air from the refrigerant circuit. Pressurized containers must be used to store refrigerant removed from the chiller during maintenance. If refrigerants that operate below atmospheric pressure are used, any air that enters the refrigerant circuit must be

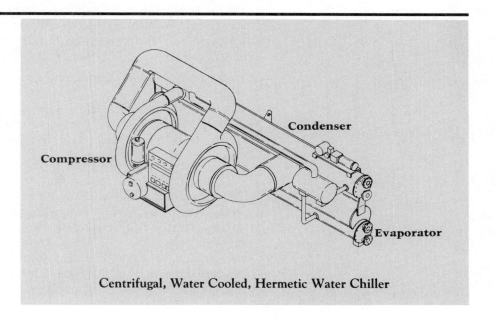

Centrifugal, Water Cooled, Hermetic Water Chiller

Figure 1.6

removed by a purge compressor. Air entering the refrigerant circuit will create surging of the compressor. However, low pressure drums can be used to store refrigerant during service.

Centrifugal chillers are available in sizes from 100-6,000 tons. Two- and four-stage chillers are more efficient than single-stage machines. It is not unusual for centrifugal chillers to operate for 25-30 years, with very little maintenance.

Installed Costs

The installed cost of a centrifugal chiller is less than that of any absorption chiller, but is higher than reciprocating and helical rotary chillers in sizes up to 200 tons. Centrifugal chillers enjoy the lowest installed cost in sizes over 200 tons. (See Figure 1.1 for the installed cost of various chillers.) The kw per ton and part load performance of single-stage, hermetic, and open centrifugal chillers are shown in Figures 1.7a & b.

Advantages:

High efficiency

Very reliable
25-30 year life

Low maintenance

Lowest installed cost over 200 tons

Good part load efficiency

Capable of operating at as low as 10% of maximum capacity

Heat recovery options available
Can heat water to 125°F.

Glycol can be cooled for low temperature applications (20°F and lower).

Disadvantages:

Licensed operating engineer often required

Smallest unit available is 100 tons.

Air-cooled package not readily available

Higher operating cost than two-stage absorption chillers in high electric energy and electrical demand areas, such as New York City
(See Figure 1.9, the comparative yearly operating cost of centrifugal, single-stage, and two-stage absorption chillers.) Because of the $30/kw demand cost, and $0.75/kwh energy cost, the energy operating cost of the centrifugal chiller is 38% more than a two-stage absorption chiller.

Higher initial cost than reciprocating and screw chillers, in smaller centrifugal chiller sizes
Figure 1.1 indicates that the installed cost of water-cooled reciprocating chillers is less than that of centrifugal chillers up through 200-ton capacity. Centrifugal chillers enjoy a lower installed cost than reciprocating chillers at capacities over 200 tons.

Absorption Chillers

Absorption refrigeration uses water as the refrigerant and lithium bromide as the absorbent. Absorption cooling takes place when heat is absorbed as water vaporizes. Instead of using a compressor to help condense and evaporate the refrigerant, the absorption cycle is a heat-operated refrigeration machine. The heat-operated generator is used to produce the pressure differential in the cycle. Lithium bromide has an affinity for water and is capable of extracting or absorbing vapor from the water. The cycle

Performance of Various Single-Stage Hermetic and Open Centrifugal Chillers

Tons	Full Load KW/Ton	Part Load KW/Ton		Dual Compressor Part Load KW/Ton	
		Percent Capacity	Percent Power	Percent Capacity	Percent Power
200	.69	60	50	60	50
300	.68	50	44	50	40
400	.65	25	28	25	25
500	.63				
500-1000	.62				

Open Type Centrifugal Chiller Performance

Tons	KW/Ton Moderate Efficiency	KW/Ton High Efficiency	Part Load Performance		
			Percent Capacity	Percent Single Compressor	Full Load Power Dual Compressors
150	.647	.607	50	45	42
200	.648	.600	25	28	23
250	.648	.583			
300	.660	.583			
400	.623	.565			
500	.600	.548			
600	.622	.565			
750	.615	.568			
950	.663	.653			

Figure 1.7a

Part Load Performance (% Power)

Capacity %	Reciprocating	Helical Rotary	Electric Drive Centrifugal	Single Stage Absorption	Two Stage Indirect-Fired Absorption	Two Stage Direct-Fired Absorption
	KW/Ton	KW/Ton	KW/Ton	Lbs. Steam per Ton	Lbs. Steam per Ton	Gal Oil per Ton
100	0.86	0.70	0.60	18	9.7	0.09
50	50%	42%	44%	50%	50%	50%
25	35%	26%	28%	28%	30%	30%

Figure 1.7b

uses lithium bromide to evaporate water (refrigerant). The water is condensed, sprayed and evaporated, cooling the chilled water circulating in the evaporator tubes.

Absorption chillers are evacuated at the factory, and sealed for proper operation of the absorption cycle. The refrigerant is cooled as it evaporates, and is absorbed by the strong lithium bromide solution. After the solution is diluted by the water, it is pumped to a container where heat is added to drive off the moisture. The concentrated lithium bromide is returned to the absorber for another cycle. The water vapor is condensed and returned to the evaporator to complete another cycle. Water is chilled as it circulates through the tubes in the evaporator.

Indirect-fired chillers use low pressure steam (8 psi), or hot water (240°F) as the heat source to separate the water from the lithium bromide solution.

Single-Stage Absorption Chillers

Single-stage absorption chillers are very inefficient; 18 pounds of steam are required to produce one ton of cooling. Waste heat generating plants, such as co-generation systems and incinerator waste heat, are good applications for single-stage absorption chillers. The exhaust gas of a gas turbine can be passed through a waste heat boiler to generate steam. The steam can be used for the absorption chiller. See Chapter 10, "Heat Recovery Systems."

Installation Costs

The installed cost of a single-stage absorption chiller, including the cooling tower is $1,060 per ton for a 100-ton chiller, and $600 per ton for a 600-ton unit. (See Figure 1.1 for installed costs of various chillers.) See Figure 1.8 for an absorption unit.

The maintenance cost of a single-stage absorption chiller is a little higher than that of a centrifugal chiller.

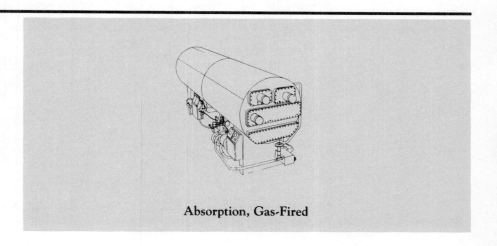

Absorption, Gas-Fired

Figure 1.8

Advantages:

Operating engineers not usually required

Smaller electric service

Lower electrical demand charge

In the event of a power failure, a smaller source of emergency power can provide stand-by service for absorption systems.

Can be solar or waste heat-powered
Co-generation will produce waste heat for operation.

Disadvantages:

Twenty percent higher installed cost than centrifugal chillers
(This is without taking into account the higher cost of the electric service required for a centrifugal chiller.) A 500-ton centrifugal chiller at 0.65 kw/ton requires 325 kw of electric service for the compressor. The absorption chiller has a solution pump and a refrigeration pump that consume approximately 4 kw of electric power.

High energy operating cost
Low pressure steam absorption chillers consume 18 pounds of steam per ton of cooling at peak load. At \$7/1,000 pounds of steam, the cost of one ton of absorption cooling at peak load is \$0.126. With an electricity cost of \$0.10/kwh, one ton of electric drive centrifugal cooling is \$0.06/ton-hour.

Low efficiency: 0.65 COP (Coefficient of Performance)
This accounts for the high steam consumption.

Very low efficiency at part-load operation
Absorption chillers consume 50% of the maximum steam supply to produce 50% of their maximum capacity, and 28% of the maximum steam supply to produce 25% of maximum capacity. For comparison, a centrifugal chiller uses 44% of maximum power to produce 50% of maximum capacity, and 28% of its maximum power to produce 25% of maximum capacity. (See Figure 1.7.)

As per Figure 1.2, the cost of one ton of absorption cooling, at 50% load, is \$0.10. At 50% load, a 0.6 kw/ton centrifugal chiller requires only 0.26 kw to produce one ton of cooling. At \$0.10/kwh, the cost of one ton of part-load centrifugal cooling is only \$0.026. With oil at \$.50 per gallon and an electricity cost of \$0.10 per kwh, the energy cost of centrifugal part-load cooling is one fourth that of single-stage absorption cooling.

Larger cooling towers and condenser water pumps are required.
At 8 psi steam pressure, 4.2 gpm per ton of condenser water is necessary. For comparison, a centrifugal chiller uses 3 gpm per ton of cooling.

Corrosive properties of lithium bromide
Lithium bromide is corrosive, and consequently, the lithium bromide inhibitor must be checked periodically to prevent serious damage to the chiller.

Crystallization of lithium bromide
This process can occur if the lithium bromide is sub-cooled in a concentrated form. Crystallization can be caused by sudden drops in condenser water temperature, low load conditions, the interruption of the heat source during a normal cycle, or air leakage into the machine. Safety controls are essential to head off conditions that may cause crystallization, which could put the chiller out of service for several days while it is decrystallized.

Higher maintenance cost as the chillers age
Normally, the maintenance on single-stage chillers is low. The only moving parts are the solution and vacuum pumps. Single-stage absorption chillers require only a little more maintenance than centrifugal chillers. It should be noted that as the chillers increase in age, maintenance costs average about $2,000/year more than centrifugal chillers.

Two-Stage Absorption Chiller

Two-stage absorption cycles result in greatly improved thermodynamic efficiency and require less energy than single-stage cycle machines. The first-effect generator receives the external heat which boils refrigerant (water) from the weak absorbent. The hot refrigerant vapor (water) goes to a second generator, supplying heat for further vaporization from the concentration from the first absorber. Two-stage units operate with a COP of 1-1.23, compared to the single-stage COP of 0.65. Steam at 40-150 psi (although 100-125 psi is common) is used as the heat source. Some models use hot water up to 400°F. Steam consumption is 9.3-12 pounds per ton-hour, compared to 18 pounds per ton-hour for single-stage absorption units.

American-made two-stage absorption chillers are manufactured in capacities of 385-1,060 tons. Foreign-made, two-stage absorption chillers are available in capacities of 100-1,500 tons.

Advantages:

Lower energy cost
Steam consumption is 9.3 pounds per ton-hour for two-stage units, versus 18 pounds per ton-hour for single-stage chillers. At a steam cost of $7 per 1,000 pounds and an electricity cost, including demand, of $0.10 per kwh, the energy cost of a two-stage absorption chiller and an electric drive centrifugal chiller are equal at $0.06 per ton-hour. In areas of high electrical cost, such as New York City, the energy cost of a two-stage absorption chiller is less than that of an electric drive chiller.(See Figure 1.9.)

Higher efficiency than single-stage absorption chiller
The two-stage chiller enjoys a COP of 1.23, compared to a single-stage COP of 0.65.

Good part-load efficiency
At 50% of cooling capacity, the two-stage chiller consumes 50% of the maximum energy input. At 25% of cooling capacity, the two-stage chiller uses 30% of the maximum energy input. A single-stage absorption chiller has the same part-load efficiency as the two-stage chiller. The two-stage chiller part-load percentage is weighed against 9.7-12 pounds of steam per ton, whereas the single-stage chiller maximum is 18 pounds per ton.

Disadvantages:

High installed cost
The installed cost of a 400-ton, two-stage indirect-fired chiller and a 600-ton unit are 23% more than that of a single-stage unit, and 93% more than an electric drive centrifugal chiller. (See Figure 1.1.)

If high pressure steam (100-125 psi) is required, a licensed operating engineer is usually mandatory.

High head room needed
American-made two-stage indirect-fired chillers are more than 8-13' in height. The units manufactured in Japan are almost 7' high for the 100-200 ton units, and up to 11' high for the 1,000-1,500 ton units.

Larger cooling tower and condenser water pumping/piping system required Two-stage chillers are rated at 4.5 gpm per ton. Single-stage units require 3-3.6 gpm per ton, and electric drive centrifugal chillers are rated at 3 gpm per ton.

Chiller Owning and Operating Costs

Electricity	energy $.075/kwh, demand $30/kw
Steam	$12 per 1,000 pounds of steam
Gas	$.60 per therm, say $.80 per 100,000 btu output.
400 tons (full load tons) (flt), 1,000 full load ton-hours per year. (flth)	
Centrifugal chiller — .65 kw/ton	
Single-stage absorption chiller — 18# of steam per ton	
Two-stage absorption chiller — 10# of steam per ton	

CENTRIFUGAL CHILLER

400 tons x .65 kw/ton x $30/ kw x 6 months/yr	=	$46,800
400 flt x .65 kw/ton x 1,000 flth/yr x $0.75/kwh	=	19,500
Yearly Energy Operating Cost		$66,300

SINGLE-STAGE ABSORPTION CHILLER

400 tons x 1,000 flth/yr x 18#/ton x $12/1,000#stm	=	$86,000

TWO-STAGE ABSORPTION CHILLER

400 tons x 1,000 flth/yr x 10#/ton x $12/1,000#stm	=	$48,000

	Electric Drive Centrifugal Chillers	Single-Stage Absorption Chillers	Two-Stage Absorption Chillers
Two 200-ton Chillers Initial Cost $400/ton	$160,000		
$500/ton		$200,000	
$900/ton			$360,000
Present Worth of Annual Operating Cost			
*$66,300 x 1/.11746	$564,447		
$86,400 x 1/.11746		$735,569	
$48,000 x 1/.11746			$408,649
Present Worth	$724,447	$935,569	$768.649

Present Worth of Operating Cost = Annual Cost x Present Worth Factor = Annual Cost x 1/Capital Recovery Factor, (CRF = .11746) Present Worth of Operating Cost = Annual Cost x 1/.11746
*Does not include the cost of licensed operating engineers.
[Based on 1990 Operating Costs.]

Figure 1.9

Limited number of manufacturers
At the present time, there are two major Japanese firms that manufacture 1,000-1,500 ton indirect-fired two-stage absorption chillers. There is one American manufacturer that produces 385-1,060 ton two-stage absorption chillers.

Tube failures
While some tube failures have occurred in the past, manufacturers have modified recent construction to eliminate this problem.

Two-Stage, Direct-Fired Absorption Chillers
Direct-fired units burn gas or number two oil as the heat source for the generator. Although not presently built in the United States, direct-fired units have been manufactured in Japan for some time.

Advantages:

High efficiency
Two-stage, direct-fired absorption chillers have a COP of 1, compared to 0.65 for single-stage units, and 1-1.23 for two-stage indirect-fired units.

Very low operating cost—a particular advantage if the electric rate is high
Two-stage, direct-fired units consume 0.09 gallons of oil, or 11.7 mbh of gas to produce one ton-hour of cooling. Figure 1.2 shows the ton-hour energy cost of various chillers. At $0.50 per gallon of oil, the cost per ton-hour for the direct-fired unit is $0.045. With gas at $0.50 per therm (100 mgh), the two-stage direct-fired ton-hour cost is $0.058. With an electrical rate of $0.10 per kwh, including demand, the ton-hour cost of a centrifugal chiller is $0.06. The ton-hour cost of a two-stage indirect-fired chiller is $0.53 with $0.50 per gallon oil, and the ton-hour cost of a single-stage absorption chiller is $0.103. In locations of very high electrical demand charges, such as New York City, the electrical centrifugal ton-hour cost can escalate to $0.135. Figure 1.10 compares the energy operating costs of electric centrifugal and direct-fired absorption chillers in a high electric rate area.

No licensed operating engineer required
Most municipalities do not require that direct-fired absorption chillers be operated by licensed operating engineers. Direct-fired chillers are fired by oil or gas burners, involving low pressure heating. Laws regulating operation of low pressure heating equipment usually apply.

Hot water heating available
Heat added to the generator to drive the refrigerant (water) from the strong absorbent is also available for hot water heating use, whenever it is not required for the chilled water cycle. Since the demand for chilled water is very low or non-existent in winter, all or most of the direct-fired chiller's heat input is available to heat the building. Heating capacity may be increased up to 200% of standard capacity by oversizing or providing multiple first-stage generators. An additional heat exchanger can be provided to produce auxiliary hot water during the summer cooling season. It is often possible to reduce the size of a separate heating boiler, or eliminate it altogether. Using a direct-fired chiller for heating not only reduces heating installation costs, but also the construction cost of building a boiler room.

Disadvantages:

Limited number of manufacturers
At the present time there are no US manufacturers of large tonnage two-stage direct-fired chillers, and only a few Japanese manufacturers.

High equipment costs
Although two-stage direct-fired chiller equipment costs are slightly greater than those of the two-stage indirect-fired chiller, the installed cost of both is almost the same. The installed cost of a 400-ton, two-stage, direct-fired chiller is $760 per ton. The indirect-fired chiller installed cost is $810 per ton, and the centrifugal chiller installed cost is $420 per ton.

Larger cooling tower and condenser water pumps
Two-stage direct-fired chillers are rated at 4.6 gpm per ton, and electric drive centrifugal chillers are rated at 3 gpm per ton. Cooling towers, condenser water pumps, and condenser water piping must be sized to handle 53% greater capacity.

Limited experience in U.S
Large capacity direct-fired chillers have been in operation in the U.S. for a relatively short time, whereas centrifugal chillers have been in use for over 50 years. Problems with tube failures in earlier models seem to have been resolved in the newer machines.

Energy Cost of 200-Ton Direct-Fired Chiller
Example of chilled water (CHW) system energy cost calculation

200-ton cooling load, 10 hours/day (h/d), 1000 full load ton-hours (flth) per year, months/year (m/y).

Direct-fired absorption chiller		
Electrical demand load:		
200-ton direct-fired chiller		7 kw
Chilled water pump (CHW) primary	0.022 kw/ton	
secondary	0.240 kw/ton	
Total CHW pumps	0.262 kw/ton x 200 tons	= 52.4
Condenser water pump	0.060 kw/ton x 200 tons	= 12
Cooling tower fan	0.13 kw/ton x 200 tons	= 26
Supply and return evaporator fans	0.40 kw/ton x 200 tons	= 80
Total electrical demand		177 kw
Operating energy cost		
Chiller 0.09 gal oil/ton x 200 tons x $0.80/gal x 1000 flth		= $14,400
Demand 177 kw x $30/kw x 6 months		= 31,860
CHW pumps 52.4 kw x 10 h/d x 20 d/mo x 6 mo x $0.75/kwh		= 4,716
Cond. water pump	12 kw	
Cooling tower fan	26.6	
	38.6 kw x 1000 flth x $0.075/kwh	= 2,895
Evaporator fans 80 kw x 10 h/d x 20 d/mo x 6 mo x $0.075/kwh		= 7,200
Total cooling season energy cost		$61,071
$61,071/200 tons x 1,000 flt = $0.31 per ton-hour		

Figure 1.10a

Self-Contained Packaged Refrigeration Units, Air- or Water-Cooled

Self-contained packaged air-conditioning units consist of compressors, condensers, and evaporators, with completely connected and tested refrigerant piping and refrigerant charge. Air-cooled units can be factory-assembled with a connected air-cooled condenser within the single cabinet, or arranged for separate remote condensers. Evaporator fans and motors, fan belts, filters, starters, and controls are all factory-assembled. The entire assembled unit is factory-wired, with all controls in a control center assembled and tested. Steam, hot water, or electric heating coils are available for installation in the unit.

These units can be equipped with supply air plenums with a grille for free-blow applications when the units are located within the conditioned space. They require only field-connected electrical power supply, condenser water piping for water-cooled units, and ductwork for ducted systems.

These units are available in capacities of from 3-60 tons. Packaged variable air volume units are available in 15-60 ton sizes, complete with prewired

Energy Cost of Centrifugal Chiller
Example of chilled water (CHW) system energy cost calculation

200-ton cooling load, 10 hours/day (h/d), 1000 full load ton-hours (flth) per year, 20 days/month (d/m), electric energy $0.075/kwh, electric demand $30/kw, oil $0.80/gal

Electric drive centrifugal chiller

Electrical demand load:		
200 tons x .65 kw/ton		130 kw
Chilled water pump primary	0.022 kw/ton	
secondary	0.240 kw/ton	
Total CHW pumps	0.262 kw/ton x 200 tons	= 52.4
Condenser water pump	0.040 kw/ton x 200 tons	= 8.0
Cooling tower fan	0.086 kw/ton x 200 tons	= 17.2
Supply and return evaporator fans	0.400 kw/ton x 200 tons	= 80.0
Total electrical demand		287.6 kw

Operating energy cost

Chiller energy 130 kw x 1000 flth x $0.075/kwh		= $15,000
Demand 287.6 kw x $30/kw x 6 months (mo)		= 51,768
CHW pumps 52.4 kw x 10 h/d x 20 d/mo x 6 mo x $0.75		= 4,716
Cond. water pump	8.0 kw	
Cooling tower fan	17.2	
	25.2 kw x 1000 flth x $0.075/kwh	= 1,890
Evaporator fans 80 kw x 10 h/d x 20 d/mo x 6 mo x $0.075/kwh		= 7,200
Total cooling season electrical energy cost		$80,574
$80,574/200 tons x 1,000 flt = $0.40 per ton-hour		

Figure 1.10b

supply air temperature controller. The supply air controller actuates a microprocessor which activates electric cylinder unloaders and hot gas bypass into the evaporator to regulate the discharge air temperature. Indoor packaged air-cooled units require field-installed condenser air intake and discharge ductwork. Fans can usually be arranged for vertical or horizontal air discharge. Although vertical packaged units are available in 3-60 ton sizes, indoor packaged horizontal units are usually manufactured in 2-5 ton sizes.

Packaged air-conditioning unit fans are not designed to serve high static pressure air distribution systems. The table in Figure 1.11 shows the maximum external static pressure capability of the units. Supply and return ductwork, supply and return air grilles, heating coils, dirty filters, variable volume units (air terminal units), and system effect factor all contribute to external static pressure.

The performance of the 3-10 ton water-cooled packaged self-contained air-conditioning units is less efficient than that of the larger packaged units. Most units contain reciprocating compressors and three-row cooling coils, and their performance is limited by the relatively low performance of that combination. Water-cooled 3-10 ton units average 1.2 kw/ton. Air-cooled

Self-Contained Air-Conditioning Unit Allowable External Static Pressure	
Indoor Air- and Water-Cooled Self-Contained Air-Conditioning Units	
Unit Capacity (tons)	Maximum Available Static Pressure (w.g.)
3	0.8
5	0.6
8	1.0
10	1.4
15	1.4
20	2.0
25	2.5
30	2.5
40	3.5
60	3.0
Indoor Horizontal Air-Cooled Packaged Air-Conditioning Unit Maximum Available Static Pressure	
Tons	Static Pressure Inches (w.g.)
2	0.50
3	0.32
5	0.28

Figure 1.11

2-20 ton units average 1.4 kw/ton. Water-cooled 15-60 ton units average 0.9 kw/ton, and air-cooled 15-60 ton units, 1.4 kw/ton.

Rooftop Single Packaged Air-Conditioning Units

These air-cooled packaged air-conditioning units are one piece and self-contained in a horizontal arrangement. Their casings are weatherproof, and the unit can be mounted on a roof curb, on a steel support elevated a few feet above the roof, or on a concrete pad on the ground. The units can be factory-equipped with electric or gas heaters, power exhaust or return air fans, or economizer controls, in constant or variable air volume applications. Units are available from 5-100 tons. The larger units are rated at 1.02-1.07 kw/ton at design conditions. The 5-ton units are rated at 1.48 kw/ton, and the 10-ton unit at 1.25 kw/ton.

Advantages:

Low installed cost
Water-cooled self-contained air-conditioning systems with the units located within the conditioned space, and roof-mounted cooling towers offer the lowest installed cost at $1,360 per ton. Figure 1.12 shows that the installed cost of other systems is almost 30% and more than the water-cooled self-contained air-conditioning system.

The installed cost of rooftop single zone self-contained systems is almost 30% more. The installed cost of an air-cooled self-contained air-conditioning system, with the unit located within the conditioned space to eliminate return air duct systems, and the air-cooled condenser mounted on the roof, is 44% more than the water-cooled self-contained unit. Direct

			Installed Cost per Ton of Various Air-Conditioning Systems (Cooling Only)					
Tons	SCAC Units Water-Cooled	SCAC Units Air-Cooled	Rooftop Air-Cooled	Reciprocating Water-Cooled CHW	Reciprocating Air-Cooled CHW	Reciprocating DX Split System	Centrifugal CHW
15	1440	1900	1640	2200	2170	1590	—
30	1360	1780	1725	1710	1610	1660	—
60	1360	1800	1720	1640	1690	1880	—
100	—	—	—	1990	1470	—	—
150	—	—	—	—	—	—	2000
200	—	—	—	—	—	—	1925
300	—	—	—	—	—	—	1840
400	—	—	—	—	—	—	1785
CHW	Chilled water						
SCAC	Self-contained air-conditioning units						
Centrif	Centrifugal chiller, water-cooled central system						
DX	Direct expansion						

Figure 1.12

expansion, chilled water, constant volume systems with terminal reheat bear an installed cost 56% higher, and a chilled water variable air volume system is 75% more than a water-cooled self-contained unit system.

Cooling costs can be metered to tenants.
Most landlords prefer not to take on the responsibility (and cost) of providing and maintaining air-conditioning systems. Self-contained packaged air-conditioning systems ideally provide a solution. Each tenant can install an independent air-conditioning system, maintain it, and pay for all of the energy for that system. The landlord can establish minimum standards for the design and installation of the systems, to protect his property. The standards might include equipment performance and impact to the building structure, such as roofs, beams, walls, water, electrical, and control systems.

Flexible zoning with multiple units
By arranging separate air-conditioning units for each floor, or for each major tenant of a building, economy of operation can be realized. Units can be turned off when the served space is unoccupied. This is the best possible energy management technique. The more diverse the occupancy schedules of the various tenants, the more cost effective the individual tenant systems become. Turning off a system when it is not needed results in less energy use than a partly loaded central system.

Staged installation and activation to match building space occupancy
As floors or sections of a building are rented or reconstructed, dedicated systems for those spaces can be installed. Up-front installation costs for a central system that will serve a future space can be postponed until a system is actually needed for that space.

Redundancy
The adverse affects of a system malfunction are confined to the individual space served by that system.

Reduced air distribution costs
Fan energy is one of the major operating cost factors of an HVAC system. A central system usually operates at 3.5″ of external static pressure. Individual dedicated units located close to the served space usually operate at 1.5″ of external static pressure. The reduced total static pressure of the individual air handling system enables it to use 43% less fan energy than a central system. The duct system installed cost is also less for individual air handling units, compared to a central system, because the main supply and return ducts are smaller.

The capacity to supplement central systems that are performing poorly
One method of providing additional cooling to spaces that are not receiving sufficient cool air to satisfy the increased cooling load is by the installation of self-contained air-conditioning units, either directly in, or near to, the space requiring the additional cooling. Horizontal units can be installed above the ceiling when floor space is not available. Typical applications are office spaces where increased lighting, people, and electronic or computer loads have been added. Conference rooms with a diverse cooling load can be temperature-regulated by operating a self-contained unit located above the ceiling, only when it is needed.

Disadvantages:

Lower cooling efficiency
The larger, self-contained, packaged air-conditioning compressors are rated at 1.02-1.07 kw/ton at design conditions. Reciprocating compressors for "split" systems are rated at 0.92-0.96 kw/ton. Reciprocating chillers

consume 0.86 kw/ton for the compressor and 0.262 kw/ton for the chilled water pumps, for a total of 1.12 kw/ton. Self-contained packaged units are more energy efficient than reciprocating chilled water systems. For comparison, helical rotary compressors are rated at 0.70 kw/ton. Helical rotary chillers (at 0.70 + 0.262 kw/ton for the chilled water pumps, for a total of 0.96 kw/ton) are more energy efficient than self-contained packaged units. In capacities over 100 tons that warrant a centrifugal chiller, the total kw/ton (0.60 + 0.262 = 0.826 kw/ton) is 15% more energy efficient than a self-contained packaged unit system.

Advantage number three, being able to turn off units that are not required to operate, can offset the peak efficiency advantage offered by central, chilled water systems.

Shorter service life

Standard, self-contained, packaged air-conditioning units have a normal life expectancy of 15 years. On the other hand, reciprocating compressors and chillers are expected to operate for 20 years, and centrifugal and absorption chillers should perform for 25 years.

High maintenance cost

Self-contained packaged air-conditioning systems are approximately 40% more costly to maintain than chilled water systems. The maintenance cost of unitary DX variable air volume systems is $41/ton. Variable air volume chilled water systems cost $28/ton to maintain. The replacement costs are not included in the above figures. The maintenance costs are for full maintenance contracts that include routine inspections, labor, and parts replacement. Since the service life of centrifugal or absorption chillers is 67% longer than self-contained packaged units, the replacement costs for self-contained units will be much higher than that of the larger chillers.

Noise in occupied space

Self-contained air-conditioning units are usually located within, or near to, the conditioned space. Although the unit sections are insulated, fan and compressor noise is generated and transmitted to occupants near the units.

Maintenance has to be performed within the occupied space.

When self-contained units are located in the occupied space (either floor-mounted or mounted in the space above the ceiling), servicing the equipment involves some disruption of the occupants. If unit placement is not properly coordinated, otherwise routine tasks, such as changing filters, may become a two-man project, as desks, filing cabinets, or equipment may have to be moved. Again, the occupant is prevented from conducting his normal work during the maintenance process.

Higher electrical requirements

The total kw requirements for multiple self-contained packaged air-conditioning units exceed that of a central system. Each packaged system must be sized to handle the peak load of the area served by that system. The total kw for the packaged air-conditioning system is the sum of all of the packaged unit kw's.

Since a central system is sized to produce only the cooling to satisfy the maximum instantaneous cooling load, the total tonnage generated and the kw to deliver the cooling are less than for the packaged system. The instantaneous cooling load is less than the sum of all of the zone peaks. The sun cannot shine on the east windows at the same time that it shines on the west windows. A building with a 200-ton instantaneous peak cooling load could have a system consisting of four 60-ton rooftop packaged systems at 1.07 kw/ton. The total compressor kw would be:

$$4 \text{ units} \times 60 \text{ tons/unit} \times 1.07 \text{ kw/ton} = 257 \text{ kw}$$

A 200-ton central centrifugal chiller would produce 200 tons at 0.60 kw per ton for a total compressor demand of 120 kw. A larger electric service, and a greater electric utility demand cost are associated with multiple packaged units.

Individual units consume valuable floor space.
If self-contained packaged air-conditioning units are floor-mounted within the conditioned space, or located in floor-by-floor or zone-by-zone mechanical rooms, more building floor space is needed. Multiple units or mechanical rooms consume more floor area than central equipment rooms. Roof-mounted equipment eliminates the need for floor-by-floor mechanical equipment rooms, mechanical room construction costs, and the loss of rental space due to the space occupied by the mechanical rooms.

Water Source Heat Pumps

Water source heat pumps are self-contained water-cooled packaged air-conditioning units with reverse refrigerant cycles. Reversing valves are installed in the refrigerant circuit to permit the condenser to act as an evaporator, and the evaporator to act as a condenser, during the heating cycle. Refrigerant hot gas is diverted to the evaporator, and refrigerant liquid is diverted to the condenser. The condenser piping circuit must be equipped with an expansion valve in order to be enabled during the heating mode. The condenser tubes must be designed to accept liquid refrigerant from the expansion valve during heating, and hot gas during cooling. The evaporator must be capable of accepting hot gas during heating, and liquid refrigerant from its expansion valve during cooling. The entire refrigeration cycle piping circuit must be designed and installed to permit oil circulation during all cycles.

During the heating cycle, when the condenser acts as an evaporator, the refrigerant absorbs heat from the water passing through the condenser. The cold water leaves the condenser (which now serves as a heating unit evaporator), and enters the cooling heat pump unit condenser, where it picks up that condenser's heat.

Heat pump units are available as console units to fit under windows, and as horizontal units for ceiling applications. The major application for water source heat pumps is as perimeter heating units. As long as there is a winter cooling load available, the rejected condenser heat of the cooling unit can be cooled by the heat pump, and the heating unit's rejected condenser heat will provide heat for the space served. See Figure 1.13 for a diagram of the water source heat pump system.

Advantages:

Low installed cost for heating and cooling
If water storage is not provided, the installed cost of water-source heat pump systems is $6.70 per square foot of building area. The installed cost of a chilled water VAV system with hydronic radiation is $8.40 per square foot.

Individual room control
Console-type heat pump units are usually installed under the windows. Each unit has its own thermostat so that every room containing a perimeter heat pump unit can have thermostatically controlled heating and cooling. Large interior spaces usually are supplied with horizontal ducted heat pump units. Each unit is controlled by a room thermostat.

Electricity for heating and cooling can be metered and paid for by the tenant. Since water source heat pumps are completely electric systems, it is very convenient to arrange for each tenant to pay for their heating and cooling energy, along with their light and power use.

Low operating cost when winter cooling loads exist and electric rates are low When buildings contain high internal heat gains, such as computer rooms, telephone rooms, electric switchgear rooms, elevator machine rooms, auditoriums, and large interior office space, a winter cooling load exists. If the electric rate is low compared to oil or gas heat, it may be desirable to heat with heat pumps, rather than cooling the internal heat gain spaces with cool outside air, and heating the perimeter spaces with hydronic radiation and a boiler plant. Figure 1.14 compares heat pump heating with oil heat.

Figure 1.14 shows that with an oil cost of $0.80 per gallon and an electrical cost, including demand, of $0.06 per kwh, the cost of oil heat is 2.42 times that of water-source heat pump heat. With an oil cost of $0.80 per gallon, the cost of heating with a water-source heat pump equals the cost of oil heat at an electric cost of $0.145 per kwh.

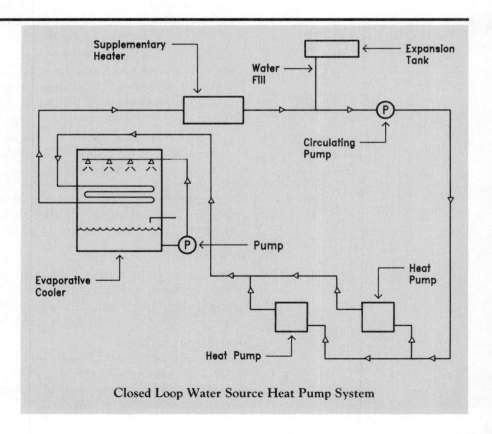

Closed Loop Water Source Heat Pump System

Figure 1.13

Disadvantages:

Noise in occupied space
Perimeter units are usually located under the windows within the occupied space. The unit fan and compressor generate noise when they operate. Although not quite as noisy as a window unit or through-the-wall air-conditioning unit, the heat pump unit is noticeable when it operates. A central system, on the other hand, with remote mechanical equipment rooms, should not transmit noise to the conditioned space.

Shorter service life expectancy
An operating life of 10-15 years can be expected for water-source heat pumps. Centrifugal and absorption chillers operate for 25-30 years, and reciprocating chillers and compressors perform for 20 years.

Higher energy cost when electric rates are high
As per Figures 1.10a & b, if the electric rate is above $0.145/kwh, the electric heat pump heating energy cost is greater than oil or gas heat. At $0.10/kwh, electric heat pump heating is equal to $0.55 per gallon of fuel oil.

Unit maintenance must be performed in the occupied space.
Heat pump units are usually installed within the occupied space, and servicing the units disturbs the room occupants.

High maintenance costs
Heat pump units operate fans and compressors 52 weeks per year, whereas air-conditioning systems employing oil or gas heat operate compressors only 25 weeks per year. Water-source heat pump system maintenance costs are

Comparison of Water-Source Heat Pump Heating with Oil Heat
Oil 140,000 btu/gal, cost of oil $0.80/gal, 60% seasonal boiler efficiency, Heat-pump one kw per ton of cooling, cost of electricity including demand $0.06/kwh
Heat-pump heating = 12,000 btu/ton + 3,400 btu/kw compressor input = 15,400 btu/kw input.
Oil heat
140,000 btu/gal x 60% seasonal boiler efficincy = 84,000 btu/gal Cost of oil heat = $0.80/gallon for 84,000 btu
Heat-pump heating
15,400 btu for every kw input
84,000 btu oil heat/15,300 btu/kw (heat-pump heat) = 5.5 kw of electric heat to equal one gallon of oil heat.
5.5 kw x $0.06/kwh = $0.33 for 84,000 btu/hr of heat pump heat.
$0.80 (oil heat)/$0.33 (heat-pump heat) = 2.42
Oil heat (at the above energy rates) is 2.42 times as expensive as heat pump heat.
At an oil cost of $0.80/gallon, and an electric rate of $0.145/kwh, the cost of heat pump heating is equal to the cost of oil heat.
$0.80 (oil heat)/5.5 = $0.145

Figure 1.14

$47 per ton. Self-contained air-conditioning unit system maintenance costs are $41 per ton. A chilled water VAV system with perimeter heat has a maintenance cost of $28 per ton.

Must have a winter cooling load to produce heat for the perimeter units
A heat pump operates as a heating unit by delivering the condenser heat to the space to be heated. In order to generate condenser heat, the heat pump must cool something with its evaporator. The "something" to be cooled in a water-source heat pump system is water or glycol that picks up heat from the condenser of the "cooling" heat pump. Whenever there is an insufficient cooling load to produce condenser heat for the loop, heat must be added to the loop by an electric, hot water, or steam source. The more outside heat is added to the loop, the less energy-efficient the heat pump system becomes. Ideally, the winter cooling load should be equal to the heating requirements of the building.

Chapter Two:

REFRIGERATION CONDENSING SYSTEMS

In the early days of air-conditioning, the heat generated by the refrigeration cycle was removed from the condenser by passing water from the municipal water system over the condenser and discharging that heated water to the sewer system as waste. Today, this method of water cooling is either prohibitive costwise or restricted by local codes or ordinances in order to conserve water. Indeed, the demand for potable water has exceeded supply in many localities, to the point that limitations on its availability have curtailed new construction.

Cooling Towers

Cooling towers reduce the amount of the potable water that is consumed to about 5% of the total circulated water. A cooling tower is a water saving device that uses heat and mass transfer to cool water. Figure 2.1 shows that the water to be cooled is distributed in the cooling tower by spray nozzles, to expose a very large water surface to the atmospheric air brought into contact with the spray water.

A fan circulates relatively cool and dry atmospheric air to the water. Some of the latent heat is taken from the water by this exposure to cool air. Part of the water is evaporated, thereby cooling the remaining water. The circulating air accepts the transfer of the heat of vaporization from the sprayed water and discharges it to the atmosphere. Some of the mass of water is transferred to the air stream. The evaporation rate is approximately 1% of the circulating water rate. Drift, the water mass transfer to the circulating air, is also about 1% of the water flow.

Cooling Tower Types

Natural Draft
Natural draft cooling towers are used primarily in large power plants where the savings of fan energy and the elimination of plume recirculation (because of the tower height) justify the high initial cost. Air movement through the tower is propelled by the difference in the entering and leaving air densities (chimney effect). Hyperbolic natural draft cooling towers are used in power plants.

Mechanical Draft
More mechanical draft cooling towers are in use than any other type. Forced draft towers push the air through the tower, whereas induced draft towers draw the air through the tower. Propeller fans are used in induced

draft towers when the air pressure loss through the tower is low. Figure 2.2 shows a forced draft and an induced draft cooling tower.

Range and Approach

The cooling tower "range" is the temperature difference between the entering and leaving water (ECWT and LCWT). Many designs are based on water entering the tower at 95°F and leaving at 85°F, for a 10°F range. The "approach" is how close the leaving water temperature is to the entering air wet bulb temperature. This is dependent on the design wet bulb temperature for the region where the tower is to be installed. If the leaving water is 85°F and the entering wet bulb is 78°F, the approach will be 7°F. Figure 2.3 is a diagram of range and approach.

Economic Cooling Tower Selection

Figure 2.4 shows that decreasing the approach from 15°F to 7.5°F doubles the size of the tower. It may be possible to use a slightly higher cooling tower cold water temperature during peak load conditions, to reduce the cooling tower size.

Before making a cooling tower selection, evaluate the cooling tower performance with a lower gpm than would be used in a "normal"

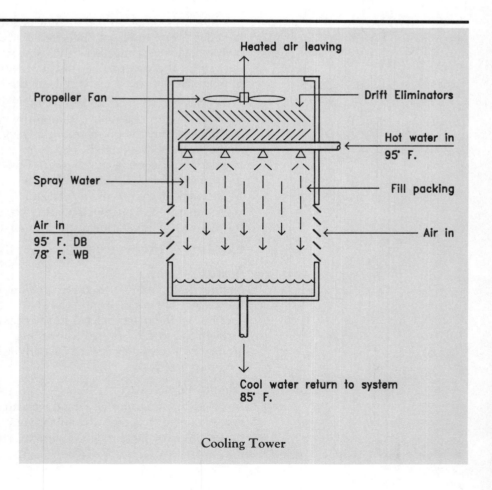

Cooling Tower

Figure 2.1

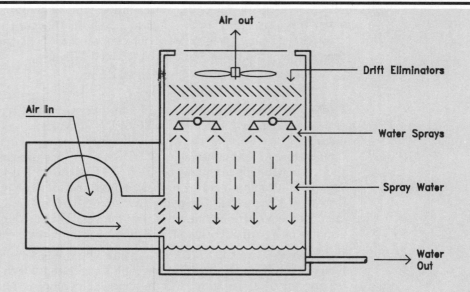

Air out

Drift Eliminators

Air in

Water Sprays

Spray Water

Water
Out

Forced Draft, Counterflow, Blower Fan Tower

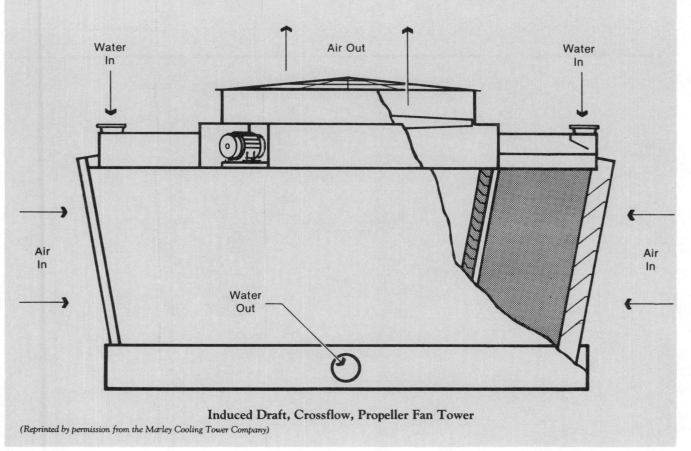

Water
In

Air Out

Water
In

Air
In

Water
Out

Air
In

Induced Draft, Crossflow, Propeller Fan Tower

(Reprinted by permission from the Marley Cooling Tower Company)

Figure 2.2

31

selection. Cooling towers for refrigeration chillers are usually selected for a water flow rate of 3 gpm/ton. Figure 2.5 shows that for the same heat transfer (capacity), wet bulb temperature, and tower leaving water temperature, the tower size can be reduced 10% by:

- decreasing the gpm from 144 to 118, or
- increasing the range by 22%.

Decreasing the gpm from 144 to 118 and increasing the range by 50% will reduce the required tower size by 20%. If 85°F water leaves the tower, circulating less water to the tower will raise the tower entering water temperature to 100°F instead of 95°F. Reducing the (design) gpm through the tower will reduce the tower size to 80% of the 95°F inlet water tower size. The installation cost is further reduced with the use of 3″ pipe size instead of 4″ pipe size. The pump size is one third smaller, and the pump energy required will be one third less, forever. The smaller cooling tower also means a permanently lower requirement for cooling tower fan energy.

If the efficiency of the towers is the same, the size requirement of a cooling tower depends on the entering wet bulb temperature for the approach, the leaving water temperature, and the range. The range is determined by the refrigeration machine selection. The cooling capacity and efficiency of the

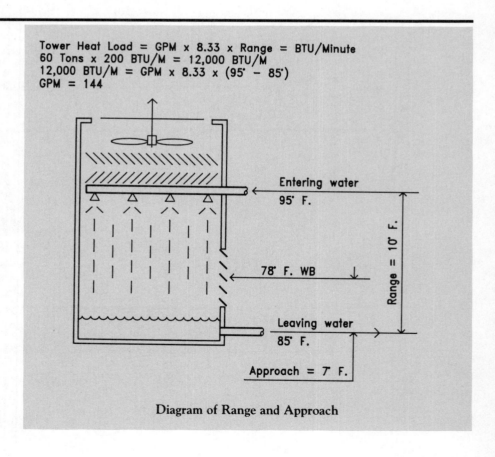

Tower Heat Load = GPM x 8.33 x Range = BTU/Minute
60 Tons x 200 BTU/M = 12,000 BTU/M
12,000 BTU/M = GPM x 8.33 x (95° − 85°)
GPM = 144

Entering water 95° F.

78° F. WB

Leaving water 85° F.

Range = 10° F.

Approach = 7° F.

Diagram of Range and Approach

Figure 2.3

refrigeration machines noted in Chapter 1 vary inversely with the head pressure. As the head pressure increases, the cooling capacity decreases and the kws increase.

Figure 2.6 shows that the condenser performance is dependent on the gpm, the difference between the condensing temperature and the entering condenser water temperature (ECWT), and the condenser size (heat exchange surface).

Figure 2.7 indicates that 60.5 tons of heat rejection, with 85°F entering condenser water, can be handled with 145 gpm through a nominal 50-ton condenser, and 118 gpm through a nominal 60-ton condenser. The total heat rejection (THR) remains the same. Only the gpm and the water temperature rise (range) change.

btu/minute = gpm x 8.33 lbs./gal. x range.
12,000 btu/m = 144 gpm x 8.33 lbs./gal. x 10°F
12,000 btu/m = 118 gpm x 8.33 lbs./gal. x (To-85°F)
To = 97.2°F

By reducing the condenser water circulation from 144°F to 118°F, the water temperature entering the tower is increased from 95°F to 97.2°F. Under

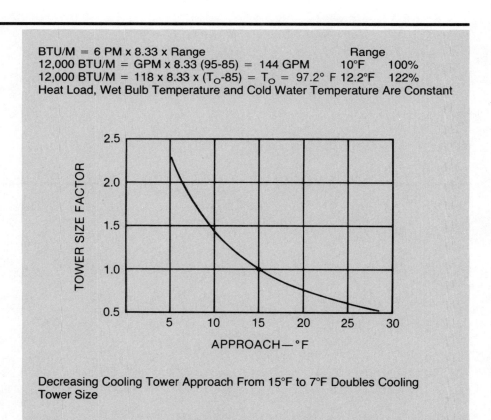

BTU/M = 6 PM x 8.33 x Range   Range
12,000 BTU/M = GPM x 8.33 (95-85) = 144 GPM   10°F   100%
12,000 BTU/M = 118 x 8.33 x (T$_O$-85) = T$_O$ = 97.2° F 12.2°F   122%
Heat Load, Wet Bulb Temperature and Cold Water Temperature Are Constant

Decreasing Cooling Tower Approach From 15°F to 7°F Doubles Cooling Tower Size

Effect of Chosen Approach on Tower Size at Fixed Heat Load, GPM, and Wet-Bulb Temperature
T$_O$ = Water Temperature Entering Cooling Tower
(Courtesy of the Marley Cooling Tower Company)

Figure 2.4

these conditions, the cooling tower size can be reduced by 10%. The reduced gpm will lower the operating cost, and could result in an additional installed cost savings, if pipe and condenser water pump sizes can be reduced.

Interference

Interference (illustrated in Figure 2.7) occurs when a portion of the saturated tower effluent contaminates the air intake of the downwind tower. Cooling towers should be installed so that the prevailing winds will carry the effluent away from the towers. If physical restraints cannot arrange cross wind airflow, the tower selection entering air wet bulb temperature should be elevated by as much as three degrees.

Walls or Enclosures

Restrictions to the air entering the tower can impede airflow through the tower and reduce the tower's performance. The distance from the wall to the induced draft tower (Figure 2.8) should be no less than the height of the tower. Induced draft tower fans cannot operate against much static pressure, so the velocity entering the tower should remain low.

Forced draft towers have centrifugal fans to overcome the tower high internal static pressure. A slightly higher static pressure imposed by a restricted air intake is not as significant as it is with an induced draft tower. The lower air discharge velocity of the forced draft tower makes recirculation of effluent air more likely than the high velocity of the

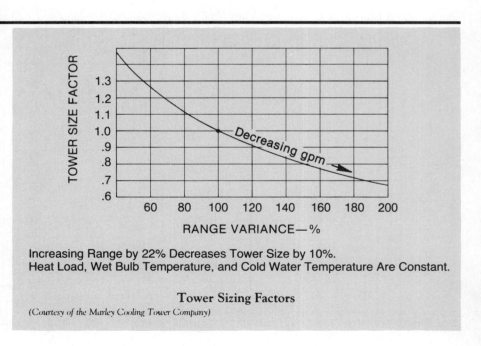

Increasing Range by 22% Decreases Tower Size by 10%.
Heat Load, Wet Bulb Temperature, and Cold Water Temperature Are Constant.

Tower Sizing Factors

(Courtesy of the Marley Cooling Tower Company)

Figure 2.5

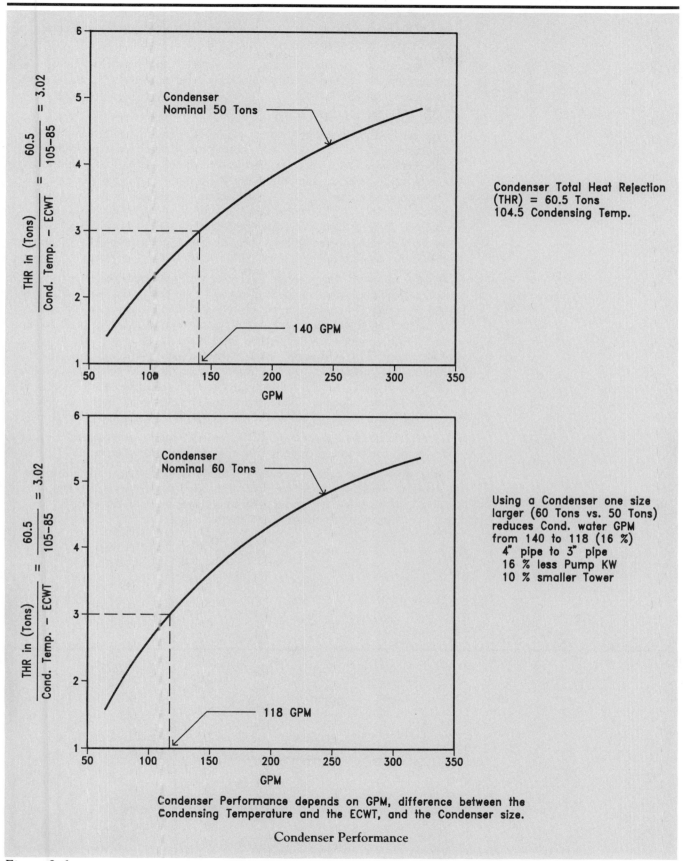

Condenser Total Heat Rejection
(THR) = 60.5 Tons
104.5 Condensing Temp.

Using a Condenser one size
larger (60 Tons vs. 50 Tons)
reduces Cond. water GPM
from 140 to 118 (16 %)
4" pipe to 3" pipe
16 % less Pump KW
10 % smaller Tower

Condenser Performance depends on GPM, difference between the
Condensing Temperature and the ECWT, and the Condenser size.

Condenser Performance

Figure 2.6

induced draft cooling tower. The wall proximity to a forced draft tower should be no less than twice the tower width, to minimize recirculation. This arrangement is shown in Figure 2.9.

Where critical recirculation cannot be avoided, the design wet bulb temperature for induced draft towers should be increased 1°F, and 2°F for forced draft towers.

Free Cooling

"Free cooling" with cooling towers is accomplished by using the cold condenser water produced by the cooling tower during low ambient wet bulb conditions, to produce chilled water. When the wet bulb temperature is sufficiently reduced, the cold water leaving the cooling tower can reduce or temporarily eliminate the use of mechanical refrigeration, to produce the chilled water. In northern latitudes, free cooling hours could be available 75% of the total yearly operating hours, but less than 20% of southern climate operating hours. Figure 2.10 shows a typical refrigeration chiller and cooling tower system.

Figure 2.11 shows the possible condenser water temperatures that can be produced by cooling towers at various wet bulb temperatures. The cooling load is usually lower when wet bulb temperatures are depressed. Process loads, such as computer rooms, are an exception. At 50% of peak cooling loads, 45°F condenser water can be produced at a 35°F wet bulb temperature. This will permit chilled water cooling without using the refrigeration chiller. At 50°F wet bulb temperature, the tower can produce 57°F at half load. A 57°F chilled water temperature can usually handle the cooling needs under part load conditions. For example, there are over 4,700 hours in Kansas City when the wet bulb temperature is below 50°F.

Figure 2.12 shows a direct cooling system in which cold condenser water is filtered and introduced directly into the chilled water system.

Direct systems are efficient because a heat exchange process is eliminated with cold condenser water circulating through the chiller. Cooling tower water is dirty, however, and the water strainer must be relied upon to remove contaminants that would otherwise interfere with the chiller heat

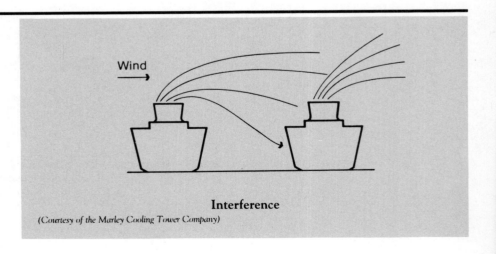

Interference

(Courtesy of the Marley Cooling Tower Company)

Figure 2.7

exchanger. Chillers and cooling coil performance are based on a 0.0005 fouling factor. Any increase in the fouling factor could wreak havoc on the cooling performance, and create tremendous problems for operating personnel.

Indirect Free Cooling System

Contamination of the chilled water system can be eliminated by installing a heat exchanger piped as per Figure 2.13, to separate the condenser and chilled water circuits. Heat exchangers are available that will produce a 2°F to 4°F temperature differential across the heat exchanger. When the compressor is not operating, the cooling tower does not have to remove the compressor heat (heat of compression), so it is able to lower the range from 10°F to 8°F. The 8°F rise in condenser water temperature enables the heat exchanger to produce a substantial 10°F temperature rise on the cooling coil side.

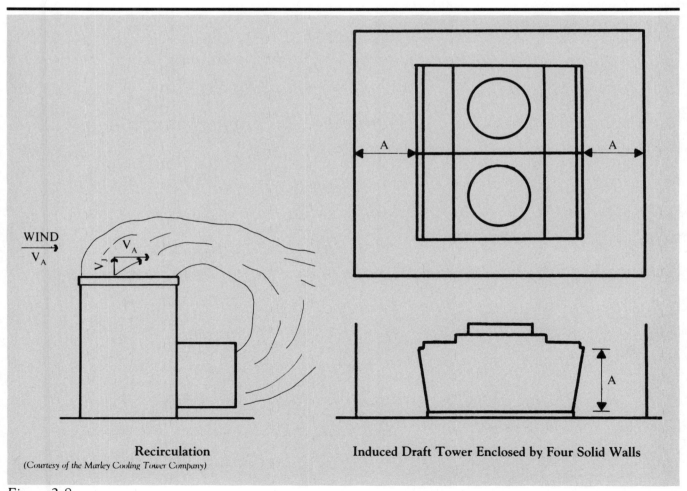

Recirculation

(Courtesy of the Marley Cooling Tower Company)

Induced Draft Tower Enclosed by Four Solid Walls

Figure 2.8

Advantages

Lower condenser water supply temperature

Condenser water supply temperatures within 5°F of the wet bulb temperature are possible. Condenser water supply temperatures of 85°F are standard with 78°F wet bulb. A refrigeration condensing temperature of 105°F is common with 85°F tower water, resulting in 0.886 kw/ton performance. Air-cooled systems operate at a 120°F condensing temperature with 95°F air on the condenser, and consume 1.10 kw/ton.

Usually lowest installed cost

Air-cooled systems under 40 tons with remote air-cooled condensers usually cost about 30% more to install than water-cooled systems with cooling

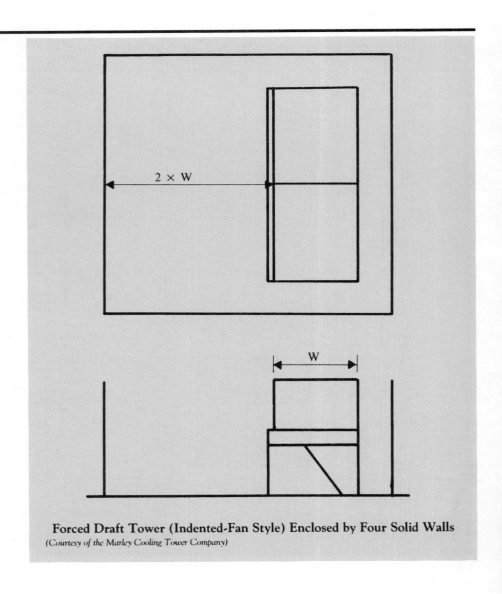

Forced Draft Tower (Indented-Fan Style) Enclosed by Four Solid Walls
(Courtesy of the Marley Cooling Tower Company)

Figure 2.9

towers. The installed cost of air-cooled systems over 40 tons with remote air-cooled condensers is approximately 40% higher than water-cooled systems with cooling towers. The installed cost of packaged air-cooled chillers is only very slightly (1-2%) higher than water-cooled systems with cooling towers.

Possible lowest operating cost

Water-cooled reciprocating chillers are 25% more efficient than air-cooled chillers (0.886 to 1.1 kw/ton). Water-cooled centrifugal chillers are 69% more efficient than reciprocating air-cooled chillers at peak load. In southern climates, even with condenser water system treatment costs, the operating cost of water-cooled systems is still less than that of air-cooled systems. The operating cost of air-cooled systems in northern climates can be less than water-cooled systems. See Chapter 5 for more on air-cooled and water-cooled systems.

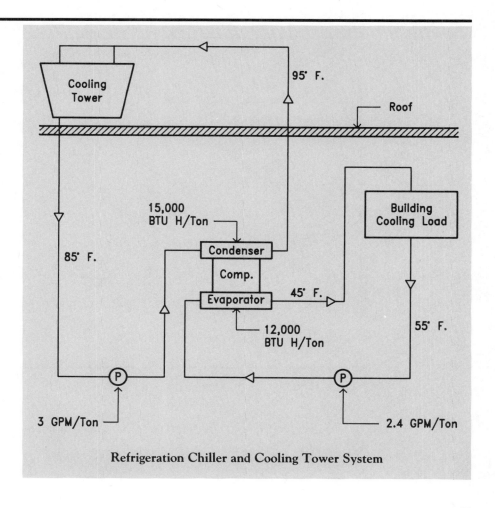

Refrigeration Chiller and Cooling Tower System

Figure 2.10

Availability of free cooling

Free cooling, described earlier in this chapter, is relatively simple to implement with cooling tower systems and closed circuit coolers. Air-cooled systems cannot provide the benefit of free cooling when outside air for cooling is not utilized.

Capable of handling very large chillers

Cooling towers can service all sizes of refrigeration systems. They are available to handle flow rates up to half a million gpm or more. Packaged air-cooled reciprocating chillers are made up to 200 tons. Although air-cooled centrifugal chillers are available, they are rarely used because of the high initial and operating costs.

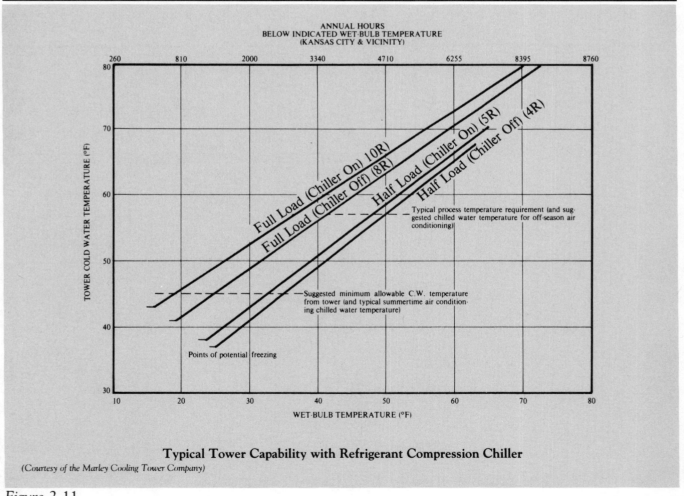

Typical Tower Capability with Refrigerant Compression Chiller

(Courtesy of the Marley Cooling Tower Company)

Figure 2.11

Heat rejection equipment can be a considerable distance from the chiller.

There is almost no limit to the distance that a cooling tower can be located away from the equipment that it serves. Condenser water piping is easier to install than refrigerant piping. Circulating pumps can move the condenser water as far as it needs to go. Long piping runs mean a greater drop in pressure for the circulating pump to overcome. Refrigerant piping pressure drop is critical. Liquid line pressure drop should not exceed the equivalent of 2°F, or 5.8 psig for R-22. Hot gas pressure drops should not exceed 2°F, because of compressor hp increases and loss of refrigeration capacity. The 2°F loss results in a total length of 200 equivalent feet for each line.

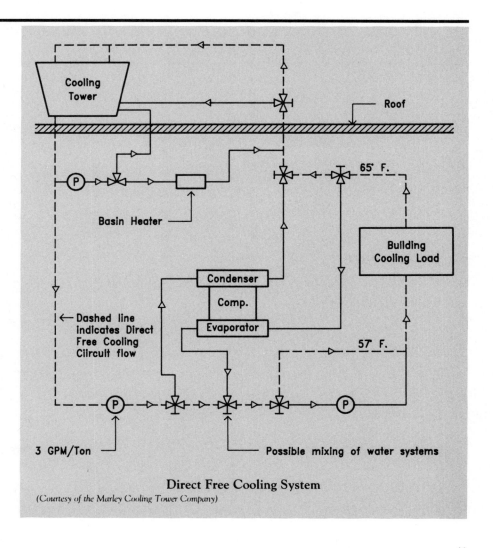

Direct Free Cooling System

(Courtesy of the Marley Cooling Tower Company)

Figure 2.12

41

Disadvantages

Some waste of water

Cooling towers waste a certain amount of water due to evaporation, drift, and the constant overflow for removal of leaves, scum or other solids. Approximately 5% of the circulated water is lost.

High maintenance cost

Maintenance of the condenser water system (including cleaning of the tower basin and condenser tubing) can be particularly expensive because it must be performed during non-business hours when the system can be shut down.

Water treatment necessary

Water treatment is absolutely necessary for cooling towers in order to eliminate the accumulation of contaminants, such as bacteria, algae, etc.

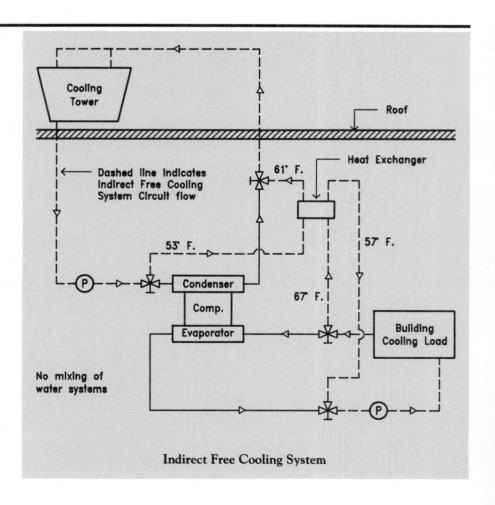

Indirect Free Cooling System

Figure 2.13

Tower cleaning and painting required
Constant exposure to the elements results in the threat of corrosion, and the need for regular cleaning and painting.

Must be winterized or drained to prevent freezing

Refilling required
The basin and piping must be refilled with water when cooling is needed again in spring (after they have been drained for winter weather to avoid freezing).

Regular inspection and maintenance required
Fill, eliminators, float valves, fans, basin headers, spray trees, etc. all require regular inspection and repair (maintenance).

Moisture carryover (drift)
Water carryover from the tower air discharge can be damaging to property, such as automobiles, windows, and building materials that are contacted by the water.

Circulating pump required

Greater weight and structural supports required

Exhaust plume during cool weather (fog)
Avoiding and coping with cooling tower disadvantages are issues addressed in Chapter 29, "Cooling Towers."

Air-Cooled Condensers

Air-cooled condensers use outside air to cool the hot refrigerant gas that is pumped by the compressor to the heat exchange coil (condenser). The outside air reduces the refrigerant temperature, which condenses the gas and subcools the liquid refrigerant.

Condenser Coils and Fans
The coils are usually copper tube bonded to aluminum fins. Outdoor air-cooled condensers are usually equipped with propeller fans, which have limited static pressure capability. The addition of only 0.3″ of static pressure will reduce the air delivery by almost 20%. The maximum available static pressure of air-cooled condenser propeller fans is 0.5″ wg, which would result in a 30% loss of cfm. Indoor units have centrifugal fans to handle the static pressure imposed by the intake and exhaust ductwork. Care must be taken with air intake and exhaust louver locations to avoid recirculation of the hot exhaust air into the intake louver.

Temperature Difference and Subcooling
Air-cooled condensers are usually selected for a 25°F temperature difference between entering air temperature (EAT) and saturated condensing temperature (SCT). With 95°F EAT at the condenser, the SCT would be 120°F. Most condenser ratings have temperature differences that range from 10°F to 40°F.

Air-cooled condensers have a higher capacity rating if no liquid subcooling takes place in the condenser. Without subcooling, the entire condenser coil is available to condense the gas. If, on the other hand, the air-cooled condenser is to perform liquid subcooling, the part of the condenser coil that is filled with the liquid to be subcooled is not available for condensing purposes. Five to ten percent of the condenser's capacity can be lost in this way.

Subcooling is usually 15°F, and results in higher system capacity at slightly higher head pressures, while allowing greater length of refrigerant run. If the refrigeration system is selected for 15°F subcooling, the air-cooled condenser should be selected accordingly. The system must contain more refrigerant when subcooling is applied.

Advantages

No water treatment required
Air-cooled condensers do not use water. Therefore, there is no water treatment required.

No winterization or spring start-up measures required
Mechanical cooling is always available, since there is no reason to drain for winterization.

Possible lower installed cost

Possible lower yearly operating cost

No condenser water pump

No water carryover (drift)

No plume

No draining or refilling
Again, with no water required, the system does not need to be drained for winter and refilled for start-up in spring.

Few mechanical parts to maintain
Other than the fan and the drive, there are very few mechanical parts to maintain.

Disadvantages

Possibly objectionable operating noise

Higher refrigerant head pressure at high ambient temperatures

Possible higher operating cost in warm climates

Higher installed cost with remote air-cooled condensers, especially in larger sizes

Not suitable for very large chillers

Field-fabricated refrigerant piping required for remote units

Requirement that the condenser be positioned close to the compressor

Larger compressor required for same load as a water-cooled unit

Deterioration of condenser coils, fans, and casings (unless coated) in harsh atmosphere

Requirement for correct refrigerant charge
Avoiding and correcting operational problems in air-cooled condensers will be further explored in Chapter 30.

Closed Circuit Evaporative Coolers

A closed circuit is not open to the atmosphere, thereby eliminating a major source of system contamination. As Figure 2.14 shows, the closed circuit cooler consists of a water cooling coil inside a cooling tower. The warm condenser water in the coil is cooled by the evaporative cooling process of the water spray and the relatively cool dry air that is moved across the coil by the fan.

Advantages

No contamination of condenser water system

The condenser water loop is a closed system not in direct contact with outside air. When inhibited glycol is used as the condenser cooling fluid, water treatment of the closed system is not required or desired. Only the water that circulates between the evaporative cooler basin and the spray nozzles needs chemical treatment.

Produces colder condenser fluid than a dry cooler

Dry coolers can cool the condenser water to within 10-20°F of the ambient dry bulb temperature. Condenser water supply temperatures during periods of 95°F outdoor temperatures will be 105°F to 115°F. Dry coolers will produce refrigeration condensing temperatures of 120°-140°F. Closed circuit evaporative coolers can cool the condenser water to within 8° of the ambient dry bulb temperature. An outdoor wet bulb temperature of 78°F will produce a 96°F condensing fluid and yield a condensing temperature of 105°F. Approximately 8% more compressor power will be used with a dry cooler than would be consumed by using a closed circuit evaporative cooler unit.

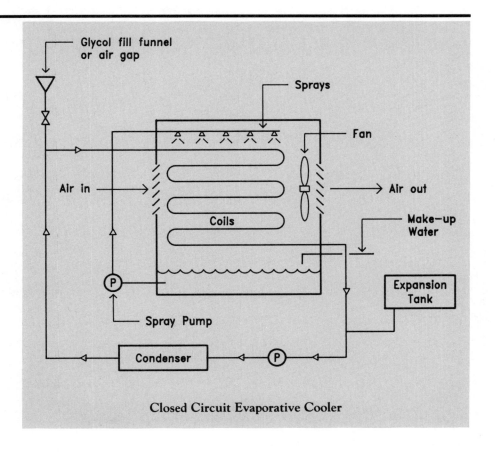

Closed Circuit Evaporative Cooler

Figure 2.14

Suitable for glycol use to eliminate condenser water freeze-up

A 30% by volume solution of glycol will form ice crystals at 7°F and still permit flow. Mechanical cooling will be available for winter use since the condenser water circuit does not have to be drained. Only the evaporative cooler basin needs to be drained or heated in preparation for freezing weather.

Can be used for winter "free chilled water cooling" operation

When the ambient dry bulb temperatures are sufficiently lower than the required chilled water temperature, the compressor can exercise now and then, or rest completely. At outdoor temperatures below 35°F, it is possible to produce 45°F glycol with the dry or evaporative cooler. Figure 2.15 demonstrates that at outdoor temperatures between 35°F and 70°F, compressor use can be minimized if a glycol pre-cooling coil is used.

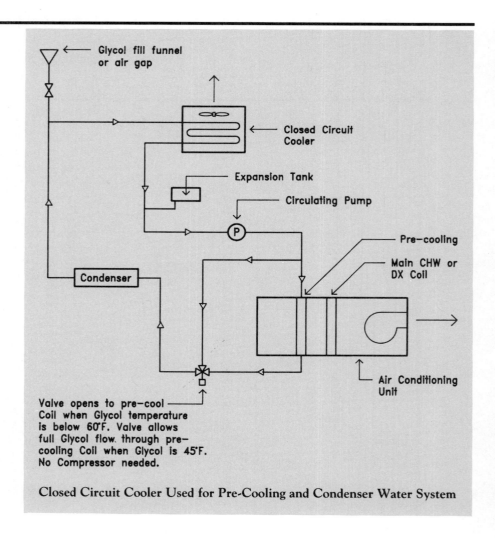

Closed Circuit Cooler Used for Pre-Cooling and Condenser Water System

Figure 2.15

Capacity to use rejected heat used for the heating system

Glycol temperatures near 120°F that are rejected from the condenser to the evaporative cooler can be diverted to the building's heating system. This system is covered in Chapter 10, "Heat Recovery Systems."

Minimal water treatment

Because only the water in the evaporative cooler basin is in direct contact with outside air, water treatment is applied only to the basin water content.

Freedom to locate closed circuit evaporative cooler a considerable distance from the compressor

The pressure drop in the condenser water piping system does not affect the refrigeration power consumption, as is the case with the remote air-cooled condenser. The evaporative cooler can be located as far away from the compressor as is necessary. It is necessary only that the condenser water pump is sized to handle the piping pressure drop.

Disadvantages

Higher condenser fluid supply temperature than cooling towers

Cooling towers produce condenser supply temperatures within 5-10°F of the ambient wet bulb temperature. Evaporative closed circuit coolers utilize water spray and forced air movement to induce evaporative cooling in order to pick up heat from the cooling coil in the evaporative cooler. The extra heat transfer from the fluid in the coil results in a slightly higher fluid temperature leaving the evaporative cooler.

Higher initial cost than cooling tower

The installed cost of an evaporative closed circuit cooler is almost twice the installed cost of a cooling tower, but slightly less than the air-cooled condenser, and about the same installed cost as a dry cooler. The evaporative cooler is much smaller than the dry cooler, permitting a competitive manufacturing cost.

Results in higher compressor condensing temperature than cooling tower or evaporative condenser

The extra heat exchange process of the closed circuit evaporative cooler results in a slightly higher compressor condensing temperature. A greater evaporative cooler heat exchange coil surface can produce condensing temperatures equal to those produced by cooling towers and evaporative condensers.

Requirement to drain or heat water in basin in winter

Water in the evaporative cooler basin will freeze in winter unless it is kept at a temperature above 32°F or drained from the sump. The basin water can be heated with electric heating coils in the basin, or by glycol or steam heating coils. Electric basin heating is the most commonly used method. An indoor tank eliminates the concern of freezing weather operations.

Some water use and water treatment

Water that is consumed through evaporation, drift, and bleed-off totals approximately 2.5% of the total circulated water. This is less than the 3% circulated water that cooling towers waste, but it still amounts to a substantial amount of water.

More maintenance than that required for air-cooled condenser or dry cooler

Normal maintenance of air-cooled condensers and dry coolers consists of possible fan motor bearing or motor replacement and yearly coil cleaning. Evaporative closed circuit coolers require the same motor maintenance, plus water treatment, spray pump, nozzles, and maintenance. Evaporative coolers need considerably more attention than air-cooled condensers or dry coolers.

Vapor plume during cool weather operation

Because warm moist air is discharged from the closed circuit evaporative cooler, a vapor plume will be created during cool weather operation.

Dry Coolers

Dry coolers are evaporative coolers without the water side basin, spray pump, nozzles, and associated piping. Dry coolers are essentially radiators. As Figure 2.16 shows, warm fluid leaves the condenser, enters the dry cooler coil, and is cooled by ambient air circulated through the coil by a fan. Larger air volumes are required to cool the fluid than those needed by evaporative-type coolers, because dry coolers rely on the ambient dry bulb temperature to remove condenser heat.

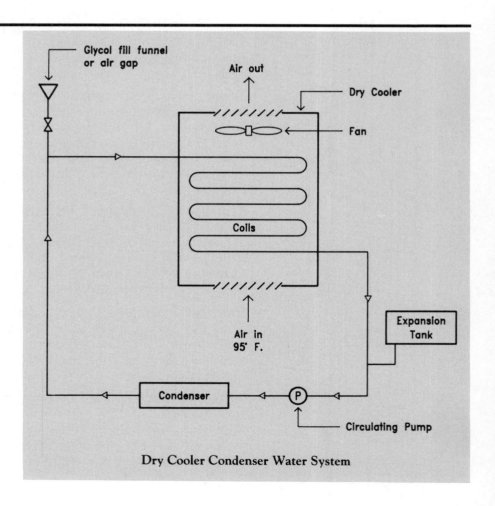

Dry Cooler Condenser Water System

Figure 2.16

Advantages

No water required
Dry coolers are used in a completely closed circuit system, and do not employ evaporative cooling. Therefore, no makeup water is used.

Little maintenance
Maintenance is reduced to lubricating bearings, possible fan and motor bearing replacement, cleaning the coil, and checking the glycol strength and operating pressure once a year.

Permits winter operation
When dry coolers are used with glycol, refrigeration compressors can be operated without the need for heaters of remote basins.

Free winter cooling
Dry coolers offer the benefit of free winter chilled water cooling.

Waste heat utilization
Rejected heat produced by the dry cooler system can be applied to the building heating system.

These advantages have been addressed in the Closed Circuit Evaporative Cooler section.

Eliminates water treatment
Since no makeup water is used, no water treatment is needed. When glycol is used as the condensing fluid, inhibitors are usually incorporated into the glycol. Water treatment is unnecessary with dry cooler systems.

Dry coolers can be located a considerable distance from the compressor.
The pressure drop in the condenser water piping system does not affect the refrigeration power consumption, as is the case with the remote air-cooled condenser. The dry cooler can be located as far away from the compressor as is necessary. The condenser water pump, of course, must be sized to handle the piping pressure drop as well as the other design parameters.

Disadvantages

Higher initial cost than cooling towers
The initial cost of dry coolers is about twice that of cooling towers.

Higher condensing temperature during design days
Since dry coolers rely on the ambient dry bulb temperature rather than the lower wet bulb temperature for the heat exchange, higher leaving fluid temperatures are produced. The higher condenser water supply temperature produces a higher condensing temperature. When a compressor is operating at a higher condensing temperature, more kw/ton are required.

More fan energy than evaporative-type coolers
Because evaporative cooling is not employed, dry coolers need more air to cool the condenser water than evaporative-type coolers. Evaporative-type coolers use approximately 300 cfm of ambient air per ton, while dry coolers use 700-1000 cfm per ton. The greater cfm means more fan horsepower.

Glycol is required for freeze protection in cold climates.
Freeze protection for dry coolers is provided by circulating inhibited glycol-based fluids through the condenser water system. The correct glycol composition must be used, based on system design temperatures, toxicity levels, thermal efficiency, and corrosion protection. Its integrity must be checked annually to ensure that freeze and corrosion protection is intact. Automotive-type glycol should never be used in HVAC work. Automotive-

type antifreeze solutions contain corrosion inhibitors that are formulated to be compatible with the materials in a car, and may not work well with HVAC materials. Automotive-type antifreeze lasts up to five years. Glycol that is made for HVAC use is made to last over twenty years.

Evaporative Condensers

The evaporative condenser is a combination of an air-cooled condenser and a cooling tower. The hot refrigerant vapor from the compressor is circulated through a condensing coil that is continually sprayed by recirculated water while ambient air is passed over the coil by a fan. Figure 2.17 pictures the system.

Advantages

Possible lower installed system costs

The installed cost of evaporative condensers is only a little more than that of air-cooled condensers, and almost twice that of a cooling tower. Since compressors served by evaporative condensers operate at a lower condensing temperature than those in air-cooled systems, the compressor could be smaller, reducing the system construction cost. When the cost of deleting

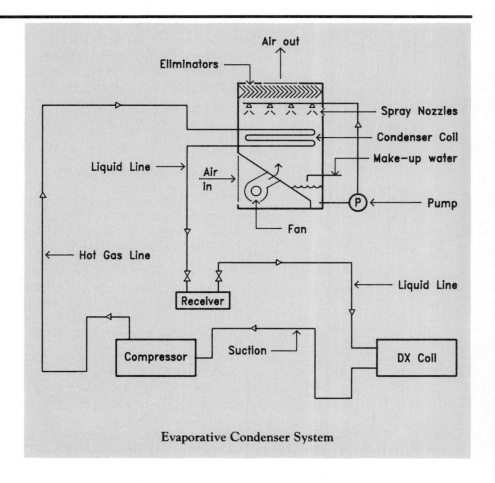

Evaporative Condenser System

Figure 2.17

the shell and tube condensers used by cooling tower systems is considered, evaporative condenser system installation cost could be lower than cooling tower systems.

Lower system operating costs

Except for well water condensing systems, evaporative condensers propagate the lowest condensing temperatures of all other condensing systems. Evaporative condensers produce these lower condensing temperatures by:

- utilizing ambient wet bulb temperatures that are 20°F cooler than the dry bulb temperature that air-cooled condensers and dry coolers use for cooling.
- using one-step heat transfer between refrigerant and ambient air, as opposed to the two-step cooling process of cooling tower systems (refrigerant-to-water plus water-to-ambient air).

Winter operation as an air-cooled condenser

During outdoor freezing conditions, it is possible to operate the evaporative condenser as an air-cooled condenser. The basin can be drained or bypassed to an indoor tank installed to avoid basin heater use.

In general, it is better to use the spray pump continuously because periodic drying of the coil generates scale accumulation.

Disadvantages

Uses water

Evaporative condensers waste about 2.5% of the gpm of water per ton of cooling.

Remote installations involve field-installed refrigerant piping.

The problems with field-installed refrigerant piping are addressed in Chapter 30.

Cannot be located very far from the compressor

The length of refrigerant piping must be kept to a minimum in order to keep the refrigerant head pressure that the compressor must produce as low as practical. Liquid line pressure drop should not exceed 1°F, and hot gas line pressure drop should not exceed 2°F.

Scale build-up on condenser coil

Maintenance of evaporative condensers is higher than that of air-cooled condensers or dry coolers. One item that requires attention is scale removal from the condenser coils. Evaporation of water sprayed on the condenser coils can leave mineral deposits that reduce the condenser's capacity for heat transfer. Scale removal is very difficult. Acid used on the coil surface can eat away at the coil, as well as the scale. Good water treatment minimizes scale accumulation.

The use of evaporative condensers is pretty much overlooked by most system designers. The possible lower installation and operational cost of the systems should overcome any additional maintenance costs that go along with water use systems.

Some air-conditioning unit manufacturers offer a packaged unit with evaporative condensers. These should be explored when evaporative condensers are considered.

Comparing Air-Cooled and Water-Cooled Systems

The decision to select an air-cooled system or a water-cooled system is usually based on the personal preference of the engineer and/or the owner rather than on technical data as to initial costs, operating costs, equipment life, and life cycle costs.

Specific applications may indicate otherwise, but in general, owner-operated systems are selected with maintenance as the primary factor. The quality of operating personnel is a major consideration in the selection of the more sophisticated energy efficient systems.

Based on the above factors, the selection of air-cooled and water-cooled systems should include consideration of these factors:

- Unless the quality of the installation crew has been established as outstanding, avoid field-installed refrigerant piping.
- In general, systems over 300 tons should be centrifugal or two-stage absorption chillers with cooling towers.
- Packaged air-cooled chillers are usually applicable to systems between 25 and 300 tons. As helical rotary and scroll chillers gain wider acceptance, their initial costs are declining. As they continue to become more affordable, packaged air-cooled chillers will be the first choice for owner-operated systems.
- Systems under 25 tons are best served by packaged air-cooled systems.

All of the above recommendations are based on "normal" comfort cooling situations with relatively unskilled operating personnel. Highly trained operators could encourage selections for "free cooling" possibilities, where climate permits that choice.

Chapter Three:

AIR HANDLING EQUIPMENT

The air handling equipment addressed in this chapter relates to the air-conditioning equipment that is remote from the refrigeration plant. Air handling units consist of the equipment that filters, heats, humidifies, cools, and circulates the air. The four major types of air handling units are:

- Factory-Fabricated Fan Coil Units
- Built-Up Units
- Customized Units
- Rooftop Units

Multizone units can be found in all of the above four categories.

Factory-Fabricated Fan Coil Units

Factory-fabricated units contain the fan and coil sections assembled in a sheet metal cabinet. Accessories such as filters, mixing boxes, and heating coils are assembled in sheet metal enclosures that can be attached to the other sections of the unit. Figure 3.1 shows a typical factory-fabricated air handling unit.

Factory-fabricated units are available with the following standard options and arrangements:

- Draw-Through
- Blow-Through
- Vertical Arrangement
- Horizontal Arrangement
- Chilled Water or Direct Expansion Cooling Coils
- Forward-Curved, Backward-Curved, and Airfoil Fans
- Hot Water, Steam, or Electric Heating Coils
- Flat or "V" Bank Filter Boxes
- Mixing Boxes
- Fan Capacity Control

Draw-Through

Draw-through units (shown in Figure 3.2) are the most common fan coil arrangement in use. The supply fan in the draw-through unit is positioned to pull the air through the filter, heating coil, and/or cooling coil, and blow the air into the supply duct.

Blow-Through

The blow-through unit design allows air to be drawn through the filter section and blown through the heating and cooling coils into the supply duct. A blow-through unit is shown in Figure 3.3.

Vertical Arrangement

Factory-fabricated units can be assembled in a vertical or horizontal arrangement. Vertical units are arranged with the coils under the fan. The vertical arrangement occupies less floor space than the horizontal arrangement. However, the coil section from the workspace above functions as the base (bottom) of the unit assembly. The fan section is mounted on top of the coil section. Note that the airflow must make a turn as it passes through the cooling coil, making it difficult to produce uniform air velocity across the cooling coil. It is possible to arrange the fan discharge in a horizontal or vertical configuration. A vertical discharge may necessitate a right angle turn at the fan discharge. This will produce an objectionable system effect factor. See Chapter 12 for the reduction of fan performance that this duct connection generates. A horizontal air discharge from a vertical air handling unit can simplify the ductwork by eliminating a 90° elbow at the fan outlet.

Horizontal Arrangement

Figure 3.2 shows the arrangement of horizontal air handling units. These units occupy more floor space than vertical air handling units. Airflow through the coils tends to be more uniform because the air does not have

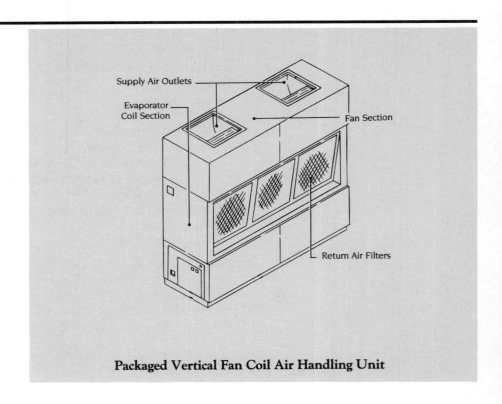

Packaged Vertical Fan Coil Air Handling Unit

Figure 3.1

to turn before it enters the coil. Duct branches can be run in several directions because of the greater available height between the fan outlet and the ceiling.

Chilled Water or Direct Expansion Cooling Coils

These units can be furnished with four-, six-, or eight-row chilled water or DX coils, with eight or fourteen fins per inch plate coils, or spiral fins. Many circuiting options are available to suit the desired thermal or operating performance. If the required coil performance can be obtained,

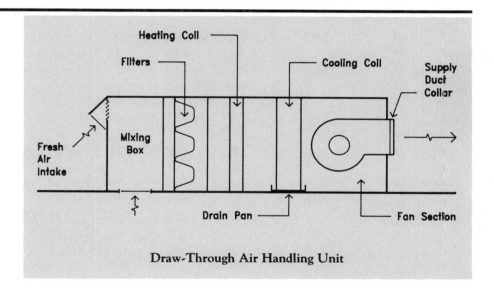

Figure 3.2

Draw-Through Air Handling Unit

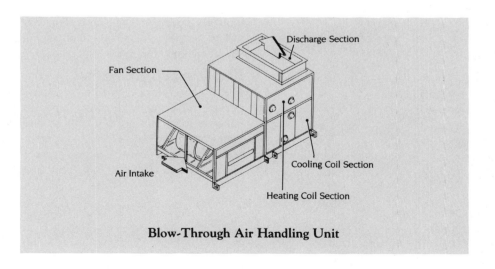

Figure 3.3

Blow-Through Air Handling Unit

coils should be selected with a minimum of six rows, and no more than eight fins per inch. This will facilitate coil cleaning.

Cooling coils require a drain pan to collect the water condensed from the air stream as the air comes into contact with the cooling that is colder than the air dewpoint temperature. This condensation is piped to an open drain via a trapped drainage piping system.

Forward-Curved, Backward-Curved, and Airfoil Fans

Standard factory-fabricated air handling units are available with these fan options. The forward-curved fan is less expensive than either the backward-curved or airfoil because it turns at a slower speed and requires lighter construction of fan wheels. Because of the lighter construction, multiple fans can be mounted on a longer shaft to produce a lower height unit. Forward-curved fans usually are available for a 4.5″ maximum total static pressure.

Backward-curved and airfoil fans turn at a higher speed than forward-curved fans, have heavier fan wheels, and cost more. Airfoil fans in factory-fabricated air handling units can operate against 7.5″ of total static pressure.

Hot Water, Steam, or Electric Heating Coils

Any of the mentioned coil sections can usually be arranged in preheat or reheat positions. Hot water heating coils are available in one or two rows and eight or fourteen fins per inch. Steam coils are usually one-row steam distributing type (tube within a tube), six, nine, or twelve fins per inch. A maximum fin spacing of six or eight fins per inch makes coil cleaning much easier.

Flat or "V" Bank Filter Boxes

Units can be provided with flat filter or angle filter sections to accommodate 2″ or 4″ thick throwaway or pleated filters. Angle filter sections accept only 2″ filters.

Bag filter sections accepting 2″ prefilters, and bag filters up to 23″ long are available for some units. Care must be taken to ensure that there is sufficient space between the filter section and the preheat coil to accept the filter extension at full airflow. A filter extension (blank) section should be provided for long bag filters.

Mixing Boxes

Figure 3.4 shows mixing box assemblies. Mixing boxes contain the filter section, a mixed air plenum, and duct connections with automatic dampers for outside air and return air ducts. The cost of a factory-fabricated section must be weighed against the cost of fabricating the mixed air plenum in a local sheet metal shop, and the field labor to install the automatic dampers.

Fan Capacity Control

Airflow modulation can be provided by the unit manufacturer for forward-curved fan units by the installation of opposed blade discharge dampers, variable inlet vanes (VIV), or by fan speed control. Units equipped with backward-curved or airfoil fans can be furnished with VIV or fan speed control. Speed control is best accomplished by using variable frequency drives. Fan modulation is further addressed in Chapter 14.

Advantages

Lower equipment cost

Because of standardization, controlled production, and lower factory labor cost than construction field labor cost, factory-fabricated units are less expensive to produce than field-erected built-up units. The cost of installing a 7,000-15,000 cfm factory-fabricated air handling unit is about 40% of the installed cost of a built-up unit, and 10% to 25% less than the installed cost of a customized unit.

Quicker delivery

Because of standardization of unit components, the manufacture and delivery of the factory-fabricated unit is usually quicker than the purchase and assembly of the built-up unit.

Disadvantages

Limited quality options

Manufacturers standardize the construction of the units to reduce costs. Some manufacturers produce units of different quality for the same capacities. Casings are usually constructed of galvanized steel, with heavy gauge formed channel support members. Access doors may or may not be double-walled, with 1.4 lb. density fiberglass insulation. Fans are forward-curved, backward-curved, or airfoil. Some manufacturers offer forward-curved fans only, while some offer airfoil fans only above a certain size. Coils may only be available with aluminum fins on copper tubes. Coils may also only be available with galvanized casings.

The following unit quality options may not be available:

- Heavier gauge casings
- Heavier gauge double-walled access doors with piano hinges

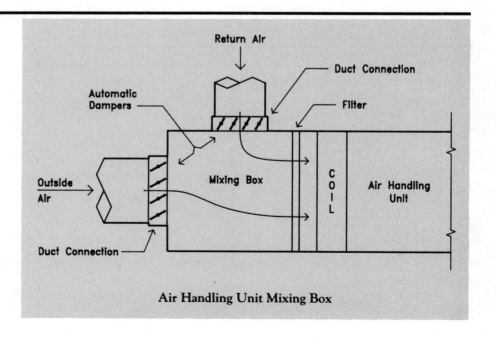

Air Handling Unit Mixing Box

Figure 3.4

- Stainless steel cooling coil casings
- Stainless steel drain pans
- Solder-coated coil fins (lead-coated copper)
- Plug (plenum) fans
- Direct drive fans
- Two fans in parallel

Limited space considerations, fixed dimensions

Standardization to reduce manufacturing costs results in a limited variety of standard dimensions. Unit sizes are determined by the cooling coil area. A 13,000 cfm unit would have a 26 S.F. coil at 500 fpm coil face velocity. A standard horizontal unit might be 4'-8" high and 8'-3" wide. Figure 3.5 shows the space required for coil removal and for servicing the unit.

The factory-fabricated unit should be installed in a 20' wide mechanical equipment room, if the unit is to be properly serviced. If the unit were configured with a 5' width and a 7'-6" height, the mechanical room width could be reduced to 13'. This built-up unit occupies 37.5% less floor space than a factory-fabricated unit.

Maximum fan motor hp

Fan cabinets are manufactured to standard dimensions for each size unit. The 13,000 cfm unit can accommodate up to a 15 hp motor. If a 20 hp motor is needed to overcome the system static, a booster fan must be provided.

Size limitations imposed by local union agreements or working conditions

Certain sheet metal local rules limit the size of factory-fabricated units in order to retain as much work for their membership as possible. In some geographic areas a 29,000 cfm unit is the largest capacity permitted in one unit. In such cases, either multiple units or built-up units must be installed.

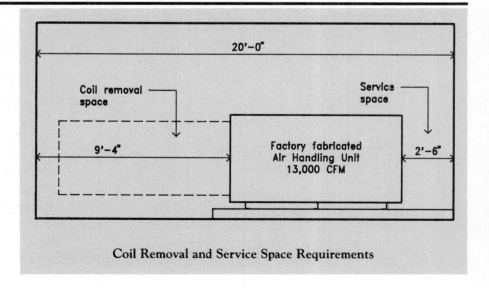

Coil Removal and Service Space Requirements

Figure 3.5

Size limitation

Most manufacturers limit their production of units to 50,000 cfm. If larger air quantities are required, multiple factory-fabricated or built-up units must be installed.

Built-Up Units

Built-up air handling units are comprised of separate casings that enclose fans, coils, drain pans, filters, mixing boxes, and plenums. Built-up unit casings are either factory-manufactured or made in a local sheet metal fabrication shop. Casings are usually double-walled, with a heavy gauge outer wall and a lighter gauge inner wall. Fiberglass insulation is sandwiched between these walls. The inner wall may be perforated to provide sound attenuation. Figure 3.6 is a built-up air handling unit.

Advantages

Space flexibility, variable dimensions

Built-up unit dimensions are limited only by the coil and fan selections. Coil selections can be customized for the area configuration desired. Standard coils can be obtained in lengths from 24″ to 144″, in 6″ increments, and heights of 18″ to 40″. The previous 13,000 cfm factory-fabricated unit was 8′-3″ wide and necessitated a 20′ wide mechanical equipment room. Two 48″ 28-tube face, 13.2 S.F. face area coils mounted two high would reduce the unit width to 5′ and the mechanical equipment room required width to 13′.

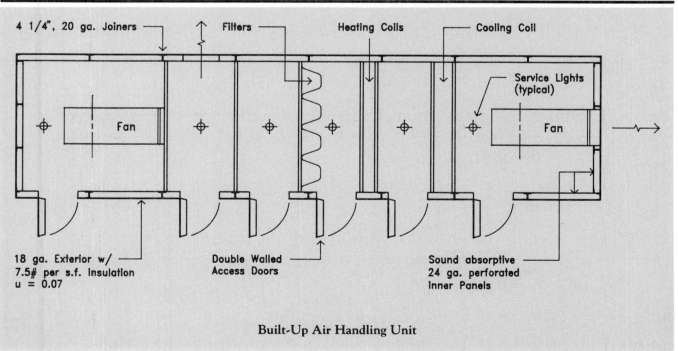

Built-Up Air Handling Unit

Figure 3.6

This represents a 35% savings of floor space! With building construction costs at $50-$100 per square foot, these could be very significant savings. Perhaps more important is the possibility of using an existing room for the air handling unit.

Possibly higher quality

Because casings, coils, fans, and filters are specified individually, a higher quality level can be obtained. Customized matching of components can also increase efficiency. Casings can be double-walled with 18-gauge exterior walls and 20-gauge inner walls, with 2″ of insulation between the walls. The 20-gauge interior wall makes it possible to enter the unit for installation and service without tearing up the unit insulation.

Cooling coils can be supplied with stainless steel casings, or the entire coil and casings can be covered with a protective coating. Condensate drain pans can be stainless steel for long life and no major maintenance. Access doors can be double-walled, internally insulated, and equipped with double hinges and sturdy handles.

Fans can be selected for the actual static pressure they must overcome, and with the precise fan motor and drive assembly. Plenum fans can be used to simplify ductwork. Vaneaxial fans are available with adjustable pitch in motion control of VAV systems. (See Chapter 20 for more on Fans.)

Multiple fans possible

For applications such as hospitals, museums, research laboratories, and auditoriums where loss of cooling or air distribution is to be avoided, fan redundancy is important. Built-up units can be configured to accommodate two fans operating in parallel. Chapter 20 on fans explains the advantages and methods of providing fan redundancy. Figure 3.7 shows a multiple fan arrangement.

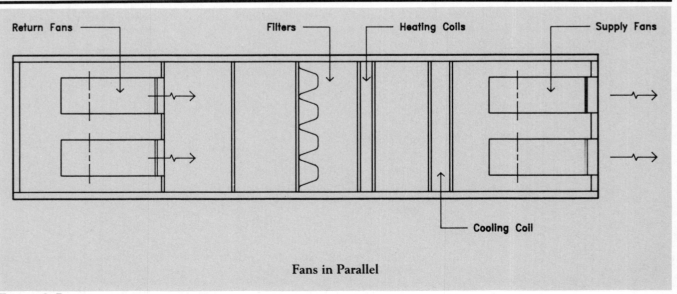

Return Fans   Filters   Heating Coils   Supply Fans

Cooling Coil

Fans in Parallel

Figure 3.7

Options for vaneaxial and plenum fans
The advantages of vaneaxial and plenum fans are addressed in detail in Chapter 20. Figure 3.8 shows a vaneaxial fan installation in a built-up unit. Plenum fans in built-up type units are shown in Figure 3.9.

Very large systems possible
There is almost no limit to the capacity of built-up units. A casing can enclose as many coils and fans as are required to handle the established load.

Long service life
Built-up units with stainless steel cooling coil casings and condensate drain pans, and epoxy or solder-coated coil fins should last over forty years.

Possible lower in-place cost in existing buildings
Factory-fabricated air handling units may not physically fit through existing doors, or on the building elevators. Installation may require use of a crane to rig the units through windows or outside air louvers, etc. If building openings are not large enough to accommodate the largest unit section, openings may have to be made. In such applications, the cost of delivering a built-up unit via the building's elevators, and erecting the unit in the mechanical equipment room, may be less than installing a factory-fabricated unit with its associated construction work. When booster fans are necessary to assist the factory-fabricated unit to overcome the system static pressure, the cost differential between the units is reduced.

Possible savings of floor space
This benefit has been addressed earlier (see "Space Flexibility, Variable Dimensions") with the example of a 35% savings of floor space.

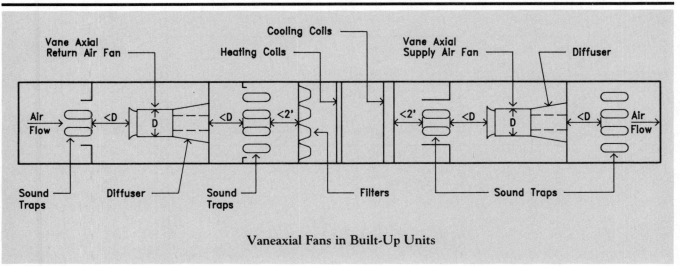

Vaneaxial Fans in Built-Up Units

Figure 3.8

Disadvantages
Higher initial cost
The normal installation cost of a factory-fabricated air handling unit is about 40% of the installed cost of a built-up unit. One should keep in mind, however, that there is nothing "normal" about installations in the construction field. One built-up unit installation could be less expensive than two or more factory-fabricated units serving the same load. Mechanical room space costs could favor the installed cost of a built-up unit.

Possible longer installation time
If the timing is right, the production run of the factory-fabricated air handling unit may coincide with the construction schedule. "Fast track" delivery at a small cost premium is available from most manufacturers. It is possible to obtain delivery of a factory-fabricated air handling unit in four to six weeks. Built-up air handling units are custom built to suit the specialized design. Delivery and erection of these units usually runs fifteen to twenty weeks.

Customized Units

Customized air handling units, not made for the mass market, usually reach a higher level of quality and offer special component configurations to meet particular design needs. When a higher quality air handling unit is desired than that which is offered by the "mass market" manufacturers, or special particulars are needed (such as plenum or direct drive fans, or extended fan sections to accommodate VAV boxes) customized units should be explored. Figure 3.9 presents customized unit applications.

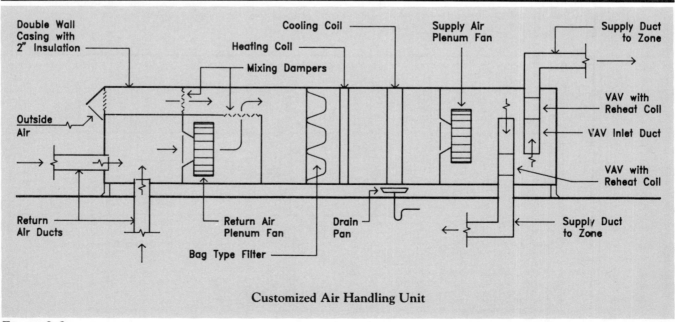

Customized Air Handling Unit

Figure 3.9

Multizone Units

Multizone units provide individual space temperature control, with one fan supplying the conditioned air to each of the zones. The multizone unit is a blow-through unit with the cooling coil and the heating coil in parallel air streams. Each zone is served by an automatic damper that regulates the volume of air passing over the cooling or heating coil to that zone. Figure 3.10 is a multizone unit system.

Advantages

Individual space temperature control

Each zone thermostat positions its automatic zone damper to adjust the flow of cool or warm air entering the common supply duct serving that zone.

Constant ventilation

The volume of air that is supplied to each zone does not vary. The supply air temperature to the zone varies as the zone thermostat positions the automatic damper to adjust the warm or cool air volume entering the zone supply duct. Since the supply air volume does not vary, the outside air volume introduced into each zone remains the same. It is not unusual for VAV systems to reduce the outside air volume to the system during reduced supply airflow. (See Chapter 23 for more on avoiding and solving problems with VAV systems.)

Disadvantages

High space humidity

The cooling and heating coils of multizone units are in parallel air paths. The air that does not pass through the cooling coil is not dehumidified. This moist air is supplied to the space and is unable to remove humidity from the space. High room humidity levels result from multizone units with fairly high outside air percentages.

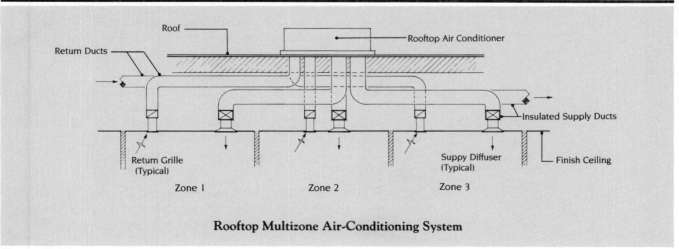

Rooftop Multizone Air-Conditioning System

Figure 3.10

Poor temperature control, damper leakage

Standard multizone units are not usually equipped with tight sealing dampers. The space between the hot and cold deck, and between each zone duct connection to the unit, is not airtight. Considerable quantities of air pass between the hot and cold deck dampers, and from one zone to another. Some manufacturers do provide tight sealing dampers or sufficient space to isolate one zone from another. Damper linkages tend to become sloppy over time. Unless regular adjustments are made, control becomes unsatisfactory.

Extensive ductwork

Each multizone unit zone is served by a duct that is connected to the unit supply section. If the unit serves twenty zones, there will be twenty different ducts, possibly running side by side, between the unit and the zones that are serviced.

Long ductwork and duct leakage

Some of the zones served by the multizone unit may have air volumes of only 500 to 1,000 cfm that may have to travel a considerable distance from the unit. If the duct fabrication or installation is less than excellent, as is often the case, the air leakage may reach a high percentage because of the high duct static pressure developed to overcome the long air distribution system. If the duct air leakage reduces the 500 cfm leaving the unit to 300 cfm, comfort conditions cannot be obtained. (See Chapter 19 for further coverage on ductwork problems.)

Higher air pressure drop with blow-through design

Multizones are constructed in a blow-through configuration. The air pressure drop through the heating circuit must equal the air pressure drop through the cooling section. Because of the proximity of the coils to the high velocity fan outlet, equalizing plates are installed in front of the coil sections to produce a uniform airflow across the coils. This arrangement produces a very high static pressure loss. This fan energy penalty is forever.

Rooftop Air Handling Units

Rooftop air handling units are similar to indoor units, except that they are located on the roof and either mounted on a roof curb or on structural supports above the roof level.

Advantages

Saves valuable building space

Since rooftop units are located on the roof, valuable space that would be used for the mechanical equipment room (MER) is saved for other purposes. In new construction projects, the construction costs of the MER can be avoided. Provisions must be made to access the roof with filter cartons, etc. for maintaining and servicing the unit.

Factory-assembled, wired, and tested

Factory-assembled units can be wired, charged, and tested before shipment to the job site. Factory prewired power and control panels save field labor and coordination. Packaged rooftop units are completely self-contained with compressor, condenser, and evaporator, and usually a gas-fired or electric heating section.

Less field labor

Rooftop units are assembled and tested at the factory. Field labor is limited to rigging and setting the unit on the roof curb, piping the coils and drains, connecting the power to the power panel or motors, and attaching the ductwork.

More singular responsibility

The manufacturer assumes responsibility for constructing the unit. Field responsibility for the unit involves only setting the unit. Built-up unit casings, fans, and filters are assembled by the sheet metal contractor, the coils are erected by the pipefitters, and the controls connected by a temperature control contractor.

Some high quality units available

Rooftop units are available with double-wall construction, 2″ of insulation between the walls, galvanized steel exterior panels with weather-resistant coatings, double-walled hinged access doors, stainless steel drain pans, stainless steel cooling coil casings, epoxy-coated coils, and plenum or direct drive fans. High quality units should give a service life comparable to indoor units.

Disadvantages

High maintenance, deterioration due to weather

The rooftop unit sits on the roof every minute of every day. It faces the sun every morning with its damaging rays, accepts the rain and freezing snow, and is exposed to contaminants from chimneys, industrial areas, and auto pollutants. Rooftop units are also climbed on and into by maintenance and repair personnel. Access doors are opened, and if not easily closed and secured, are left open. Unless some provision is made to protect the equipment with heavier construction materials, strong and easily operated access doors, and epoxy coatings, they cannot survive under these conditions for very long.

Shorter service life

Many standard rooftop units are in very bad condition in ten to fifteen years. High quality units with weather-resistant coatings, on the other hand, should last as long as indoor units.

Structural requirements to support the unit

The weight of curb-mounted rooftop units must be carried by the roof structure. Some rooftop units are installed on steel supports above the roof. In either case, structural work must be designed and installed as required. This is part of the installation cost of a rooftop unit evaluation.

May require weatherpoof ductwork

When air handling units are installed on the roof, ductwork may also have to be installed above the roof. When exposed to weather, ductwork must be sealed watertight to prevent water from entering the building. Supply and return air ductwork must be insulated to avoid heat transfer to and from the outdoors and to prevent condensation. The insulation must be waterproofed, as wet insulation loses its thermal benefit. The quality of the outdoor duct installation is dependent on the installation personnel. Excessive duct leakage will cause ballooning of the exterior duct insulation and waterproof jacket, with the resultant water leaks. If the waterproof covering is less than perfect, water penetration into the insulation will result. It should be noted that all roof penetrations are sources for water entry. These include the curbs where the ducts enter the building and every duct or piping support. See Figure 3.11.

Possible damage to the roof surface

Rooftop units are located on the roof. Installation of the unit, electrical, piping, ductwork, insulation, and controls, as well as start-up and service all take place on the roof. The traffic of these tradesmen reduces the life of the roof. Screws may be dropped on the roof, stepped on, and penetrate the roof membrane. Filters must be changed, bearings lubricated, and controls adjusted. All of this activity takes place on the roof and ultimately contributes to roof leaks.

Must be serviced in inclement weather

It is the nature of the service world that equipment tends to fail during periods of terrible weather. Either the sun has heated the unit's metal surface to the point that touching it burns the skin, or an electrical problem has developed during a torrential downpour, or the heating control has malfunctioned and it is 10 degrees below zero, with a 40 mph wind.

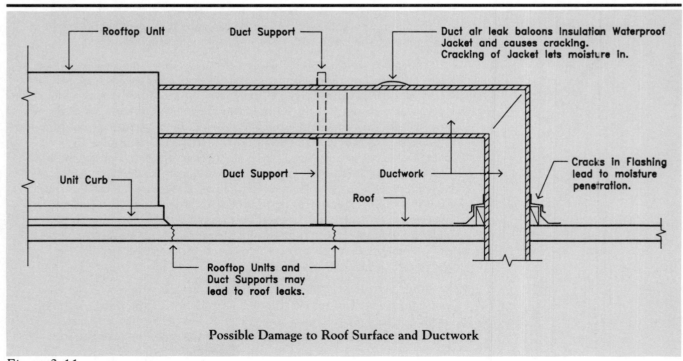

Possible Damage to Roof Surface and Ductwork

Figure 3.11

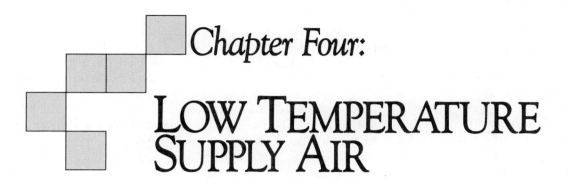

Chapter Four:

LOW TEMPERATURE SUPPLY AIR

The dramatic increase in the cost of energy, especially electricity, has prompted many changes in the HVAC industry. Variable air volume and the microprocessor are now the foundation of standard HVAC design. The latest advancements have been inspired by the electric power shortage, as electric utilities struggle to meet peak demands for power.

One method of easing the burden of generating sufficient power to meet the peak demand is to produce and store chilled water or ice during the off-peak hours of electric demand, for use during the peak hours of electric energy usage. Low temperature supply air is a cousin to this type of cooling storage (ice bank). Generating and storing ice makes it possible to create air temperatures as low as 40°F.

For most air-conditioning systems, the fan energy required to transport the supply and return air represents about 40% of the cooling energy. Figure 4.1 is an example that demonstrates these relationships.

Reduction in Fan Energy

If the quantity of supply and return air circulated to conditioned spaces is reduced, the fan energy necessary to move that air is also reduced. As Figure 4.2 describes, the volume of air needed to cool a space depends on the temperature difference between the supply air and the space temperature.

The sensible heat gain of the space to be cooled warms the cold air supply to the space to the desired room temperature. The latent (moisture) heat generated inside the conditioned space adds moisture to the supply air. Most comfort air-conditioning systems are designed for a 55°F supply air temperature. Heat from the supply fan and from the conditioned space warms the supply air up to 75°F.

The following calculations determine the supply air quantities for a 216,000 btuh internal sensible heat load.

$$cfm = \frac{btuh \ (internal \ sensible \ heat)}{1.08 \times td \ (temperature \ difference \ between \ room \ air \ and \ supply \ air \ temperatures)}$$

For 55°F supply air: $cfm = \frac{216,000}{1.08 \times (75-55)} = 10,000 \ cfm$

For 45°F supply air: cfm $= \dfrac{216{,}000}{1.08 \times (75-45)} = 6{,}667$ cfm

If 45°F rather than 55°F supply air is delivered to a space, a one-third reduction of the 55°F air volume will produce the same cooling benefit. The 45°F supply air reduces the relative humidity of the space, so some additional latent heat is removed at the cooling coil. The 45°F supply air system actually requires less than two-thirds of the 55°F air, because less supply fan heat is added to the supply air.

The fan brake horsepower adds heat to the fan supply air. The fan brake horsepower (bhp) is equal to:

$$\frac{\text{cfm} \times \text{sp}}{6356 \times \text{efficiency}}$$

Assuming a duct sp of 1″ and a fan efficiency of 75%, the fan bhp due to the fan moving 10,000 cfm through the 1″ static pressure duct system is:

$$\frac{10{,}000 \text{ cfm} \times 1 \text{ sp}}{6356 \times 0.75} = 2.1 \text{ bhp}$$

The fan bhp due to the fan moving 6667 cfm through the 1″ static pressure duct systems is:

$$\frac{6667 \text{ cfm} \times 1 \text{ sp}}{6356 \times 0.75} = 1.39 \text{ bhp}$$

Two-thirds of the fan energy is required for the 45°F supply air system, as compared to a 55°F supply air system.

$$\frac{1.39}{2.1} = 0.66$$

Comparison of Fan Energy to Total Cooling Energy

Example: A 100-ton chiller in use 10 hours per day, 20 days per month, over a 6-month cooling season, at 1,000 full load ton-hours (flth) of cooling per year, 12 month fan operation for ventilation and intermediate season cooling, $0.10 per kwh.

Chiller:	100 kw x 1,000 flth =	100,000 kwh
Chw pump:	26.2 kw x 10 h/d x 20 d/m x 6 m/y	31,440 kwh
Cwp:	8.6 kw	
Ct fan:	4.0	
Total	12.6 kw x 1,000 flth =	12,600 kwh
Evaporator fans:	40 kw x 10 h/d x 20 d/m x 12 m/y =	96,000
	Total	240,040 kwh

Total yearly cooling and ventilation energy use: 240,040 kwh
Evaporator fan percentage of total cooling and ventilation yearly energy consumption is 40%. (96,000/240,040).

Figure 4.1

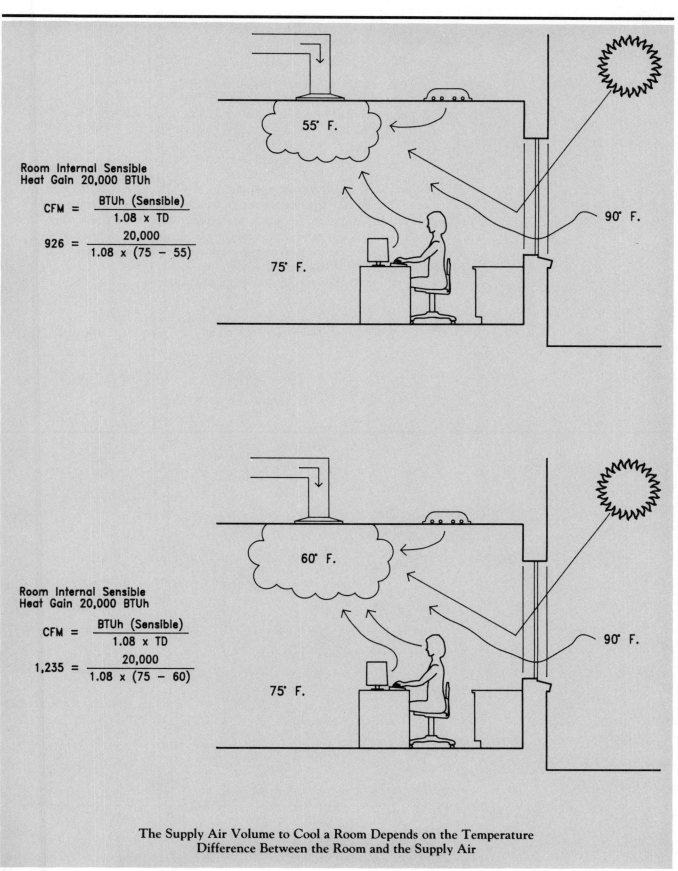

Room Internal Sensible
Heat Gain 20,000 BTUh

$$CFM = \frac{BTUh\ (Sensible)}{1.08 \times TD}$$

$$926 = \frac{20,000}{1.08 \times (75 - 55)}$$

55° F.

90° F.

75° F.

Room Internal Sensible
Heat Gain 20,000 BTUh

$$CFM = \frac{BTUh\ (Sensible)}{1.08 \times TD}$$

$$1,235 = \frac{20,000}{1.08 \times (75 - 60)}$$

60° F.

90° F.

75° F.

**The Supply Air Volume to Cool a Room Depends on the Temperature
Difference Between the Room and the Supply Air**

Figure 4.2

The amount of fan heat that is added to the supply air is determined by the fan brake horsepower.

$$1 \text{ bhp} = 2545 \text{ btuh}$$

$$\text{Temperature difference (td)} = \frac{\text{btuh}}{\text{cfm} \times 1.08}$$

1.08 is the constant to convert cfm to lbs./hr. x specific heat.

$$\text{ft}^3 = 0.24 \times 60 \text{ min./hr.}/13.34 \text{ ft}^3/\text{lb.} = 1.08$$

$$\text{td} = \frac{1.39 \text{ bhp} \times 2545 \text{ btuh}}{6667 \text{ cfm} \times 1.08} = 0.49 \text{ F}$$

Fan heat raises the supply air temperature 0.49°F for every inch of static pressure that the fan operates against. The total air supply moved by the supply fan is the sum of the supply air required to remove the sensible heat from the conditioned space, plus the supply air required to remove the fan heat, plus the supply air necessary to remove the duct heat gain. The psychrometric chart in Figure 4.3 diagrams the air condition leaving the cooling coil until it warms up to the room temperature.

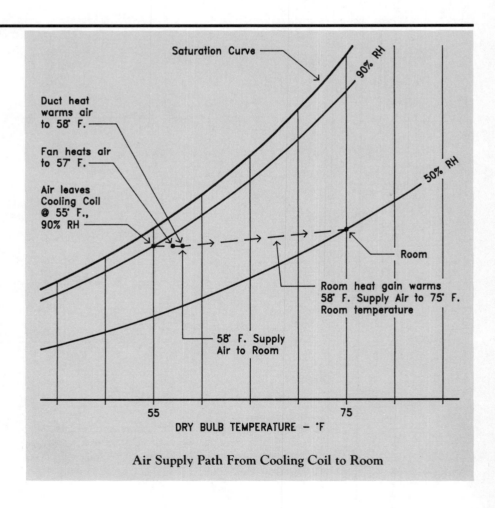

Air Supply Path From Cooling Coil to Room

Figure 4.3

The supply air quantity for the above 55°F system =

$$\frac{216{,}000 \text{ btuh} + (2.1 \text{ bhp} \times 2545 \text{ btuh per bhp})}{1.08 \times (75-55)} = 10{,}247$$

The air supply volume must be increased from 10,000 cfm to 10,247 cfm, to compensate for the heat added by the supply fan.

$$\frac{10{,}274}{10{,}247} = 1.025$$

2-1/2% more supply air is needed to offset the supply fan heat.

The supply air quantity for the above 45°F supply air system =

$$\frac{216{,}000 \text{ btuh} + (1.39 \text{ bhp} \times 2545)}{1.08 \times (75-45)} = 6775 \text{ cfm}$$

$$\frac{6775}{6667} = 1.016$$

Only 1.6% more supply air is needed to offset the 45°F supply air fan heat.

The 45°F supply air system requires only 66% of the supply air of a 55°F supply air system.

$$\frac{6667 \text{ cfm}}{10{,}247 \text{ cfm}} = 0.66$$

Advantages

Reduced supply air quantity
As detailed above, a 45°F supply air system uses 34% less primary air than a 55°F system. It is not desirable to circulate less than four air changes of supply air within the occupied space. When VAV systems reduce the airflow below four air changes per hour, air circulation does not take place below the office modular screens. The outgassing from fabrics, floor coverings, furniture and building materials never gets diffused. This creates an uncomfortable environment. In order to maintain a satisfactory air change rate, supply air booster fans are used to mix the 45°F primary air with return air. Room air at 75°F is mixed with the 45°F primary air to deliver 55°F supply air to the conditioned space. The room thermostat controls the amount of 45°F primary air to be mixed with the room air. The conditioned space always experiences four air changes per hour of air circulation, regardless of the internal load.

Smaller air handling units, return air fans, primary supply air and return air duct systems, and less duct insulation
The installation cost savings for these items could amount to $1.80/sf.

Smaller supply and return air fans
As shown above, 34% less air is delivered, and 34% less fan energy required to deliver the primary air. The supply air booster fans add fan horsepower to the system, but at only 25% of the resistance of the primary air system.

Less chilled water gpm, piping, insulation, and smaller chilled water pumps
The water quantity is determined by the following:

8.3 lb./gal. x 60 min./h = 500

$$\text{gpm} = \frac{\text{btuh}}{500 \times \text{td}}$$

$$\text{gpm (45°F water)} = \frac{750{,}000 \text{ btuh}}{500 \times (60-45)} = 100 \text{ gpm}$$

$$\text{gpm (40°F water)} = \frac{750{,}000 \text{ btuh}}{500 \times (60-40)} = 75 \text{ gpm}$$

$$\text{gpm (35°F water)} = \frac{750{,}000 \text{ btuh}}{500 \times (60-35)} = 60 \text{ gpm}$$

Lower chilled water temperatures result in 25% to 40% less gpm and pump energy, as well as a savings of 15-25% of the installation cost of chilled water pumps, piping, and insulation.

Possible lower installed cost

Figure 4.4 is an example of comparing the installed cost of different components of various 400-ton systems. The installation cost of a 45°F air system is 18.8% less than the installed cost of a 55°F air system.

Fan-powered air terminal unit costs are not included in the Figure 4.4 items. Fan-powered or induction type air terminal units are essential for 45°F supply air systems. Mixing of return air with 45°F primary air is required to maintain the 4-6 air changes required for comfort.

The installed cost of fan-powered air terminal units is $0.80 per cfm. Fan-powered air terminal units would add $0.80/cfm x 108,000 cfm = $86,400 to the cost of the 45°F air system, for a total initial cost of $591,000. This is slightly less than the $599,850 initial cost of the 55°F supply air system.

Although fan-powered air terminal units are not essential for 55°F variable air volume systems, they do avoid the common VAV complaints of stuffiness and drafts. Fan-powered air terminal units provide the proper air changes at all times.

Initial Cost Comparison of 45°F and 55°F Supply Air Systems		
Item	55°F air	45°F air
	400 tons	435 tons
Refrigeration plant ($430/ton)	$173,000	$176,000
Air handlers and primary duct ($1.80/cfm)	$282,000 (157,000 cfm)	$194,000 (108,000 cfm)
Cooling coils	$28,000	$42,000
Chilled water pumps ($25/gpm)	$16,250 (654 gpm)	$13,000 (520 gpm)
Chilled water piping and insulation, 2,000 ft. 5 inch, $50/ft. 4 inch, $40/ft.	$100,000	$80,000
Total	$599,850	$505,000

(Adapted from Heating/Piping/Air Conditioning, January 1988, Robert T. Tamblyn, P.E.)

Figure 4.4

Lower operating cost for electric energy and electric demand charges

See Figure 4.5 for the comparison of electric demand for 55°F supply air and 45°F supply air systems. Both of the 400-ton centrifugal chilled water VAV systems include fan-powered air terminals. The 45°F supply air system has a 4% lower electric demand than the 55°F supply air system.

Figure 4.6 exhibits the yearly electrical energy costs of a 400-ton chilled water VAV air-conditioning system with 55°F supply air. The energy consumption is based on 1,000 full load ton-hours (flth), 12 hours of operation per day (h/d), 5 days per week (d/w), and 25 cooling weeks per year (w/y). The VAV fans operate 52 weeks per year and use 50% of the comparable constant volume yearly fan energy. The cost of electricity is averaged at $0.10 per kwh. The electric energy cost of the 562,700 kwh per year is $56,270.

Figure 4.7 displays the yearly electrical energy cost of operating the above facility as a 45°F supply air system. The yearly electric energy cost for the 519,000 kwh/yr. is $51,900. This is 8.4% or $4,370 per year less than a 55°F supply air system.

Lower life cycle cost

Figure 4.8 compares the life cycle cost of a 55°F and a 45°F supply air system. The combination of a lower installed cost and the lower operating cost make the value (as of 1990) of owning and operating the 400-ton 45°F supply air system lower than that of the 55°F supply air system.

Please note that the $86,400 cost of the fan-powered air terminal units has not been added to the cost of either system. If the $86,400 cost of the fan-powered air terminal units was added to the 45°F supply air system and not to the 55°F supply air system, the present value of the 45°F air system would be $1,033,653, which is about 4% less than the present value of the 55°F supply air system.

Comparison of Electric Demand of 55°F and 45°F Supply Air Systems		
Electrical Item	**55°F air**	**45°F air**
	400 tons	435 tons
Chiller	260 kw 0.65 kw/ton (size 73)	282 kw 0.65 kw/ton (size 78)
Cooling tower fan	20 kw	21 kw
Condenser water pump	21 kw	22 kw
Supply air fan	147.5 kw	107.5 kw
Chilled water pump	21 kw	18.2 kw
Total	469.6 kw	450.8 kw

(Adapted from Heating/Piping/Air Conditioning, January 1988, Robert T. Tamblyn, P.E.)

Figure 4.5

Possible up-front utility payment for kw saved

Many electric utilities are offering customers $200 to $500 per kw saved by the customer. The utility payment for the 45°F air system above could be:

469.5 kw − 450.8 kw = 18.7 kw saved x $500/kw = $9,350

This sum should reduce the initial cost of the 45°F supply air system in the life cycle analysis.

Figure 4.6

Yearly Electrical Energy Use of 400-Ton 55°F Supply Air VAV Chilled Water Air-Conditioning System

Chiller	260 kw x 1,000 flth =	260,000 kwh/y
Cooling tower fan	20 kw x 1,000 flth =	20,000
Condenser water pump	21 kw x 1,000 flth =	21,000
Chilled water pump	21 kw x 12 h/d x 5 d/w x 25 w/y =	31,500
Fans	147.5 kw x 12 h/d x 5 d/w x 52 w/y x 50% =	230,000
Total kwh per year		562,700

Total yearly electrical energy cost of operating a 400-ton air-conditioning system.

562,700 kwh x $0.10/kwh = $56,270 per year

(Adapted from Heating/Piping/Air Conditioning, January 1988 Robert T. Tamblyn, P.E.)

Figure 4.7

Yearly Electrical Energy Use of 400-Ton 45°F Supply Air VAV Chilled Water Air-Conditioning System

45°F Supply Air System ($0.10/kwh), 1,000 Full Load Ton-Hours, 12 Hours/Day, 25-Week Cooling Season, Fans Operate 52 Weeks/Year, VAV, Fan Energy is 50% of Constant Volume System.

Chiller	282 kw x 1,000 flth =	282,000 kwh/y
Cooling tower fan	21 kw x 1,000 flth =	21,000
Condenser water pump	22 kw x 1,000 flth =	22,000
Chilled water pump	18.2 kw x 12 h/d x 5 d/w x 25 w/y =	27,300
Fans	107.5 kw x 12 h/d x 5 d/w x 52 w/y x 50% =	167,700
Total kwh per year		519,000

Total yearly electrical energy cost of operating a 400-ton air-conditioning system.

519,000 kwh x $0.10/kwh = $51,900 per year

(Adapted from Heating/Piping/Air Conditioning, January 1988 Robert T. Tamblyn, P.E.)

Disadvantages

Quality of the duct and chilled water pipe insulation application must be improved to avoid condensation on the duct, pipe, and insulation surfaces.

Condensation occurs when 55°F air is supplied in the air distribution system, and 45°F chilled water is circulated in the piping system. Unless the quality of the insulation application is better than average, operational problems will surface with low temperature air systems. The 45°F supply air and the 35°F water are so much farther below the dewpoint temperature of the surrounding air than the 55°F supply air system, that the ceilings will become saturated with water if average insulation procedures prevail.

Provisions must be made to supply 55°F air to the space.

A 45°F supply air temperature will create problems with standard diffusers. Heavy condensation will take place on the diffusers, especially during the morning system start-up. VAV modulation of supply air to the room will reduce air circulation to an unacceptable level.

Although there are diffusers made specifically for low temperature air, most low temperature supply air systems employ fan-powered VAV terminals to deliver 55°F air to the space. The fan-powered boxes add additional maintenance in the occupied areas of the building.

Existing chillers cannot produce water temperatures less than 40°F.

Special ordered medium temperature brine units must be used to produce 35°F brine. These are available on a retrofit basis.

Volume of outside air for intermediate season cooling is reduced.

The total volume of 45°F air delivered to the space is two-thirds that of a 55°F air system. There is one-third less outside air available for "free cooling."

Life Cycle Cost of 55°F and 45°F Supply Air 400-Ton Chilled Water VAV Systems		
10% interest, 20-year period, 0.11746 capital recovery factor, no fan-powered air terminal units for either system.		
Present Value	55°F air system	45°F air system
Initial cost	$599,850	$505,400
Operating Cost		
$56,270/yr x 1/.11746	$479,057	
$51,900/yr x 1/.11746		$441,853
Present Value	$1,078,907	$947,253

Figure 4.8

Ice and Chilled Water Storage

Thermal storage is the generation and storage of a mass of cooling for future use. The stored chilled water or melted ice is circulated to the cooling coils during the peak electrical demand hours. Figure 4.9 shows the curve for a daily cooling demand. The height of each block is the cooling tonnage for that hour.

In the example in Figure 4.9, four hundred tons of cooling are needed for the one hour between 2:00 p.m. and 3:00 p.m. This represents 400 ton-hours of cooling. Between 8:00 a.m. and 9:00 a.m., 160 tons of cooling are required. That equals 160 ton-hours of cooling for that period. The total amount of ton-hours necessary to cool the building on that day

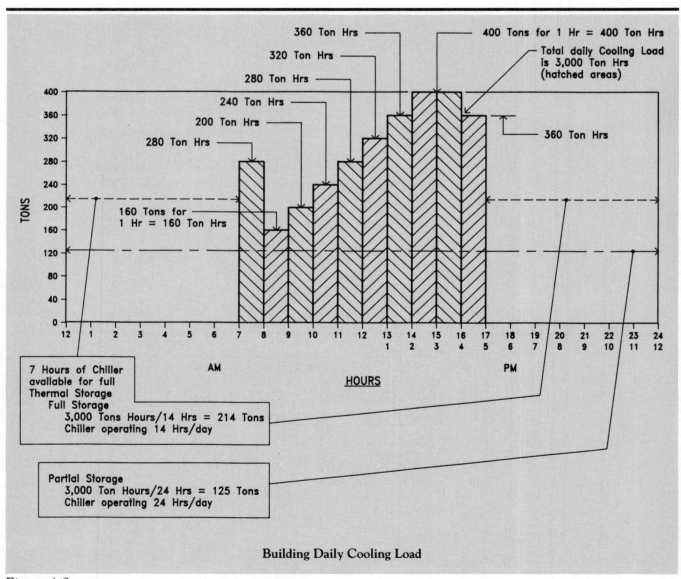

Building Daily Cooling Load

Figure 4.9

is the area under the curve. Adding all of the individual hourly ton-hours will give the total ton-hours of cooling under the curve. A total of 3,000 ton-hours is shown in Figure 4.9. All of the area outside of the cooling load profile is available for making ice or chilled water, to be used for the cooling demand that is under the curve. If the chiller were to run 14 hours per day to store the 3,000 ton-hours of cooling for use during the 10-hour cooling period, it would have to produce 3,000 ton-hours/14 hours = 214 tons.

A 400-ton cooling load can be handled by a chiller producing only 214 tons. The chiller size can be further reduced by running the chiller during the peak cooling period. The chiller will charge the ice storage unit at night and cool the load directly, along with the stored cooling. With the chiller running 24 hours instead of 14 hours, the chiller capacity will be 3,000 ton-hours/24 hours = 125 tons. A 400-ton cooling load can be handled by a chiller producing 125 tons.

Thermal storage is becoming increasingly popular because many of the utilities have difficulty meeting peak kw demand. In order to avoid building new generating plants, some electric companies are providing incentives as high as $500 per kw for every kw that can be avoided. This inducement makes thermal storage even more attractive to many building owners.

A full storage system is one in which the chiller is not operated during the period of peak electrical demand. All of the cooling load is handled by the stored water or ice. The chilled water that is generated and stored during the evening hours is circulated during the day to supply all of the cooling. Ice systems rely on the ice that is made during the non-peak evening hours to either melt into chilled water during the day, or to have the ice cool the brine that is circulating to the cooling coils during the day. The chiller produces chilled water or ice during the off-peak period, when no cooling load exists.

With a partial storage system, the chiller operates during the non-peak hours to produce chilled water or ice for storage. The chiller also operates during the peak hours to supply cooling. A smaller chiller is used in the partial storage system than in the full storage system because it operates more hours during the day to produce cooling.

Ice Storage

Ice storage is more popular than chilled water storage because of the much smaller storage space required. A pound of water can provide 144 btu's of cooling when it becomes ice. A pound of chilled water provides only 10-15 btu/lb.

Chilled water requires almost thirteen cubic feet to store one ton-hour of cooling, whereas ice needs only a little over three cubic feet to store the same cooling.

There are several methods of providing ice storage. Figure 4.10 depicts the open type storage tank, with an air pump to agitate the water. Ice builds up on the refrigerant tubes to a predetermined thickness and melts from the outside toward the refrigerant tube. Because this is an open system, extra pump head is necessary to accommodate the static head. Water treatment is also required.

Figure 4.11 shows a closed loop system that uses a chiller to produce 26°F glycol during the off-peak hours. The cold 26°F glycol is circulated in plastic tubes immersed in a container of water. The glycol freezes the water in the container into a solid block of ice. During the peak load period, the system coolant circulates in tubes and melts the ice around the tubes. The cooled glycol is circulated to the cooling coil to pick up heat.

Figure 4.12 shows the type of ice maker that uses plastic containers filled with deionized water (D.I.). During the non-peak charging period, glycol is cooled to 26°F by a chiller, and circulated through a tank containing the containers of D.I. water. The glycol freezes the water in the containers. During the discharge cycle, the warm glycol that is returned from the cooling coils gives up some of its heat to melt the ice. The cooled glycol is circulated to the cooling coils to pick up heat.

Advantages

All of the advantages of low temperature air apply to ice storage.

Uses about one-fourth the volume of chilled water storage.

For smaller size systems (under 400 tons), a lower installed cost than chilled water storage.
See Figure 4.13 for a cost comparison.

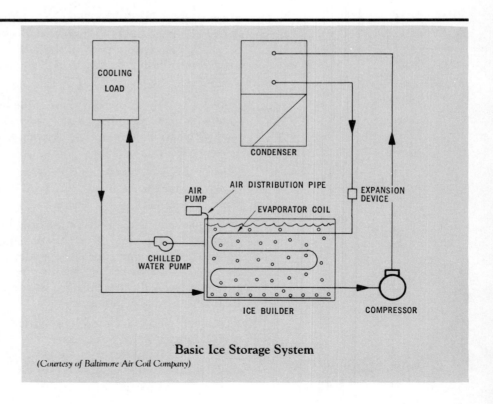

Basic Ice Storage System
(Courtesy of Baltimore Air Coil Company)

Figure 4.10

For smaller size systems (under 400 tons), lower life cycle cost than chilled water storage.

See Figure 4.14 for life cycle comparison.

Disadvantages

Same disadvantages as low temperature air.

DX systems may require field-fabricated refrigeration piping.

Higher construction cost than chilled water storage in larger sizes (over 400 tons).

Higher life cycle cost than chilled water storage in larger sizes (over 400 tons).

Requires refrigeration systems capable of operating at low temperatures. Standard condensing units and chillers cannot be used.

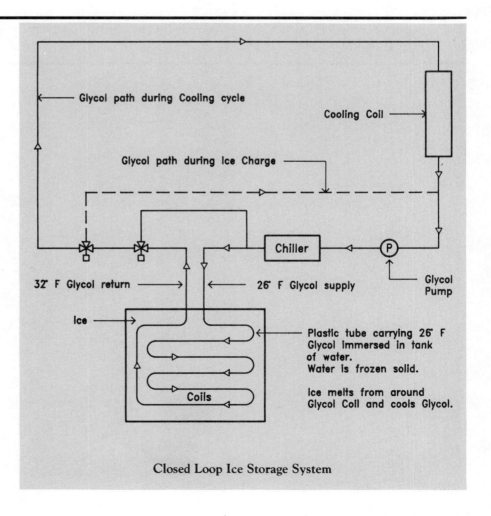

Closed Loop Ice Storage System

Figure 4.11

Chilled Water Storage

Advantages

Possible to use standard chillers for 40°F chilled water storage.

Lower installed cost over 400 tons.

Lower life cycle cost over 400 tons.

Ice storage has become a fact of HVAC life and is likely to be with us until electrical power is plentiful and inexpensive.

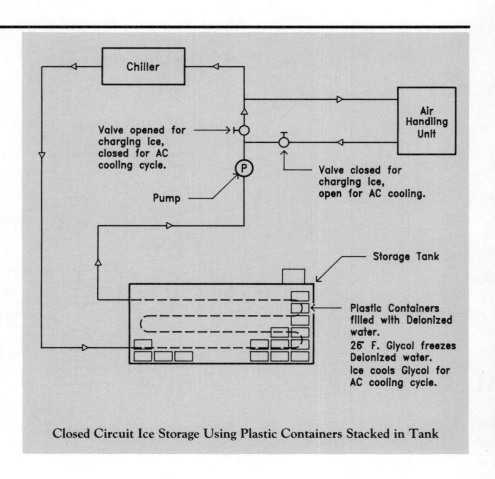

Valve opened for charging ice, closed for AC cooling cycle.

Pump

Valve closed for charging ice, open for AC cooling.

Chiller

Air Handling Unit

Storage Tank

Plastic Containers filled with Deionized water.

26° F. Glycol freezes Deionized water. Ice cools Glycol for AC cooling cycle.

Closed Circuit Ice Storage Using Plastic Containers Stacked in Tank

Figure 4.12

Comparison of Installed Cost of Chilled Water and Ice Storage Systems

	45°F Air Centrifugal Chiller 205 Tons	40°F Air 33°F Chiller 225 Tons	40°F Air DX Coil Ice Maker 225 Tons
Refrigeration cycle including condensing system	$115,000	$173,000	$160,000
Storage	$80,000 185,000 gal. uninsulated buried concrete tank (2580 ton-hr.)	$44,000 145,000 gal. insulated metal tank (2500 ton-hr.)	$96,000 45,000 gal. insulated metal tank and DX coil (2500 ton-hr.)
Storage interface	$37,000	$37,000	$37,000
Air handling system and primary ductwork between air unit and air terminal units	$162,000 (108,000 cfm)	$124,500 (83,000 cfm)	$124,500 (83,000 cfm)
Cooling coils	$41,300	$45,300	$45,300
System chilled water pumps at $10 per gpm including spare pump	$10,400 (520 gpm)	$8,200 (406 gpm)	$8,200 (406 gpm)
2000 ft. chilled water piping at $10 per ft. in.	$70,000 3-1/2 in. diameter	$60,000 3 in. average diameter	$60,000 3 in. average diameter
Total	$515,700	$492,000	$531,700

(Courtesy of Heating/Piping/Air Conditioning)

Figure 4.13

Comparison of Life Cycle Costs of Electric Demand and Items of Different Initial Cost

	45°F Air Centrifugal Chiller 40°F Water	40°F Air 33°F Chiller 35°F Water	40°F Air DX Coil Chiller 35°F Water from Ice
Electric demand cost of refrigeration cycle for 6 months per year at $6/kw	$6,840	$6,156	$7,524
Electric demand cost of fan and pump for 12 months per yr. at $6/kw	$5,062	$3,881	$3,701
Total cost of electric demand, $ per yr.	$11,902	$10,037	$11,225
Life cycle cost of electric demand, assume 10 x 1 yr. cost	$119,020	$100,370	$112,250
Initial Cost	$575,700	$492,000	$531,700
Life cycle cost	$634,720	$592,370	$643,950

(Courtesy of Heating/Piping/Air Conditioning)

Figure 4.14

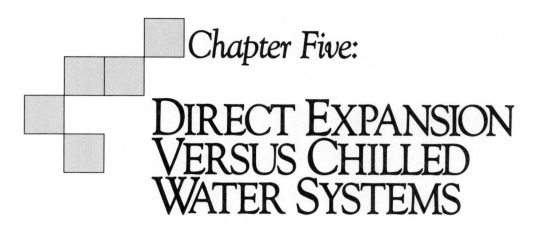

Chapter Five:

DIRECT EXPANSION VERSUS CHILLED WATER SYSTEMS

The great majority of air-conditioning installations are direct expansion (DX) systems. Almost every air conditioned home contains a DX system. DX systems are usually limited to the smaller systems, and seldom exceed 200 tons.

Almost all very large air-conditioning systems are chilled water systems. However, there are also many 10-50 ton chilled water systems.

Direct Expansion Systems

Direct expansion systems are those in which the refrigerant evaporator is in direct contact with the load. The evaporator could be a cooling coil or a heat exchanger, such as a chiller. Figure 5.1 is a schematic example of a DX system.

This analysis will address DX systems for air-conditioning employing field-fabricated refrigerant piping. Condensing units with remote condensers and/or evaporators are used with refrigerant piping connecting the compressor, condenser, cooling coils, or chillers.

Air-cooled DX systems are primarily used for small tonnage applications, such as residences, stores, small office buildings, and computer rooms.

Advantages

Low first cost
DX systems under 20 tons are 15-20% less expensive to install than chilled water systems.

Suitable for air-cooled condensers
The advantages of air-cooled condensers were addressed in Chapter 2.

Availability of equipment, parts and service technicians
There are more small DX units with reciprocating compressors installed than any other type of air-conditioning equipment. As a result, there are more technicians to service reciprocating DX, and more replacement parts available than for any other air-conditioning equipment. Service schools use reciprocating compressors and DX systems in their training programs.

Suitable for supplementary cooling systems
DX systems are ideal for supplying additional cooling to handle increased loads. These would include spaces such as conference rooms, offices where electrical equipment and people have been added to the space load, and computer rooms with added equipment. Air-cooled DX systems are simple

to install and can provide the separate temperature control required by those applications. In the same vein, DX systems are ideal for handling after-hours cooling for spaces such as private offices, box offices, and other areas that require cooling when the main plant is shut down for the evening.

Higher operating suction temperatures

Because only one heat exchange is involved to produce cold air, the compressor can operate at a higher suction pressure. Figure 5.2 shows that as the suction pressure increases, the compressor power decreases to produce the same cooling.

DX systems can operate at a 45°F suction temperature to provide 55°F supply air. Compressors used in chilled water systems must operate at an 8°F to 10°F lower suction temperature to produce 45°F chilled water.

No chilled water pump or refrigerant to water heat exchanger

DX systems employ one refrigerant to cooling load heat exchanger. The DX coil shown in Figure 5.1 directly cools the air.

Chilled water systems require two heat transfer processes. The first is a refrigerant-to-water heat exchange to produce the 45°F chilled water that will be circulated to a cooling coil. The second heat exchange occurs when heat is transferred from the air to be cooled to the chilled water in the cooling coil.

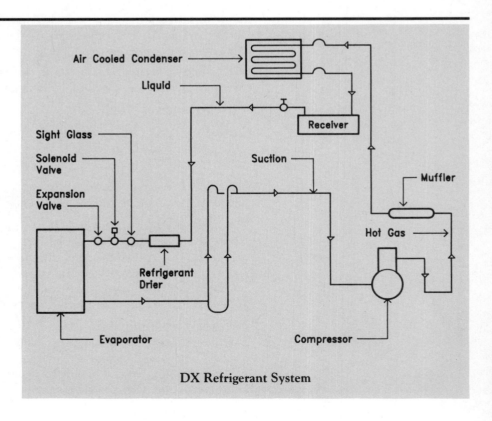

Figure 5.1

DX Refrigerant System

Suitable for tenant-operated systems

Small local DX systems can be operated and maintained by the tenants of apartments, shopping centers, and office complexes.

Building managers can assign the total responsibility and expense of operating the air-conditioning system to others.

Convenient heat pump cycle

DX systems can provide the benefits of heat pump heating using the reverse refrigerant cycle. Figure 5.3 describes an air-to-air reverse cycle heat pump.

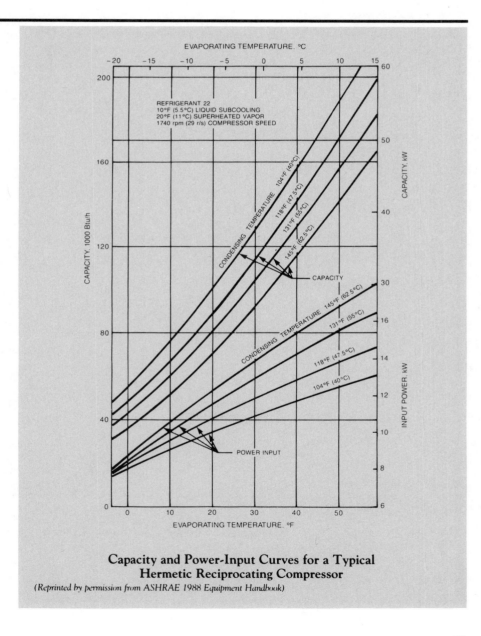

Capacity and Power-Input Curves for a Typical Hermetic Reciprocating Compressor

(Reprinted by permission from ASHRAE 1988 Equipment Handbook)

Figure 5.2

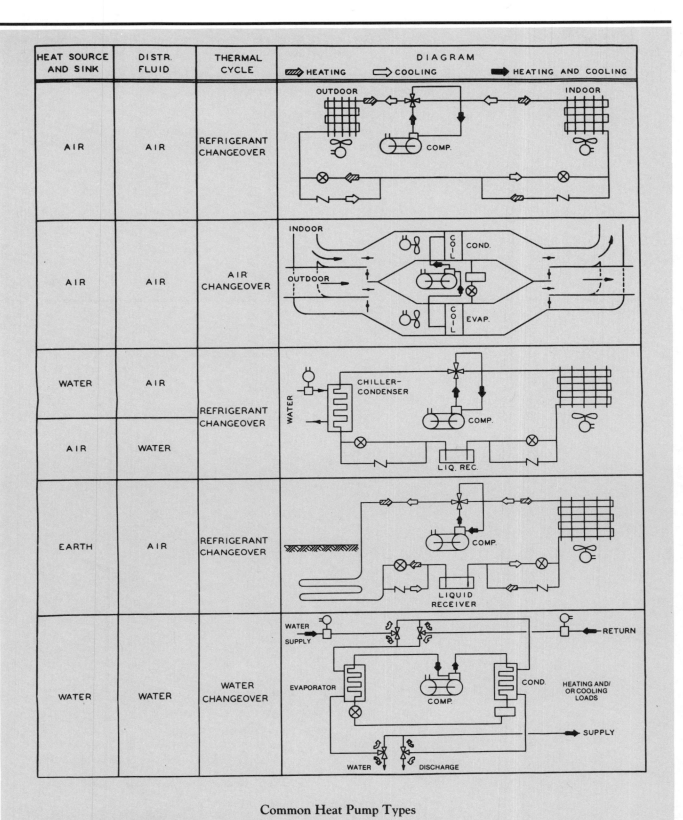

HEAT SOURCE AND SINK	DISTR. FLUID	THERMAL CYCLE	DIAGRAM
AIR	AIR	REFRIGERANT CHANGEOVER	
AIR	AIR	AIR CHANGEOVER	
WATER	AIR	REFRIGERANT CHANGEOVER	
AIR	WATER		
EARTH	AIR	REFRIGERANT CHANGEOVER	
WATER	WATER	WATER CHANGEOVER	

Common Heat Pump Types

(Reprinted by permission from ASHRAE 1987 HVAC Handbook)

Figure 5.3

Air-to-air heat pumps produce two to three times as much heat as a resistance heater with the same electrical input.

Disadvantages

Field-installed refrigeration piping
Unless well trained designers and installers of refrigerant piping are handling the project, the air-conditioning system can cause a lot of problems. The problems of field-installed refrigerant piping are explored in Chapter 28, "Refrigerant Piping."

Limited temperature control
DX systems that serve multiple areas and employ a one-room thermostat to control the compressor, cannot satisfy all areas with variable loads. DX with VAV is the alternative, but is not practical in small systems.

Little capacity for diversity
Units under ten tons do not normally offer compressor unloading capacity control devices. If one 7-1/2 ton compressor is installed, compressor short cycling occurs when the load decreases.

Limited redundancy
Redundancy is provided only with multiple compressors. The installed cost of systems with multiple DX systems will be substantially increased if multiple compressors are used.

Increased maintenance
Systems that employ multiple DX units require more maintenance. More operating equipment leads to more equipment failures for one reason or another. Equipment in different locations will be more difficult to service than equipment in a centralized plant.

Increased electrical demand
Multiple DX units produce a greater electrical demand than a central system. Individual DX units must be sized for the individual peak load of each system, and cannot take advantage of load diversity.

All zones do not normally peak at the same time. A central system can be sized for the instantaneous heat gain of the area to be served. This means that a central system can be about 25% smaller than the sum of individual units, and save 25% of the electrical demand cost.

The quantity of refrigerant in a system is limited by code.
ASHRAE 15 Safety Code For Mechanical Refrigeration limits the maximum pounds of a refrigerant that can be used in a direct DX air-conditioning system. The code stipulates that no more than 22 pounds of R-22 can be used in a direct refrigeration system for every 1,000 cubic feet of space served by that system. The code is more restrictive in that if the airflow of any enclosed space can be shut off or reduced to one-fourth of its maximum, the cubical contents of the smallest enclosed space occupied by human beings shall be used to determine the permissible quantity of refrigerant to be used in the system. Institutional and public assembly occupancies are further restricted by reducing the amount of allowable refrigerant to 50%, or 11 pounds of R-22 per 1,000 cubic feet of occupied space.

This creates a problem for 100% shut-off VAV systems. Provisions must be made for limiting the closure of all balancing and control dampers in the air distribution system to 25% of the maximum. VAV devices must be physically restrained from delivering less than 25% of their maximum design airflow.

Less suitable for very large systems

Refrigerant piping becomes economically prohibitive when long piping runs connect the compressor, condenser, and evaporator. Refrigerant pressure drops must be kept to a minimum. Higher pressure losses increase compressor power requirements in order to produce the same cooling.

Chilled water piping can be run almost any distance to circulate chilled water to the cooling coils.

Chilled Water Systems

Chilled water systems are indirect cooling systems. The refrigeration unit cools the water to, say, 45°F. A pump circulates the 45°F water to the cooling coils to pick up heat from the space, and returns the warm 55°F water to the chiller to be cooled again. Figure 5.4 shows a typical chilled water system.

Advantages

Suitable for very large systems

This item is covered under preceding "Disadvantage No. 8" in the previous "DX Systems" section.

Better control

Any number of chilled water branches serving individual zones can be connected to a chilled water circuit. Temperature control can be provided for as many rooms as desired.

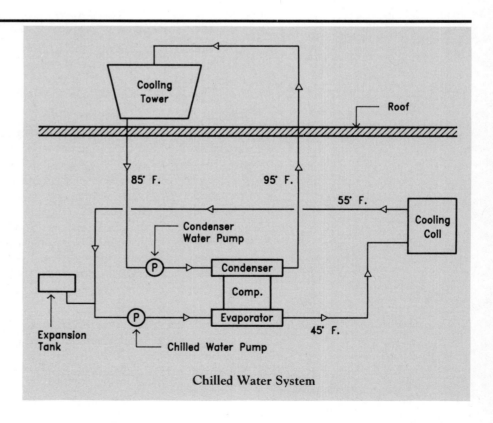

Chilled Water System

Figure 5.4

Connecting multiple controlled cooling coils to a single DX system could be difficult to implement and maintain. Figure 5.5 pictures the installers' and operators' dilemma.

Centralized maintenance

Central chilled water plants are easier to maintain than multiple DX systems. Central chilled water plants are usually located in a single area restricted for operating personnel. The maintenance staff can work in one location to service chillers, circulating pumps, water treatment, and controls.

Diversity

The chilled water plant can be sized for the instantaneous cooling load. Individual DX systems must be sized for each peak load. Since all loads do

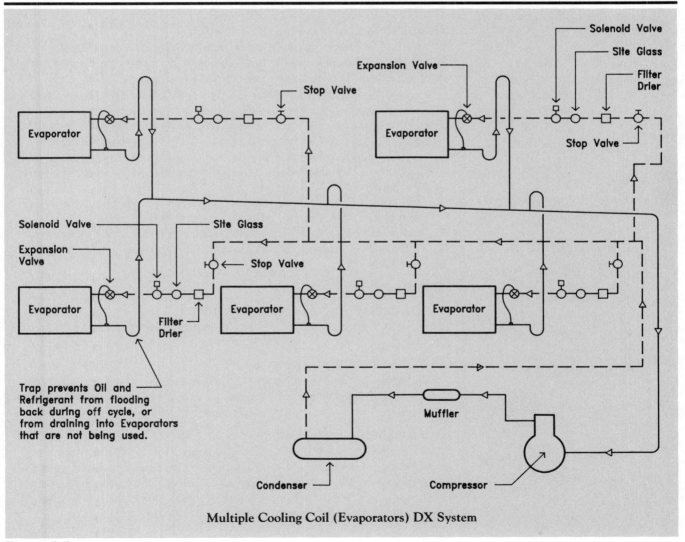

Multiple Cooling Coil (Evaporators) DX System

Figure 5.5

not peak at the same time, central chiller sizes are usually 15-30% smaller than the sum of all of the individual peak units.

Redundancy

Back-up capacity is easily accomplished with central chilled water systems. The simplest method of providing redundancy is to use two chillers in parallel, with each sized for one-half of the load. If a chiller fails on a less than design day, cooling can be provided. Less essential areas can be shut off to deliver the chilled water to more important zones. Installing three or more chillers to share the load yields better redundancy.

Flexibility

Additions and modifications to a chilled water system are relatively simple to implement. When a building is remodeled, connections into the chilled water mains can be made to serve the added or replaced cooling units. If an increase in cooling load occurs due to a renovation or addition, an additional chiller and pump can be added to the chilled water circuit.

Possible lower installed cost for larger systems

Central chilled water systems are 10-20% less expensive to install than DX systems with refrigerant piping. (See *Means Mechanical Cost Data 1990*, pages 319 and 329. Note that self-contained rooftop units are not included in the DX systems listed in that publication.)

Possible lower operating cost

Although DX condensing units are more efficient than reciprocating chillers because they operate at higher suction temperatures, central chilled water systems can produce lower operating costs because:

- Centrifugal chillers are often selected at 0.65 kw/ton, whereas 100-ton reciprocating DX units operate at 0.82 kw/ton.
- DX systems enjoy a lower cooling coil air pressure drop. A four-row DX cooling coil air pressure drop of 0.60 is lower than a six-row chilled water coil air pressure drop of 0.91 for the same coil face velocity. DX systems do not need to operate a chilled water pump at 0.57 kw/ton.
- Even with lower cooling coil air pressure drop and the operating cost of a chilled water pump, when centrifugal chillers are used in chilled water systems, the chilled water system electrical energy demand is 0.09 kw/ton less than the DX system (0.95 kw/ton − 0.86 kw/ton). See Figure 5.6.

Disadvantages

Higher initial cost for smaller systems

Chilled water systems are more costly to install than DX systems below 20 tons.

Chilled water pump required

Chilled water systems need a pump and flow control piping specialties to circulate the chilled water. DX systems do not use chilled water or a chilled water pump.

Difficulty assigning cooling costs to tenants

Electric metering of a tenant's use of the central system cannot be accomplished. Chilled water btu meters that measure the chilled water gpm and temperature rise can be installed to calculate the cooling used by a tenant who uses chilled water. It is not practical to measure cooling use from an air system at this time.

When the performance of the refrigeration piping installation contractor is known to be good, DX systems should be considered for systems up to 200 tons. In most cases, DX systems will be an attractive option in systems under 30 tons.

Central chilled water systems offer many more advantages than DX systems for all large systems. Chilled water systems can be economically justified for most systems over 50 tons.

Comparison of Central Chilled Water System and DX System Electrical Demand					
Central Chilled Plant kw/ton					
	Centrifugal .65 kw/ton chiller	Water-cooled reciprocating chiller 15-100 tons	Water-cooled reciprocating chiller 180-280 tons	Air-cooled reciprocating chiller 20-30 tons	Air-cooled reciprocating chiller 40-60 tons
Chiller	0.650	0.900	1.000	1.060	1.140
Chilled water pump	0.057	0.057	0.057	0.057	0.057
Condenser water pump	0.050	0.050	0.050	—	—
Cooling tower fan	0.040	0.040	0.040	—	—
Condenser fans	—	—	—	0.075	0.120
6-row 14 f/i cooling coil	0.060	0.060	0.060	0.060	0.060
Total kw/ton	0.857	1.107	1.207	1.252	1.377

DX Split Systems				
	Air-Cooled Units			
	3-8 tons	20-80 tons	100-120 tons	Water-cooled with tower
Compressors	1.300	1.150	1.070	0.820
Condenser fans	0.100	0.100	0.100	—
Cwp, tower fan	—	—	—	0.090
4-row DX coil 14 f/i	0.040	0.040	0.040	0.040
Total kw/ton	1.440	1.290	1210	0.950

Figure 5.6

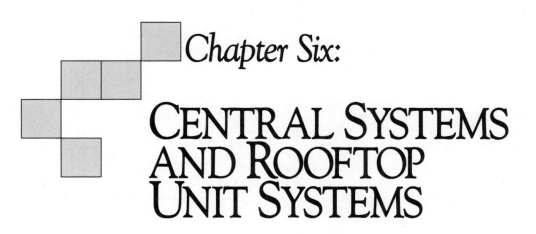

Chapter Six:

CENTRAL SYSTEMS AND ROOFTOP UNIT SYSTEMS

Central Systems

Central systems are comprised of a single chilled water or DX refrigeration plant. Multiple chillers or compressors may be included for redundancy. The refrigeration plant could be air-cooled, or water-cooled with a cooling tower. The central plant could serve air handlers for each of several zones, air handlers for each floor, or larger air handlers serving an entire building or section of the building. Figure 6.1 shows a simplified central system.

Advantages

Suitable for very large systems

Central chilled water systems employ chilled water as the cooling medium to remove heat via the various cooling coils. Water is easily transported to the coils, thereby allowing refrigeration machinery to be located remotely from the cooling coils. Central chilling plants of many thousands of tons of cooling are not uncommon. Locating the large prime movers away from the occupied space is desirable. The rooftop DX unit must, however, be located close enough to the space to minimize the ductwork requirements.

Better control

Central systems have chilled water and hot water mains circulating water to the cooling and heating coils. Any number of branches can be connected to these mains to supply cooling and heating elements with water controlled by automatic valves. It is possible to install an automatically controlled heating and cooling unit in every room in the building, if so desired. Small zones that operate at times when the central air handling unit is off can be served by a fan coil unit or small air handling unit, and can be connected to the central heating and cooling mains.

Rooftop units are usually single zone systems and rely on VAV boxes and/or reheat coils to produce the zone temperature control. Small individual zones that may require cooling when the main rooftop unit is off must be served as a separate rooftop unit.

Centralized maintenance

Because a central chilled water main supplies cooling throughout the building, the chillers can be located in a central mechanical room away from the occupied areas.

Equipment that is within a central mechanical equipment room is more likely to be well-maintained than equipment that is located on the roof, where it may have to be serviced in inclement weather.

Service contracts for rooftop systems are approximately 30% more costly than central chilled water systems.

Diversity

The central chilled water system can be sized for the instantaneous peak heat gain. Rooftop units must be sized for the peak load of the areas they serve. The fact that all areas of a building do not peak at the same time makes it possible to size the central unit for a diversity that could range between 70% and 85% of the sum of all of the peak loads.

Redundancy

Central systems provide redundancy by installing multiple chillers and circulating pumps. When a chiller is out of service, the operating chiller or chillers continue to produce cooling. The extent of redundancy depends on the number of chillers that are installed. A rooftop unit with a single compressor cannot provide cooling for the entire area served, when the compressor fails.

Flexibility

Adding new chilled water circuits to an existing chilled water system is a relatively simple operation. Modifying existing chilled water piping to suit remodeling changes is not a difficult task. An additional chiller and pump can be added to an existing chilled water system to produce the extra cooling load imposed by another structure added to the building or by increased equipment and occupancy cooling loads.

Possible lower installed cost for larger systems

For systems larger than 300 tons, the installed cost of central chilled water systems with four-pipe fan coil units is less than the installed cost of a rooftop unit system with hydronic perimeter heat. VAV rooftop unit systems enjoy a lower installed cost than central chilled water VAV systems.

Possible lower operating cost

Centrifugal chillers are more efficient systems than rooftop DX, when they operate the same number of hours. A 0.65 kw per ton centrifugal chiller

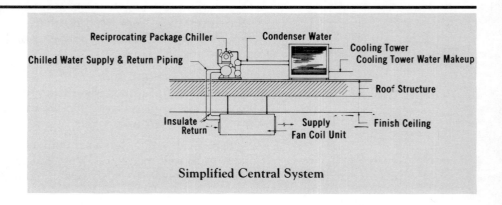

Simplified Central System

Figure 6.1

system electric demand (including a six-row cooling coil, a chilled water pump, condenser water pump and cooling tower fan) is 0.84 kw per ton, while air-cooled rooftop units (with four-row cooling coils and condenser fan) is 1.19 kw per ton. See Figure 6.2.

In areas where the electric utility cost is very high, two-stage absorption chillers are cost-effective. Central systems can employ those chillers to further reduce the energy cost of operating central chiller systems. Waste heat from cogeneration systems and incinerators can be used to fuel absorption chillers.

Disadvantages

Higher installed cost for smaller systems
The installed cost of central chilled water systems is higher than rooftop unit systems for most applications. The installed cost of two pipe central chilled water fan coil systems can be competitive with rooftop systems in sizes over 400 tons.

Possible use of building floor space
Unless the central mechanical room is located on the roof, floor space that could be used for other purposes is necessary to house the cooling plant and air handling equipment. When central systems are located on the roof, a penthouse structure must be built to house the equipment.

Comparison of Central Plant and Rooftop System Electrical Demand					
Central Plant kw/ton					
	Centrifugal .65 kw/ton chiller	Water-cooled reciprocating chiller 15-100 tons	Water-cooled reciprocating chiller 180-280 tons	Air-cooled reciprocating chiller 20-30 tons	Air-cooled reciprocating chiller 40-60 tons
Chiller	0.650	0.900	1.000	1.060	1.140
Chilled water pump	0.057	0.057	0.057	0.057	0.057
Condenser water pump	0.050	0.050	0.050	—	—
Cooling tower fan	0.040	0.040	0.040	—	—
Condenser fans	—	—	—	0.075	0.120
6-row 14f/i cooling coil	0.060	0.060	0.060	0.060	0.060
Total kw/ton	0.857	1.107	1.207	1.252	1.377
Air-Cooled Rooftop Units					
	3-18 tons	20-60 tons	80-100 tons		
Compressors	1.190	1.100	1.050		
Condenser fans	0.150	0.100	0.100		
3- or 4-row cooling coil	0.025	0.029	0.040		
Total kw/ton	1.365	1.229	1.190		

Figure 6.2

The central heating and cooling plant for a typical commercial building with 250,000 square feet of gross occupied area consumes 5,000 square feet, which is 2% of the gross occupied area. The fan room requires 15,000 square feet, which is 6% of the gross occupied area. The total construction cost of the 20,000 square feet of mechanical space at $40 per square foot is $80,000. If the 15,000 square feet of fan room space must be located within the building, the loss of rent, at $10 per square foot, would be $150,000 per year.

A fan coil unit system does not normally occupy floor space, as fan coil units can be located above the ceilings. When fan coil units are located under the windows, they do consume a little more space than direct free-standing or wall-hung radiation.

Air-cooled central systems could have a higher operating cost.
Figure 6.2 shows that air-cooled rooftop units could have a lower electric demand than air-cooled central reciprocating chillers, particularly in the 80-100 ton range.

More qualified operating personnel required
Central systems tend to be more complex than rooftop systems. Operators must be well-versed in the operation of chilled water or large DX plants and hydronic pumping systems.

More complex installation
Central systems involve the installation of large chilled water or refrigerant piping systems, with pump, water treatment and possibly cooling towers.

Rooftop Units

Rooftop units have become increasingly popular in recent years. Their popularity cannot be attributed to their good looks, but rather their saving of valuable floor space and the fact that they often generate the lowest installed system cost.

The rooftop unit can be a factory-fabricated air handling unit complete with fans, cooling coils, heating coils and filters, installed in a weatherproof cabinet. The rooftop unit can also be assembled as a complete air-conditioning unit with the air handling unit, air-cooled DX condensing unit, controls, and optional gas-fired furnace or electric heating coils, all assembled in one casing, completely wired and tested.

The rooftop unit can be provided with a prefabricated roof-mounting curb, mounted on structural steel supports above the roof, or installed on a concrete slab-on-grade.

Multiple rooftop units are often installed for zoning and to minimize ductwork. Figure 6.3 shows a rooftop system.

Advantages

Building floor space saved
Since rooftop units are located on the roof, mechanical equipment rooms are not required within the building. The rentable space is increased, or the cost of constructing mechanical rooms is avoided. See item 2 in the previous list of central system disadvantages.

Lower installed cost
Except for larger two-pipe fan coil unit systems, the installed cost of rooftop unit systems is less than that of central systems.

Possible lower operating costs in certain sizes

Figure 6.2 shows that air-cooled rooftop units have an electric demand very close to that of the water-cooled reciprocating chiller.

The seasonal electrical cost of air-cooled systems could be less than water-cooled systems with cooling towers when water treatment costs and lower refrigeration condensing temperatures (due to lower ambient air temperatures entering the condenser) become effective.

Simpler installation

The units are assembled, wired, and tested at the factory. The system can be placed into operation once the contractors set the one-piece units and provide them with power.

More singular responsibility

Because the units are completely assembled and tested at the factory, the responsibility of the construction of an entire air-conditioning plant rests with the manufacturer.

Disadvantages

Higher operating costs

The electrical energy costs of rooftop units are greater than the electrical energy costs of centrifugal chilled water systems.

The electrical energy costs of rooftop units are greater than those of water-cooled chillers. Although the kw demand of the rooftop unit is slightly higher than that of a reciprocating water-cooled chiller system, the seasonal operating cost of an air-cooled rooftop system is less than that of a central water-cooled reciprocating chilled water system. This is due to the water

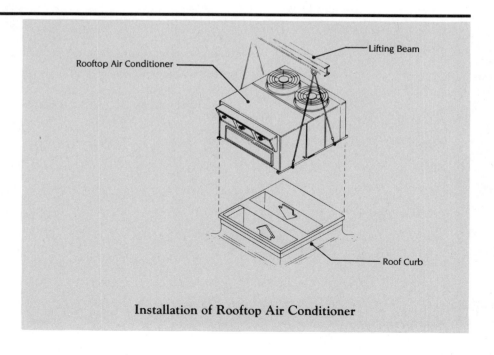

Figure 6.3

Lifting Beam

Rooftop Air Conditioner

Roof Curb

Installation of Rooftop Air Conditioner

treatment costs involved with a water-cooled system and the lower operation costs of the air-cooled system during lower ambient temperatures. Figure 6.2 compares the energy usage of the various systems.

Higher maintenance

Units that are continuously subjected to the elements will deteriorate much faster than equipment that is indoors. Rooftop units are available with heavier gauge double-walled construction, better insulation, and epoxy coatings to protect the casing, condenser fan and coil, and evaporator coil. The initial cost of these units is roughly 25% more than that of standard units.

Roof-mounted equipment is more difficult to maintain than indoor units. The absence of stair access to the roof means climbing exterior ladders while carrying tools, filters, or replacement parts to service the units. It seems as though Mother Nature insists that we must repair a failed rooftop unit during a heavy downpour, or when it is 105°F or 5°F below zero, or snowing. Some manufacturers make service vestibules for their rooftop units. It is possible to install the service vestibules so that they are accessible from inside the building.

Shorter life expectancy

Because rooftop units are continually exposed to weather, their service life is less than that of a central plant. Standard rooftop units usually have a 15-20 year service life. The custom-built units should last 30 years. It is not uncommon for central centrifugal chilled water plants to be in use for 40 years.

Structural work may be required to support the units.

An 80-ton rooftop unit with a return air fan weighs approximately 17,000 pounds.

Weatherproof ductwork may be required.

Space conditions between the ceiling and the rooftop unit may make it impossible to arrange for all of the supply and return ductwork to connect to the unit at the unit supply and return duct locations. Ductwork installed above the roof must be protected from the weather. Weatherproof ductwork is not only expensive to install, but it is difficult to prevent water intrusion.

Possible damage to the roof

Roofs can be damaged by workmen installing and maintaining the rooftop units. Every duct support and roof curb is a potential roof leak.

Rooftop units will continue to be attractive options, especially in the sizes that exclude centrifugal chillers. Owner-operated facilities may opt for the custom or semi-custom units for durability and ease of maintenance.

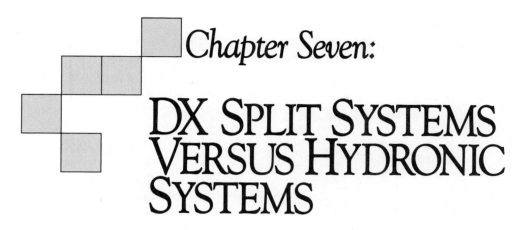

Chapter Seven:

DX SPLIT SYSTEMS VERSUS HYDRONIC SYSTEMS

Fan coil units are small air handling units with either chilled water or DX cooling coils. Where heating is required for these units, they may have either an electric or hydronic heating element.

Hydronic (Chilled Water/Hot Water) Fan Coil Systems

The fan coil units in this comparison will be the chilled water/hot water type. Chilled (or hot) water is circulated from a central plant to the fan coil units. Fan coil systems may be used in two-pipe, three-pipe, or four-pipe hydronic systems.

Two-Pipe Systems

In a two-pipe system, either chilled water or hot water is supplied through the same piping system to a single terminal coil. There is no simultaneous heating and cooling capability for the intermediate seasons. Zoning for exposure is helpful, but it need not be done on a room-by-room basis. The southern exposure should be on a separate zone because it needs cooling during the winter months. Chilled water should be available for circulation to the south zone when the rest of the building needs heating. Changeover from heating to cooling may have to be made several times a year.

Use of electric strip heaters in the fan coil units permits the circulation of chilled water during mild weather to satisfy the cooling requirements. When the temperature drops to a point where heating is needed, the electric strip heaters satisfy the space needs. Heating may be needed on the shaded lower floors of high-rise buildings, while at the same time there is a demand for cooling for the sunny upper levels.

Three-Pipe Systems

Three-pipe systems are not too common. A chilled water main and a hot water main supply the single coil and a single common line returns to the chiller or boiler. Control malfunction can make this system a nightmare.

Four-Pipe Systems

In four-pipe fan coil units, each unit contains a separate cooling and a separate heating coil. Individual chilled water lines serve the cooling coil and individual heating lines serve the heating coil.

Four-pipe fan coil systems can provide heating or cooling at any time to satisfy the individual (fan coil) requirements of any room. No summer/winter changeover is required. The operation is very simple because there are no zone valves or minimum chilled water return temperature controls involved.

The installed cost of a four-pipe system is approximately 15% higher than the installed cost of a two-pipe system.

Installation Considerations

Fan coil units can be installed:

- Under the windows—to circulate warm air over the windows during the heating season, and to deliver cooled air in summer.
- Above the ceiling—with interconnecting ductwork to distribute the air to the rooms served by the unit.
- Mounted against the ceiling or walls—to blow directly into the room.

Figure 7.1 shows the various locations of the fan coil unit.

Advantages

Better control

A thermostat and fan speed switch located in each room allow control of the room temperature and the air quantity to suit each occupant's needs. Unoccupied rooms can operate at reduced airflows and higher space temperatures to save energy.

Centralized refrigeration plant maintenance

The accessibility of a single chiller plant makes servicing easier for maintenance personnel. Multiple condensing units and air handlers create more maintenance problems.

Closer match of output to load requirements

Direct expansion systems must serve a single load. Water chillers can serve any number of individual loads. Since all loads will not peak at the same time, the central chilled water system can be sized for the simultaneous peak, rather than the sum of the individual peaks. This usually means a 25-30% reduction in equipment savings, as well as lower power demand costs.

System flexibility

Chilled water distribution systems can readily be tapped into for the addition or rearrangement of loads. Any remodeling or expansion of the facility can be accommodated by disconnecting or adding water piping. Direct expansion system modifications involve ductwork changes, condensing units, and refrigerant piping. These may require a major shutdown of the affected system.

The central chilled water loop permits shifting of loads from one location to another. When occupants are in the cafeteria, the chilled water can be automatically shifted from offices to the cafeteria units.

Redundancy

With a direct expansion system, compressor, refrigerant circuit, or air handling failure can put that entire system out of service. A central chilled water plant consists of multiple refrigeration circuits. Should one compressor fail, the second machine will serve the entire building during non-peak conditions. Should a fan coil unit fail, only the room would be affected. Spare fan coil unit fan assemblies are usually maintained in reserve.

If one chiller compressor were to go down during peak conditions, individual units can be shut off to permit full cooling where it is needed.

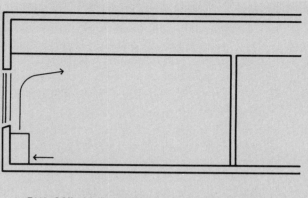

FAN COIL UNIT LOCATED UNDER THE WINDOW

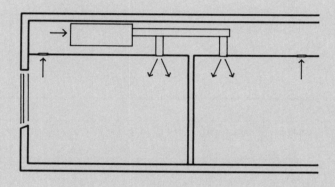

FAN COIL UNIT MOUNTED ABOVE THE CEILING
WITH DUCTWORK TO ROOMS SERVED BY THE UNIT

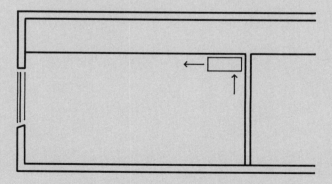

FAN COIL UNIT MOUNTED BELOW THE CEILING

Various Locations of Fan Coil Units

Figure 7.1

Longer expected life

Equipment life is extended by concentrating the load in fewer machines, which can be maintained more effectively. When the required cooling capacity exceeds 100 tons, centrifugal chillers are recommended. Centrifugal chillers, having a longer life expectancy than reciprocating chillers, could extend the chiller service life to over 30 years. When redundancy is important, two chillers should be installed.

Lower operating cost

Fan coil units with a central ventilation supply system use approximately 38% less fan energy than a split system air system. The total electric demand of an air-cooled chilled water fan coil unit system is 9% less than that of a 20-30 ton air-cooled split system (of equal capacity), and 3% less than a 40-60 ton air-cooled split system (of the same capacity). Water-cooled centrifugal chilled water fan coil unit systems enjoy 19% less electrical demand than a DX split system. See Figure 7.2 for the comparison of the electric demand of the systems.

Comparison of DX Split Systems and Central Chilled Water Fan Coil Unit Systems					
Central Plant Fan Coil Unit kw/ton					
	Centrifugal .65 kw/ton chiller	Water-cooled reciprocating chiller 15-100 tons	Water-cooled reciprocating chiller 180-280 tons	Air-cooled reciprocating chiller 20-30 tons	Air-cooled reciprocating chiller 40-60 tons
Chiller	0.650	0.900	1.000	1.060	1.140
Chilled water pump	0.057	0.057	0.057	0.057	0.057
Condenser water pump	0.050	0.050	0.050	—	—
Cooling tower fan	0.040	0.040	0.040	—	—
Condenser fans	—	—	—	0.075	0.120
Fan coil unit fans	0.135	0.135	0.135	0.135	0.135
Ventilating AHU fan	0.060	0.060	0.060	0.060	0.060
Total kw/ton	0.992	1.242	1.342	1.393	1.473
DX Split Systems					
Air-Cooled Units					
		3-8 tons	20-80 tons	100-120 tons	Water-cooled with tower
Compressors		1.300	1.150	1.070	0.820
Condenser fans		0.100	0.100	0.100	—
Cwp, tower fan		—	—	—	0.090
Air handling unit fan		0.270	0.270	0.270	0.270
Total kw/ton		1.670	1.520	1.440	1.180

Figure 7.2

More reliable operation

A central packaged air-cooled chilled water fan coil unit system is more reliable than a DX split system. The packaged air-cooled chiller is assembled and tested at the factory under controlled conditions. Refrigerant piping is brazed at the factory with dry nitrogen passing through the piping to prevent oxidation and scale from accumulating on the pipe interior. All internal wiring, including controls, is connected and tested at the factory. Field labor requires no more than connecting electrical power and water piping to the chiller. A DX split system, on the other hand, requires field-connected refrigerant piping. (See "Disadvantage No. 1" under "DX Split Systems" later in this chapter.)

Disadvantages

Possible higher initial cost

A chilled water fan coil unit system with a central ventilation supply air system has a 6% higher installed cost than a split DX system. However, a central DX system with variable volume and/or reheat will have a 10% higher installed cost than the chilled water fan coil system with central ventilation.

Ventilation air for central fan coil systems is best provided by a central ventilation system. The system is comprised of air handling units that filter, cool, or heat the outside ventilation air, and deliver it to the fan coil units.

Reduced economizer cycle cooling

The ventilation requirement of 15 cfm of outside air per person dictates that 15% of a comfort air-conditioning system's total air supply must consist of outside air. Since the ventilation system will deliver only 15% of the total air supply as outside air, the benefit of using outside air for cooling is reduced. Outside air switch-over for economizer operation is usually set at 68°F. High design wet bulb areas should have lower temperature dry bulb economizer set points. The penalty for limited cooling with outside air is dependent on the number of hours during the cooling season when the ambient conditions are below 62°F wet bulb temperature.

Maintenance in occupied areas

Since fan coil units are located within the conditioned space, maintenance must take place within spaces that are occupied. Fan coil unit maintenance involves replacing filters several times a year, and possible fan motor replacement. Locating most of the fan coil units above corridor ceilings will minimize occupant disturbance.

Extensive condensate disposal system

Each individual fan coil unit has a condensate drain pan to collect moisture that is condensed on the cooling coil. Each drain pan must be piped to an open drain. This usually involves running a condensate return main along with the chilled water piping. Special care must be taken to ensure that sufficient pitch is given to the condensate drain piping to permit gravity water flow. This is not always easy to accomplish when the fan coil units are installed close to the top of hung ceilings. Some local codes might require that an auxiliary drain pan be installed under fan coil units installed above a ceiling.

Low efficiency filters

Fan coil units are usually equipped with 1″ thick throwaway or cleanable type aluminum filters. Both filters are 10% efficient, but 1″ textile media filters are 20-30% efficient. Two-inch filters in air handling units are 20% efficient. (40% two-inch filters are available for air handling units at additional cost.)

DX Split Systems

A "split system" is one in which the condensing unit is physically separated from the DX air handling unit. Field-installed refrigerant piping transports the refrigerant between the two units. Split systems can utilize either air-cooled condensing units or water-cooled condensing units with cooling towers. Figure 7.3 shows the air-cooled split system.

The most common applications of split systems are found in residential air-conditioning where the air-cooled condensing unit is located outdoors and the air handling unit is located in the attic with connecting ductwork to the various rooms. Smaller residential systems are "single zone" systems with a single thermostat cycling the compressor. Larger split systems serving office buildings, banks, libraries, restaurants, supermarkets, department stores, etc. could be VAV systems. VAV DX systems must be equipped with capacity unloaders, hot gas bypass, and special controls to prevent icing of the DX cooling coil during reduced airflow. Split systems can be used as reverse cycle air-to-air or water-to-air heat pumps. Figure 7.4 shows the air-to-air heat pump method.

Advantages

Individual tenants can be responsible for their own system.
DX systems offer a major advantage to landlords, because the individual system is all electric and dedicated to a particular tenant. This makes it relatively simple to arrange for each tenant to pay all electric and maintenance costs for the system. When the project is to be for the use of, and operated by, the owner, this advantage does not apply.

Initial cost may be lower.
See fan coil systems "Disadvantages."

Compressor can be more energy efficient at peak load.
DX systems have one heat exchange process. Warm air passing through the cooling coil surrenders its heat to the cold refrigerant. A 20-ton air-cooled DX split system operating at 40°F refrigerant temperature, uses 1.15 kw/ton.

A water chiller must cool water to 45°F, and circulate the cold water to a cooling coil. The water extracts heat from the warm air passing through the coil. Since the water chiller compressor must operate at a lower refrigerant temperature, 35°F (and pressure) to produce 45°F water, more energy is required. A 20-ton air-cooled chiller uses 1.393 kw to produce one ton of

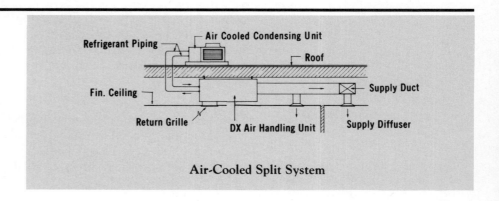

Air-Cooled Split System

Figure 7.3

Common Heat Pump Types

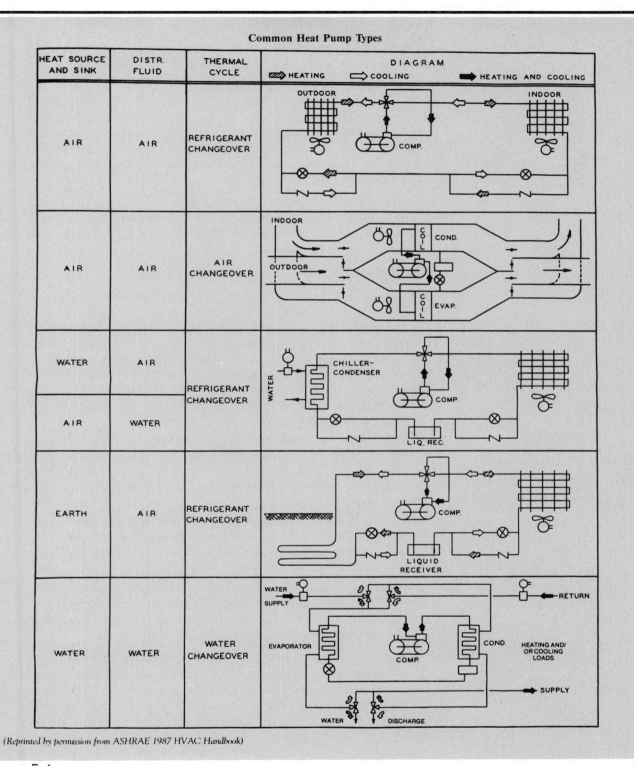

HEAT SOURCE AND SINK	DISTR. FLUID	THERMAL CYCLE	DIAGRAM
AIR	AIR	REFRIGERANT CHANGEOVER	
AIR	AIR	AIR CHANGEOVER	
WATER AIR	AIR WATER	REFRIGERANT CHANGEOVER	
EARTH	AIR	REFRIGERANT CHANGEOVER	
WATER	WATER	WATER CHANGEOVER	

(Reprinted by permission from ASHRAE 1987 HVAC Handbook)

Figure 7.4

cooling. A centrifugal chiller uses only 0.65 kw/ton. The electric demand of the various systems was previously covered. See fan coil unit advantage, "Lower Operating Cost," which shows how the total operating cost of a chilled water fan coil system is less than a DX split system.

Full economizer cycle is available.

As explained previously in "Reduced Economizer Cycle Disadvantages" in the fan coil systems section, this can be a significant factor with systems installed in a cooler climate. It is not a major consideration for systems in a warm climate or those systems with higher percentages of outside air for ventilation. These might include nursing homes, hospitals, and research laboratories.

Higher efficiency filters are possible.

Air handling units can be furnished with 40% efficient filters as a standard option. Filter efficiency can go as high as 99.9% if the high efficiency filters are installed in the supply duct.

Disadvantages

Significant field labor requirement for installation

A DX split system requires field-connected refrigerant piping. The refrigerant tubing must be sealed and nitrogen-charged to prevent dirt and moisture from entering the piping system prior to, and during installation. Dry nitrogen should be passed through the piping during brazing. Refrigerant piping must be properly sized and installed to minimize pressure drop, produce the correct refrigerant velocity at minimum load, and to permit oil return to the compressor. This may involve the design and installation of double hot gas or suction risers.

After the piping is connected, the entire system must have all air and moisture removed by a two-stage vacuum pump. The system is then charged with dry nitrogen, and evacuated again to 200 microns. Finally, the system is charged with sufficient refrigerant to permit proper operation during mild outdoor temperatures.

Failure to perform correct design and installation of refrigerant piping leads to compressor failure, due to the presence of moisture and other contaminants in the system, as well as lack of sufficient oil in the compressor.

Refrigerant expansion valves must be properly adjusted during start-up to prevent liquid refrigerant from entering the compressor. Refrigerant accessories, such as driers, sight glasses, flexible pipe vibration isolators, and pressure switches, must be specified and installed.

Code limitations on amount of refrigerant permitted in system

Codes limitations on the amount of refrigerant permitted in the entire system are based on the volume of the smallest room served by the system. A "direct" refrigeration system is one in which the DX cooling coil is located in the air circulating path to the spaces served by that system. ASHRAE's "Safety Code For Mechanical Refrigeration" permits only 11 pounds of refrigerant 22 per 1,000 cubic feet in the smallest enclosed space occupied by human beings in institutional occupancies, and 22 pounds of R-22 in other occupancies. If the airflow to any enclosed space served by the air duct system *cannot* be shut off or reduced to below one fourth of its maximum, the cubical contents of the entire space served by the air duct system may be used to determine the permissible quantity of refrigerant in the system.

Balancing dampers, fire dampers, smoke dampers, and variable air volume dampers can be shut off, placing the system in the 11 pounds of refrigerant limitation. A 10-ton DX system operates with 22 pounds of refrigerant, which exceeds the 11 pounds of refrigerant limit.

Indirect refrigeration systems (in which water or brine is cooled by the refrigerant and then circulated to the cooling coil to cool the air) are acceptable under the Safety Code for Mechanical Refrigeration, without any refrigerant limitations.

Individual room control difficult to accomplish without reheat

Since 15% of the total air supplied to the rooms is outside ventilation air, reducing the air quantity to satisfy a reduction in cooling load reduces the outside air volume to the room. This can be overcome by using air measuring stations in the outside air system. (See Chapter 23.) It is desirable to maintain a minimum of four air changes of supply air in order to maintain a feeling of air motion and to reduce odors in the room. In order to maintain acceptable temperatures in the rooms during reduced cooling load conditions, the temperature of air delivered to the room must be increased. This is accomplished in a central system by reheating the air, which increases operating costs.

A chilled water fan coil system raises the supply air temperature to the room by reducing the quantity of chilled water to the cooling coil. The supply air is cooled only as much as is necessary to satisfy the cooling load.

Increases in the cost of operation due to reheat

Energy is expended to cool the supply air to satisfy the space with the greatest cooling load. Additional energy must be expended to reheat the supply air to those spaces that do not require maximum cooling.

Diversity is lost.

A constant volume reheat system must cool the entire system air supply to a temperature that will satisfy the space with the maximum cooling needs. The system must be sized for the sum of all of the building's individual peak loads. If the sum of each space's individual maximum peak load is 100 tons, a 100-ton air handling unit and condensing unit must be installed. If the maximum instantaneous cooling load is 75 tons, the chilled water fan coil units for each space will be sized for the maximum peak cooling load for that space, but the chiller and chilled water circulating pump can be sized for 75 tons. That is the maximum cooling load that the chilled water plant must deliver at any one time.

Inactive areas difficult to serve without hot gas bypass control

Unused rooms of a facility served by a central DX system are difficult to control. Supply air to the inactive rooms can be reduced by means of an automatic damper. A reduction in airflow through the DX cooling coil will lower the refrigerant pressure and temperature. Should the temperature drop below 32°F, ice will accumulate on the cooling coil, further reducing the airflow and the refrigerant temperature, thereby producing even more ice.

Unloaders to reduce compressor capacity may not be capable of satisfying minimum load. Hot gas bypass serves as the final stage of unloading by injecting hot refrigerant gas into the cooling coil to raise the refrigerant pressure and temperature. Setting proper operation of the control and arranging for oil transport in the hot gas bypass line require special expertise.

Direct expansion split systems are not appropriate for institutional applications, without making provisions to satisfy both the refrigeration and ventilation codes. The application for DX split systems seems to fall in the residential and small commercial projects. Field installation of hot gas bypass systems should be avoided.

Chapter Eight:

BOILERS AND FURNACES

Boilers

Boilers are pressure vessels designed to generate and transfer heat to water or to make steam. Boilers are grouped into classes based on working pressure and temperature. Low pressure boilers are constructed for a maximum working pressure of 15 psi for steam and up to 160 psi and 250°F for water.

Medium and high pressure boilers are designed to operate at pressures above the low pressure maximums. When the fluid being heated is air, the unit is called a *furnace*.

The basic boiler classifications are:
- Fire Tube Boilers
- Water Tube Boilers
- Pulse
- Cast Iron Boilers
- Steel Boilers

Fire Tube Boilers

Fire tube boilers are made of steel shells with steel tube inside the shells. The water surrounds the tubes, and the hot gases from the furnace travel through the tubes and give up some of their heat to the water. These boilers are usually limited to steam pressure of 350 psig and 160 psig, and 250°F for water. Fire tube boilers are designed as scotch marine, and firebox type for use in HVAC systems. Figure 8.1 shows a packaged modified scotch marine fire tube heating boiler.

The most popular fire tube heating boiler is the packaged modified scotch marine. In this type of boiler, the combustion chamber is cylindrical, and the boiler is suitable for pressure firing. This means that the combustion chamber and flue gas passages can be above atmospheric pressure, so that no induced draft or forced draft fan is necessary to remove the flue gases. Stack or chimney height can thereby be reduced. The boilers are made in up to 35,000 MBtuh (thousands of btuh) capacity and can be oil-, gas-, or combination oil- and gas-fired. The boiler burner unit can be pre-wired and test-fired at the factory. Special units can be made for coal firing. These boilers are used for both steam and hot water heating systems.

Water Tube Boilers

Water tube boilers are composed of drums with connecting tubes. Water is contained inside the drums and tubes, and the hot furnace gases flow around the tubes to transfer their heat to the water inside the tubes.

Water tube boilers are furnished for steam or hot water up to pressures of 3,500 psi. Utilities and large facilities that employ steam turbines as prime movers use water tube boilers. Otherwise, water tube boilers are not found in common usage in HVAC systems at the present time. They were extremely popular in the coal-fired era.

Smaller packaged water tube boilers are available from a few manufacturers in capacities up to 25,000 Mbtuh. Shop-assembled water tube boilers are made up to 1,000 psi in sizes from 10,000-250,000 Mbtuh. The boilers can be equipped with integral superheaters to raise the steam temperature as much as 300°F above saturation temperature. The boilers can be pressure-fired. Figure 8.2 is an illustration.

Cast Iron Sectional Boilers

All of the pressure parts of these boilers are made of cast iron. (See Figure 8.3 for an illustration.) Water fills the hollow space within the cast iron shell. The hot furnace gases flow around the outside of the cast iron shell within the combustion chamber and flue passages. The maximum operating pressure of cast iron boilers is 15 psig for steam and 160 psig and 250°F for water. Boilers are available in sizes up to 15,000 MBH.

Cast iron sectional boilers can either be shipped in sections for job site assembly or they can be delivered pre-assembled with an insulated jacket as a one-piece unit. A major advantage of the cast iron boiler is the fact that replacement sections will fit through a standard door. This feature makes these boilers ideal for retrofit projects. They can be energized with gas, oil, combination gas and oil, or coal firing. They are designed for natural draft, forced or induced draft, or pressure firing.

Pulse Boilers

The relatively new pulse boiler is probably the most efficient boiler made today. It is made with a sealed combustion system that eliminates standby losses resulting from cool room air entering the combustion chamber and flowing through the flue passages and up the chimney. In conventional boiler systems, the room air robs the boiler of heat stored in the boiler hot water and metal surfaces. The warm chimney creates a stack effect that draws air in through the combustion chamber, flue passages, and up the a

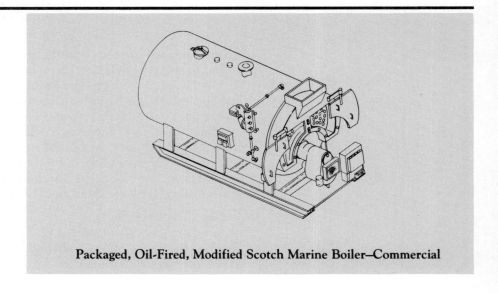

Packaged, Oil-Fired, Modified Scotch Marine Boiler—Commercial

Figure 8.1

boiler when the boiler is not firing. These problems do not occur with a pulse boiler, primarily because it does not require vast quantities of combustion air or a chimney. Figure 8.4 shows how the pulse boiler works.

The pulse boiler draws air from the outside through a small diameter PVC pipe into the combustion chamber where it is mixed with gas and ignited by a spark plug on the initial firing only. Each burn mixture that follows is ignited by the heat produced from the previous cycle. The pressure created by the combustion process forces the hot gases into the heat exchanger tubes where the heat is transferred to the water surrounding the tubes.

As the gases are cooled below their dewpoint temperature, condensation of the water vapor in the flue gases occurs; and the latent heat of vaporization is released, adding another 9% efficiency to the process. The condensate is removed by a drain at the boiler base. The low temperature exhaust is vented to the outside by a PVC pipe (chimney not required). Boiler efficiency is remarkable at over 90%. Sizes range from 47 MBH to 150 MBH.

Figure 8.2

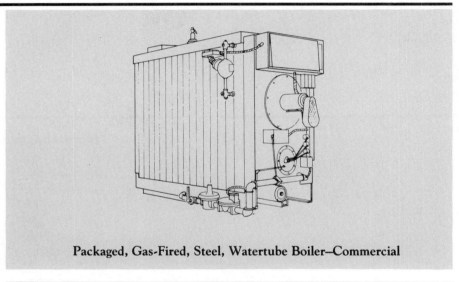

Packaged, Gas-Fired, Steel, Watertube Boiler—Commercial

Figure 8.3

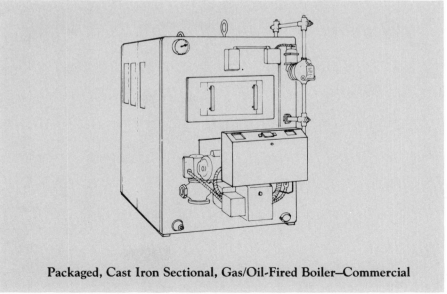

Packaged, Cast Iron Sectional, Gas/Oil-Fired Boiler—Commercial

Low Water Content Boilers

Small capacity steel and cast iron boilers are designed with very small boiler water volumes. The low boiler water content reduces the standby losses because there is not as much hot water stored in the boiler to give off heat to the surroundings and to the chimney. Boilers of 500 Mbtuh input contain as little as 12 gallons of water.

Wall-Mounted Boilers

Small wall-mounted boilers are made in sizes from 50-100 Mbtuh, designed to minimize installation costs. The boiler dimensions are 14" x 15" x 32". Boilers are shipped completely assembled with circulating pump, relief valve, expansion tank, automatic fill valve, and all controls completely wired. The sealed combustion chamber permits the flue gases to exhaust from the rear of the boiler directly to the outside, as well as providing for air intake directly into the combustion chamber, eliminating a chimney. These boilers operate at 80% efficiency.

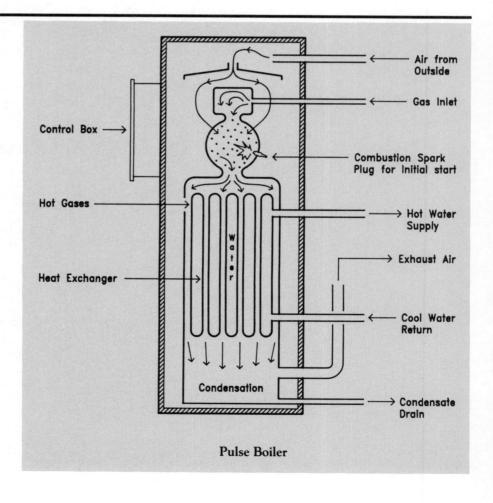

Pulse Boiler

Figure 8.4

Furnaces

Indirect-Fired Furnaces

Furnaces provide heated air for warm air heating systems. The hot gases from combustion travel through a heat exchanger to the chimney. Air is forced around the outside of the heat exchanger to accept the heat from the gases. The heated air never comes into contact with the flue gases. A fan switch starts the furnace blower when the air in the heat exchanger is sufficiently heated to a set point that will not cause an objectionable draft to the building's occupants (125°F). After the burner is shut off, the blower continues to move air through the furnace until a low air temperature is reached (90°F), at which time the blower is stopped. Continuous air circulation is also available, where the fan remains in operation and the space thermostat cycles the burner on and off.

Direct-Fired Furnaces

The direct-fired furnace is the most efficient gas-fired heating unit made. It is 100% efficient. As Figure 8.5 shows, the burner fires directly into the forced circulation air to be heated, with the flame impinging on mantles placed in the air stream. The products of combustion are mixed with the heated air and distributed to the space to be warmed. These units are only to be used with 100% outside air systems where the small amount of combustion products are diluted with the 100% outside air, so as to be within the allowable limits.

Airflow switches must be provided to shut the burner off if the outside air volume is insufficient. The direct-fired furnace is appropriate for systems utilizing large volumes of outside air. These would include garages, bus depots, research laboratories, laundries, boiler room combustion air supply, paint shops, woodworking shops, gymnasiums, etc.

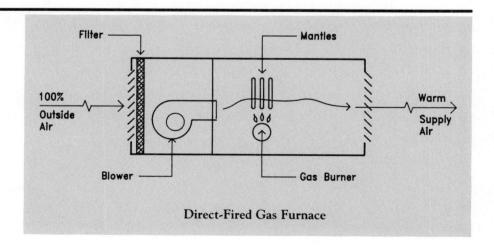

Direct-Fired Gas Furnace

Figure 8.5

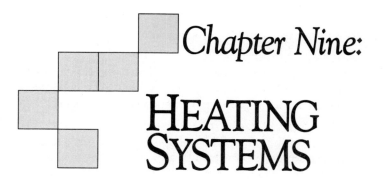

Chapter Nine:

HEATING SYSTEMS

Heating systems are installed to replace the heat that leaves a building with heated air, and to temper the outside ventilation air that is introduced into the building. The basic types of heating systems are:

- Steam Systems
- Hot Water Systems
- Radiant (Infrared or Panel) Systems
- Warm Air Systems

Steam Systems

Steam heating systems are popular in areas where district steam is available to be supplied to buildings, and where steam is used for purposes other than building heating. District steam is provided in large cities, such as New York City, Chicago and Boston. Facilities that generate steam for purposes other than heating buildings include kitchens (for preparing and warming food); domestic hot water; hospitals, prisons, and mental institutions (for sterilizers, kitchens, and laundries); cogeneration systems and electric generating plants.

District steam is usually supplied to the building at 100 psi and reduced to 15 psi for building heating use by pressure reducing valves (prv). Figure 9.1 shows a typical steam heating system.

The Steam Cycle

The two-pipe open mechanical return steam system is the most common method of steam distribution. Steam is generated in a low pressure boiler (#1 in Figure 9.1). The steam generating pressure transports the steam through the supply piping system to the heating elements (#2). The steam transfers its heat to the heating elements by releasing its latent heat of approximately 1,000 btu per pound. The condensed steam is released by the steam trap (#3) to the dry return main (#4), returned by gravity to the vented condensate pump receiver (#5), and pumped into the boiler.

Advantages

Provides steam where required for humidification; also for use in kitchens, laundries, sterilizers, domestic hot water, etc.
Steam humidification is the most common method of adding humidity to large facilities. Many kitchens utilize steam equipment for preparing and warming food. Laundries are large steam consumers, using it for heating wash water, driers, ironers, and pressing machines. Sterilizers and

autoclaves are often steam-operated as well. Facilities that use steam for any of the above are candidates for steam boilers and heating systems.

Relatively simple operation

Steam systems are relatively simple. The cycle is described above. The condensate return pump and traps are the only mechanical machines in the system, since steam pressure transports the steam to the heating elements.

Smaller pumps are required.

The condensate pump is the only pump needed in a steam system. Because the heat transfer in the steam system is latent heat, a much smaller quantity of water is circulated than in a hot water system. As Figure 9.2 illustrates, a hot water system uses 44 times the pumping energy of a steam system.

Coil freezing is less likely.

If preheat coils are piped correctly with vacuum breakers and proper trap selection, condensate will drain from the coil, and condensate freezing will not occur. (See Chapter 22.)

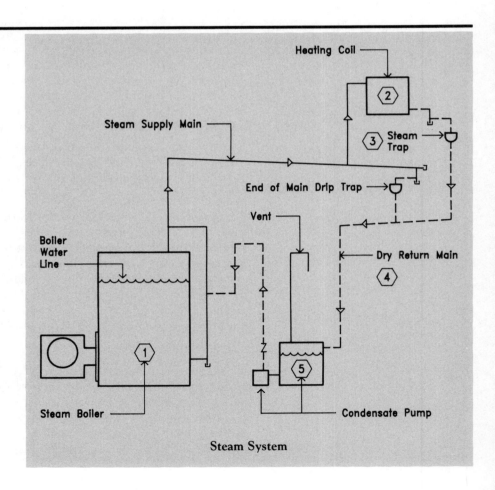

Figure 9.1

Disadvantages

Higher installed cost

The installed cost of steam systems is approximately 25% more than that of hot water heating systems. The higher initial cost of steam systems is due primarily to the cost of installing the larger steam supply piping, as compared to the smaller hot water heating piping.

An 8″ steam supply pipe and a 3″ condensate return pipe are required for a 6,000 MBtuh low pressure steam system. The same capacity hot water heating system would use 5″ supply and return heating mains. The above steam piping installed cost is 30% more than the hot water heating piping cost.

The installed cost of high pressure steam plants is more than a third higher than the cost of a hot water heating system. High pressure boilers, 300 psi valves and fittings, additional insulation, and aerators (to remove oxygen from the water entering the boiler) account for the cost difference.

Trap and water treatment maintenance required

Steam traps are self-actuating valves that close when steam tries to enter, and open when condensate enters the trap. Traps can only operate so many times before they fail. Most traps cycle ten times per minute and lose control after two or three years of operation. Traps must be checked periodically to monitor their operation, and if they are found to leak or pass steam, they must be scheduled for repair. It is not unusual to find fifty percent of the steam traps in a system performing poorly.

Steam boiler water treatment is necessary to reduce corrosion and pitting problems, to which condensate returns are extremely susceptible. In fact, most steam piping failures involve the condensate return mains.

Pumping Energy Required for Steam Versus Hot Water Heating Systems

1,000,000 btuh heating system, 15 psi steam, latent heat of evaporation is 960 btu per pound, 20 psi condensate pump discharge.

Hot water heating system, 30°F temperature difference, 60 feet head.

Steam
1,000,000 btuh/960 btu/pound = 1,042 pounds of water per hour
1,042 lbs./hr./8.33 lbs./gal. = 125 gal./hr./60 min./hr. = 2.1 gpm
20 psi x 2.3 ft/psi = 46 ft
bhp = gpm x head (ft)/3960 x eff
bhp = 2.1 gpm x 46 ft/3960 x .70 = 0.035 bhp

Hot Water
gpm = btuh/500 x t.d.
gpm = 1,000,000 btuh/500 x 30°F = 67 gpm
bhp = gpm x ft/3960 x eff
bhp = 67 x 60 ft/3960 x .70 = 1.4 bhp
1.4 bhp/0.035 bhp = Hot water heating pumps use 40 times the pump energy of a steam system condensate pump.

Figure 9.2

When steam is used for humidification or cooking, special care must be taken with the application of corrosion inhibitors. Sodium nitrite-borax or sodium chromate corrosion inhibitors must not be used. There are products available for system protection for these applications, and they must be explored with a reliable water treatment company.

Vented condensate receiver flash steam loss
When hot condensate under positive pressure enters a vented receiver, a small amount of condensate will *flash* into steam and some of the steam will be lost through the receiver vent. The vent is necessary to act as a vacuum breaker to facilitate condensate pump performance.

Sufficient vertical space must be available to accommodate steam and condensate piping pitch.
Figure 9.3 demonstrates that approximately two feet of vertical space is required to allow for the pipe pitch and trap installation.

Hot Water Systems

Figure 9.4 shows a hot water heating system that consists of a boiler or heat exchanger. The heat exchanger (#1 in Figure 9.4) heats the water to be circulated by a pump (#2) to the heating elements (#3). An expansion tank (#4) accommodates the increase of water volume in the closed system due to the increased temperature. A pressure regulating valve (#5) controls the system pressure.

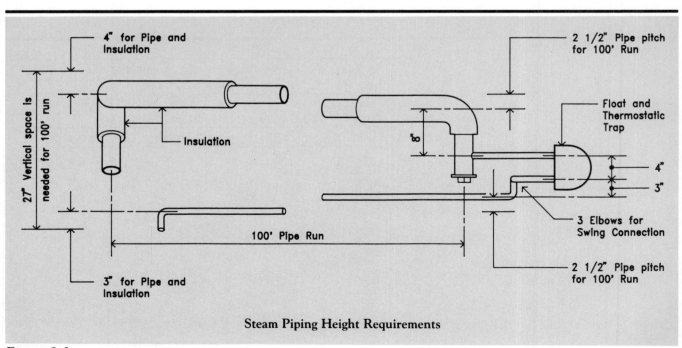

Steam Piping Height Requirements

Figure 9.3

118

Advantages

Lower installed cost
See "Higher installed cost," a disadvantage listed in the previous "Steam Systems" section.

Less corrosion in boiler and return piping
Hot water heating systems are closed systems which involve very little make-up water. Oxygen control is not as extensive as in steam systems. Water treatment should be employed to control corrosion.

Smaller pipe sizes
This has been addressed in the "Steam Systems Disadvantages" section.

Variable heating temperature for control
Except for vacuum supply systems that are now obsolete, steam systems operate at a constant steam enthalpy. The supply temperature of a hot water heating system can be reduced as the load decreases to improve control performance. (See Chapter 16, "Controls.")

Simpler piping installation
Hot water heating piping is easier to install because pipe pitch is not as pronounced and trap assemblies are not required. Vertical direction changes involve air vent valves at the high points and drain valves at the low points. As Figure 9.5 describes, steam lines must be dripped with a trap at every elevation change from horizontal to vertical.

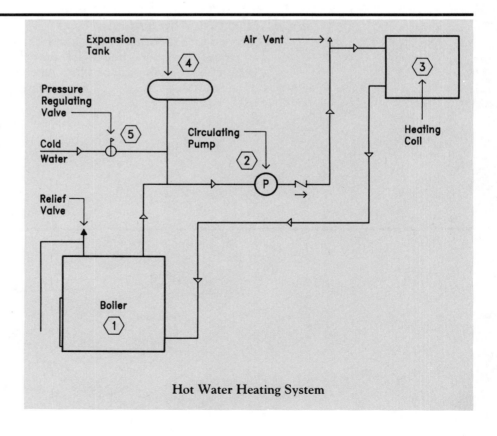

Hot Water Heating System

Figure 9.4

Suitable for glycol use

Hot water systems can be charged with a water and glycol mixture to suit the freeze protection requirements of the area. The use of glycol eliminates the water treatment system and essentially eliminates the possibility of freezing the system. Glycol slightly reduces the effective capacity of the heat exchanger, pump and expansion tank, thereby possibly requiring an increase in the size of this equipment. (See Chapter 22 "Preheat Coil Problems.")

Disadvantages

Possible freezing of pipes or coils

Hot water heating pipes are prone to freezing when circulation is impeded. Water flow in a pipe that is inside an exterior wall can be reduced or stopped by the closing of an automatic valve or air in the pipe. Once the water in the pipe is cooled below 32°F, freezing takes place and the expansion to ice could rupture the pipe or fitting wall.

Preheat coil problems and glycol freeze protection of hot water heating systems have been addressed in Chapter 22.

Greater distribution energy

The greater pumping requirements of hot water heating systems have been covered in the "Advantages" section of "Steam Systems."

Terminal Units (for Steam and Hot Water)

Terminal units are necessary to convert the steam or hot water supplied into heat for the building. Some of the common terminal units are:

- Perimeter radiation
- Fan coil units
- Cabinet, projection, and horizontal unit heaters
- Unit ventilators
- Radiant (panel) heating

Perimeter Radiation

Finned tube radiation or cast iron radiators located under windows has been the most popular method of delivering heat to buildings for some time. Figure 9.6 is an illustration of fin tube perimeter radiation.

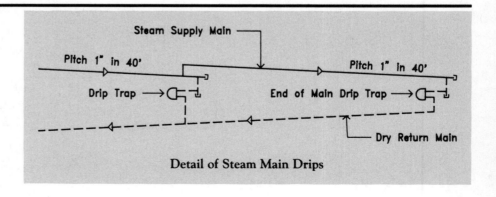

Detail of Steam Main Drips

Figure 9.5

Perimeter radiation is effective because it adds heat to the building at the location where heat leaves the building. The building's major heat loss is at the windows and exterior walls. Perimeter radiation is installed under the windows to warm the inside surface of windows and walls, producing greater indoor comfort. People become uncomfortable when their warm bodies radiate heat to cold windows and walls.

Fan Coil Units

Fan coil units are combination heating and cooling units that contain the coils, drain pan, blower, filter, and fan control, all in one cabinet. Figure 9.7 is an illustration of a small, basic type of fan coil unit.

For effective heating in cold climates, the fan coil unit should be located under the window, to warm the window by blowing warm air across it. Fan coil units located at the ceiling are effective under mild winter conditions only.

Cabinet and Unit Heaters

Cabinet unit heaters are fan coil units without cooling coils or drain pans. They produce considerably more heat than a radiator. Cabinet heaters are usually installed at locations requiring high concentrations of heating, such as entrance vestibules.

Projection and horizontal unit heaters can be either horizontal-type or projection (down-blow). They are basically enclosures with a heating coil and a propeller fan. Unit heaters also generate substantial quantities of heat for their size. Unit heaters are installed in areas where close comfort conditions are not required. It is very difficult to heat effectively using only hot air from ceiling air discharges. As Figure 9.8 demonstrates, the buoyancy of warm air keeps the warmth at the ceiling level and, therefore, the floor remains at sub-comfort zone temperatures. The high discharge

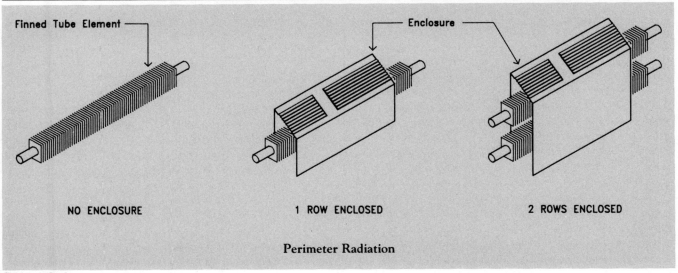

Finned Tube Element

Enclosure

NO ENCLOSURE   1 ROW ENCLOSED   2 ROWS ENCLOSED

Perimeter Radiation

Figure 9.6

velocity (such as that produced by projection or "down blow" units) that is essential to get the warm air down to the floor, may be objectionable to occupants.

Unit heaters are installed in garages, warehouses, aircraft hangers, mechanical equipment rooms, and building overhang soffits where drafts and colder floors can be tolerated. High bay areas would benefit from projection-type unit heaters.

Unit Ventilators

Unit ventilators are cabinet heaters with automatic dampers and controls for introducing outside ventilation air into the unit. Unit ventilators are usually installed against the outside wall so that a louvered outside air intake can conveniently bring outside air directly into the mixed air plenum at the bottom of the unit. The proximity of the outside air intake to the heating coil makes coil freezing a realistic hazard. Unit ventilators are installed in projects requiring large volumes of outside air. The applications for unit ventilators include school or college classrooms, auditoriums, and gymnasiums.

Radiant Heat

Radiant Panel Systems

Heat transfer by radiation takes place as wave motion similar to light waves. Heat energy is transmitted from warm surfaces to cooler bodies

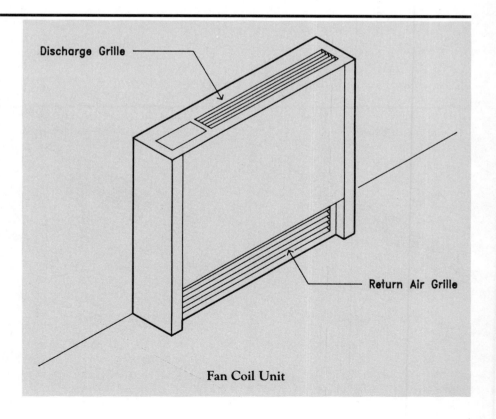

Discharge Grille

Return Air Grille

Fan Coil Unit

Figure 9.7

without the need for intervening matter. Most of us have experienced the effect of the sun's radiant heat warming us on a cool, but sunny day. When a cloud passes, we feel cold even though the air around us remains at the same temperature. The advantage of radiant heat is that it can produce comfort independent of air temperature. We can be more comfortable in a radiant-heated room at a lower temperature than we can in an air-heated room at a higher temperature.

Floor Radiant Panels

Floor radiant panels are comprised of hot water heating pipes or electric heating cables imbedded in the concrete floor slab. The principal application for this method of heating is in slab on grade construction in northern climates. The major benefit of this system is the warm floor. Because warm air rises, a large percentage of the heat produced from a floor radiant panel is convection heat.

The disadvantage of the floor radiant panel system is the thermal lag. It takes a long time to heat the mass of concrete and it takes a long time for the warm concrete to stop transmitting heat. It is not unusual for a room thermostat set at 72°F to be satisfied and stop supplying warm water to the floor panel, have the sun shine on the window, and three hours later find that the room temperature is over 80°F.

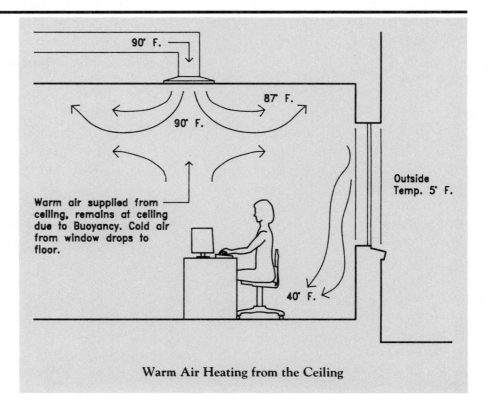

Warm Air Heating from the Ceiling

Figure 9.8

Ceiling Radiant Panels

Ceiling radiant panels are true radiant heating devices because convection does not take place at the warm ceiling. When heat is needed, the results are excellent. When ceiling radiant panels are constructed with coils imbedded in cement plaster ceilings, the same thermal lag just described in the radiant floor panel section also occurs.

Figure 9.9 shows a ceiling radiant panel constructed of a metal ceiling panel with the heating lines bonded to its top. Insulation is placed over the entire panel to contain and concentrate the heat downward. The panels are available for two by four lay-in ceiling construction. Ceiling radiant panels are installed in the ceiling near the exterior wall so that the exterior wall will be warmed by the radiant heat. At least 50% of the exterior wall heat loss should be handled by the radiant panel installed within 5' of the exterior wall. Hot water radiant ceiling panels produce about 170 btu/SF with 180°F water. The heating performance of ceiling panels is excellent because thermal lag is minimal.

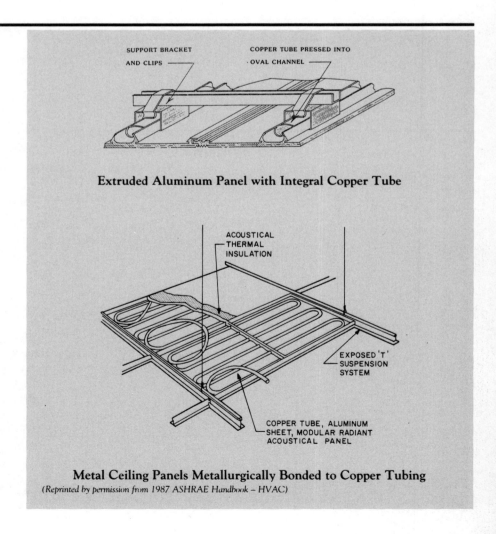

Extruded Aluminum Panel with Integral Copper Tube

Metal Ceiling Panels Metallurgically Bonded to Copper Tubing
(Reprinted by permission from 1987 ASHRAE Handbook – HVAC)

Figure 9.9

124

Electric Radiant Panels

Electric radiant panels are constructed with electric heating elements attached to the metal ceiling panels with insulation above the radiant panel. These too are made to fit into a 2' x 4' ceiling tile space. Again, electric ceiling radiant panels are installed close to the exterior wall. Electric radiant panels have a higher surface temperature than hot water radiant panels. As a result, fewer electric panels are required. One drawback is the discomfort their higher surface temperature may cause to anyone within 3' of the panels. Ceiling electric heating panels perform better when installed in 9' or higher ceilings.

The installed cost of electric ceiling panels is approximately 20% less than hydronic baseboard radiation and could be a good system choice in locations where rates are under $0.06/kwh, and demand is under $6/kw.

Gas- and Oil-Fired Radiant Panels

Fuel-fired radiant panels similar to the one shown in Figure 9.10 are ideal for buildings with high ceilings. The fuel is burned in a 4" round tube, and the hot gases are pulled through the tube by a small fan and vented to the exterior. A metal shield is installed over the tube to reflect the heat downward. Heating sections as long as 80' can be attached to the heater housing. Since these units are radiant heaters, comfort conditions can be attained at the floor level with lower room temperatures. Bus depot and repair facilities, garages, aircraft hangars, and gymnasiums can benefit from this type of heating system, which makes it possible to enjoy comfortable working conditions with 50°F room temperatures. It is important, however, not to have any objects within 4' of the heater. (Such a heater installed 2' from the top of an automobile blisters the vehicle's roof paint.)

Warm Air Systems

Warm air is produced in a furnace or by heating coils in the air distribution system. The advantages of warm air heating systems are similar to those associated with air-conditioning, and are described in the following paragraphs.

Advantages

Use of the cooling air distribution system

Many cooling systems employ an air distribution system to deliver the cool air to the conditioned space. Heating can be provided by adding a heating coil to the air handling unit or to the duct branches at the perimeter areas. Ceiling-supplied warm air systems are not efficient for northern climates; nevertheless, they are a popular choice for reasons of economy. Cooling air distribution systems that supply the air from the floor or from a low wall diffuser are suitable for heating as well as for cooling.

Humidification easily provided

Humidification can be added to an air distribution system by installing a humidifier in the air handling unit or in the ductwork.

Warm air heating from a furnace limits possible system freezing.

Heating with a furnace does not involve water, and no water is used in an all-air system (except for the humidifier). Therefore, system freezing is not possible. For this reason, warm air heating systems may be selected for northern vacation homes that are unoccupied for long periods of time in winter. Air systems have no heating pipes in exterior walls to freeze.

Disadvantages

Comfortable temperatures more difficult to attain
Air terminals must be properly located. Correct air velocities are essential for good air distribution.

Warm air supplied from high wall or ceiling air terminals cannot effectively reach the floor level to produce the desired level of comfort.
This is especially true in northern climates where cold inside surface temperatures are common.

Objectionable drafts.
If the system is not properly balanced, uncomfortable drafts will result.

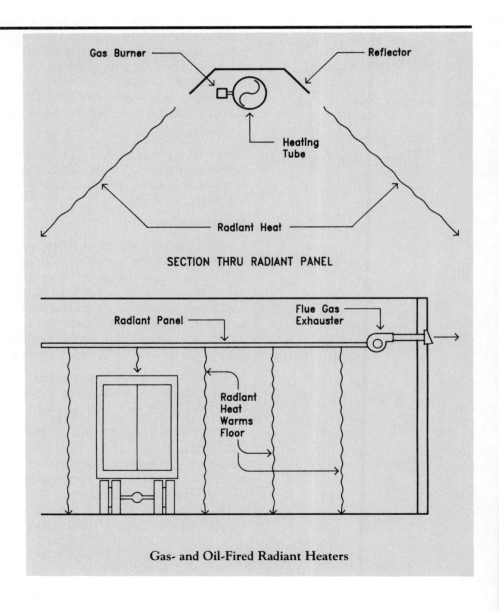

SECTION THRU RADIANT PANEL

Gas- and Oil-Fired Radiant Heaters

Figure 9.10

Dust and dirt.
The system must be regularly and properly maintained. If filters are not cleaned or replaced according to the required schedule, dust and dirt will be transmitted through the ductwork and into the living space.

Chapter Ten:

HEAT RECOVERY SYSTEMS

Heat recovery is the reclamation and use of heat energy that otherwise would be wasted. The waste heat is substituted for a portion of the heating or cooling energy that would normally be required for heating the building, outside ventilation, makeup air, or domestic hot water. Cooling energy can also be saved by transferring a portion of the warm outside ventilation air to the cool building exhaust air. It is important to be sure that the energy expended in the reclamation process does not exceed the value of the reclaimed energy over a period of time.

In this chapter, we will describe and compare four different types of heat recovery systems:

- Air-to-Air Runaround System
- Heat Pipe System
- Fixed Plate Heat Exchanger
- Waste Heat Boiler

Air-to-Air Runaround System

The air-to-air runaround system is a means of extracting a portion of the heat from the warm exhaust air and transferring that heat to the outside ventilation air for the building. The exhaust air heat source may be toilet exhaust air, general exhaust air, or heat from the boiler and incinerator flue gases entering the chimney. Figure 10.1 is an example of an air-to-air runaround system that reclaims part of an animal facility's building exhaust air, and transfers that heat to the outside ventilation air.

In a runaround heat recovery system, a pump circulates the 40% glycol solution through the piping system that connects the coils. The glycol temperature leaving the outside air coil could be below 30°F. The cold glycol could condense and freeze the moisture on the exhaust air coil. A freeze control thermostat must sense the glycol temperature leaving the outdoor air coil and divert the warm glycol leaving the exhaust air coil to mix with the cold glycol, to keep the glycol entering the exhaust air coil above 30°F.

The quality of the exhaust air must be evaluated to determine if filters should be installed to protect the coils. If filters are not installed, frequent coil cleaning will be required. Coil selections should not exceed six rows and eight fins per inch to facilitate coil cleaning.

The fan energy expended by the added resistance of the filters and coils, plus the glycol pump energy, must be calculated for an entire year's operation. This is an added operating expense that must be compared to the heat recovery energy savings. Figure 10.2 is an example of a heat recovery energy study. Runaround systems are about 50% efficient. The installed cost of a runaround system is about the same as a fixed plate heat exchanger (discussed later in this chapter) and half the installed cost of a heat pipe.

Note that the additional fan and pump energy costs of the example in Figure 10.2 exceed the preheat steam cost savings. In this case, the installation of an air-to-air runaround system is not justified.

Heat Pipe

The heat pipe is an air-to-air heat reclamation device. As Figure 10.3 shows, the refrigeration tube is continuous from the exhaust duct to the outside air duct. Each tube contains liquid refrigerant that evaporates when it absorbs heat from the warm exhaust air. The refrigerant migrates as a gas to the cold end of the tube, where it condenses and releases heat to the cold outside air. The condensed liquid then runs back down the pipe to the warm end of the tube to start another cycle. Tilt controls are available to control the heat exchange by regulating the refrigerant flow (by tilting the tubes to slow or speed up refrigerant flow).

Heat pipes are about 60% efficient, but cost about twice as much to install as runaround coils. Since the continuous heat pipe tubes must be situated in both the outside air and the exhaust ducts, the two air streams must be brought close to each other at the heat pipe. This is not usually easy to accomplish, since outside air and exhaust louvers are not normally close to each other.

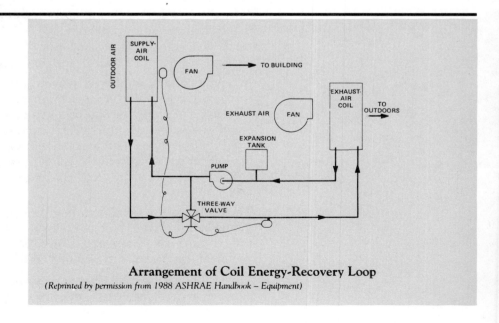

Arrangement of Coil Energy-Recovery Loop

(Reprinted by permission from 1988 ASHRAE Handbook – Equipment)

Figure 10.1

Fixed Plate Heat Exchangers

Fixed plate heat exchangers are static devices that are configured in a cross-flow arrangement. The primary heat exchange plate separates the air streams. This essentially eliminates leakage between the air streams. There is no secondary heat exchange fluid involved, so that a broad temperature range is possible. Figure 10.4 shows four types of basic plate heat exchangers.

The efficiency of the fixed plate heat exchanger is about 60%. The installed cost of the plate heat exchanger is approximately the same as the runaround coils and one-half that of a heat pipe. The duct connections for a fixed plate heat exchanger are similar to the heat pipe arrangement. The exhaust and outside air ducts must be brought to the common heat exchange device.

Waste Heat Boilers

Exhaust combustion gases are passed through waste heat boilers so that the 800°F turbine or 1,800°F incinerator flue gas can convert the water in the

---

Heat Recovery Energy Study

New York City, using an electric rate of $0.15/kwh, steam cost $10 per 1,000 #. 80,000 cfm of outside air at 0°F, 104,000 cfm exhaust air at 75°F, 8,760 hours per year. Glycol pump operates 4,000 hours per year.

Present Cost of Heating
80,000 cfm x 1.08 x 117,811 degree hours to 65°F = 10,0178,870 MBH
10,0178,870 # steam x $10/1,000 # = $100,179 per year.

Energy saved by heat recovery is 50% of present usage.
Cost of energy saved is $100,179 per year x 0.50 = $50,090 per year.

Energy penalty of runaround system.
Outside air coil static pressure, 1.0 in wg.

80,000 cfm x 1.0 in wg x 0.75 kw/hp / 6356 x 0.7 eff = 13.5 kw

Exhaust air coil static pressure, 1.72 in wg
Exhaust air coil filter static pressure, 0.80 in wg
Total additional exhaust static pressure, 2.52 in wg

104,000 cfm x 2.52 in wg x 0.75 kw/hp / 6356 x 0.7 eff. = 44 kw

Glycol pump 340 gpm, 60 ft. head, 40% glycol factor of 2.

340 gpm x 60 ft. x 2 x 0.75 kw/hp / 3969 x 0.7 eff. = 11 kw.

Total fan kw penalty = 13.5 + 44 = 57.5 kw
Cost of additional fan energy = 57.5 kw x 8,760 hrs./yr. x $0.15/kwh = $75,555 per year

Cost of pump energy
    11 kw x 4,000 hrs./yr. x $0.15/kwh = $6,600/yr.

Total electric penalty cost = $75,555 + $6,600 = $82,155/year

Energy penalty $82,155 - energy saved $50,000 = $32,155 additional cost every year

Cost of installing runaround system is $200,000.

Save $200,000 capital outlay and $32,155 per year and do nothing.

Heat recovery involving additional coil and filter pressure drops 8,760 hours per year. In high electric rate areas, may not be advisable.

Figure 10.2

waste heat boiler heat exchanger to high pressure steam. A 2,800 kw gas turbine can deliver 17,500 lbs. of 150 psi steam per hour by passing 800°F exhaust gases through the waste heat boiler. To prevent condensation, the flue gas temperature must not be allowed to drop below 300°F. Figure 10.5 shows a waste heat boiler and turbine generator arrangement.

Boiler Economizer

A boiler economizer extracts heat from the 500°F boiler flue gas and preheats the boiler feed water. Controls must be installed to divert the water flow from the economizer, to prevent extracting heat from the flue gas to the point that the flue gas dewpoint temperature is reached. Sulfurous acid will otherwise form and will attack the breeching and any other steel surface that it can locate.

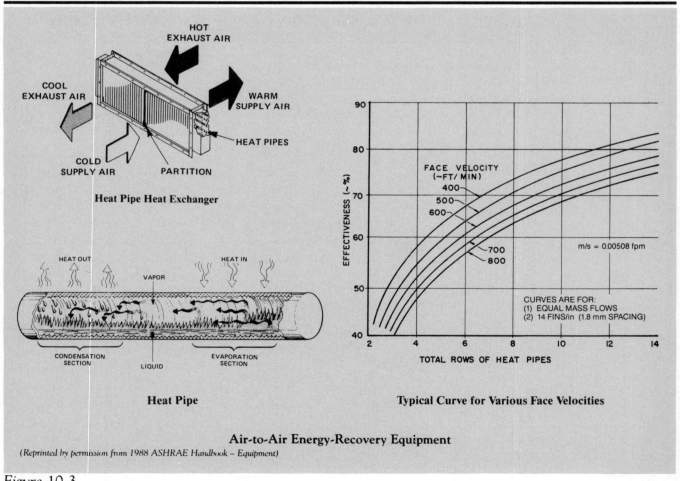

Heat Pipe Heat Exchanger

Heat Pipe

Typical Curve for Various Face Velocities

Air-to-Air Energy-Recovery Equipment

(Reprinted by permission from 1988 ASHRAE Handbook – Equipment)

Figure 10.3

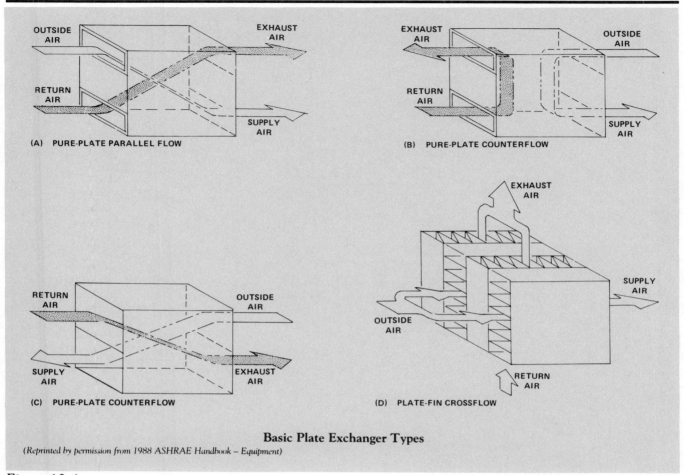

Basic Plate Exchanger Types

(Reprinted by permission from 1988 ASHRAE Handbook – Equipment)

Figure 10.4

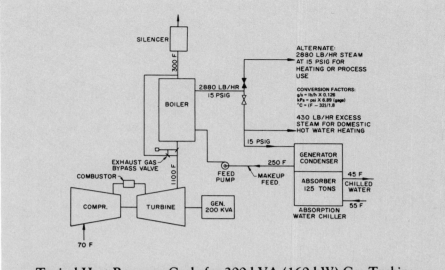

Typical Heat Recovery Cycle for 200 kVA (160 kW) Gas Turbine
(Reprinted by permission from 1987 ASHRAE Handbook – HVAC)

Figure 10.5

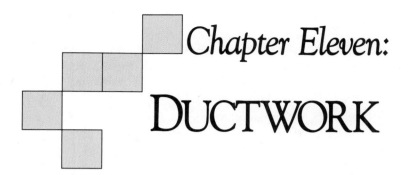

Chapter Eleven:

DUCTWORK

Ductwork represents approximately one third of the installation cost of an air-conditioning system and one half of the heartache endured by owners, engineers, and mechanical contractors. It should be noted that duct design influences the construction cost, fan operating cost, and noise generation of the duct system. The importance of properly reviewing design considerations cannot be overemphasized. Chapter 19 examines some common duct installation and operational problems, and explains how to avoid and resolve them.

Construction Cost Factors

The following considerations influence the installed cost of ductwork. Some are expressed as "rules of thumb," or guidelines that should be followed in designing a ductwork system.

Minimize the Number of Fittings.

The cost of a vaned elbow is approximately ten times the cost of five feet of straight duct. Try to eliminate as many fittings as possible. Rather than installing transitions to make a small reduction in the duct size after branch takeoffs, continue the same duct size.

Keep the Aspect Ratio Low.

Aspect ratio is the ratio of the width of a duct to its height. A 48" x 12" duct has an aspect ratio of 4. As Figure 11.1 demonstrates, the installed cost of a duct with an aspect ratio of 4 is almost 150% that of a square duct.

Use as Much Round Duct as Possible.

As Figure 11.1 points out, the installed cost of a round duct is almost 80% less than that of a duct with an aspect ratio of 4.

Try to Keep Duct Pressure Low.

Figure 11.2 shows that the installed cost of a 3"-4" wg duct class is 45% more than a duct with a 1/2"-1" wg pressure class.

Seal Ductwork.

The leakage due to unsealed ductwork means that ductwork and air handling equipment must be increased to compensate for the air loss. Avoiding and correcting duct leakage is addressed in Chapter 19, "Ductwork."

Analyze Operating Costs.

The static pressure loss of a duct system affects fan energy. A life cycle evaluation of the duct systems will determine the basis of duct design. It may be cost efficient to use smaller duct sizes to reduce the installed cost, if the cost of electricity is low enough to result in a lower life cycle cost. Figure 11.3 shows the life cycle analysis of a proposed duct system.

Note that in this example, the duct design with the highest pressure loss produced the lowest present worth. The 0.20" per 100' duct system is less costly to own and operate over a 20-year period. The 2,325 fpm duct velocity may be too high unless sound attenuators and acoustic lining were installed. Those items would, however, increase the installed and

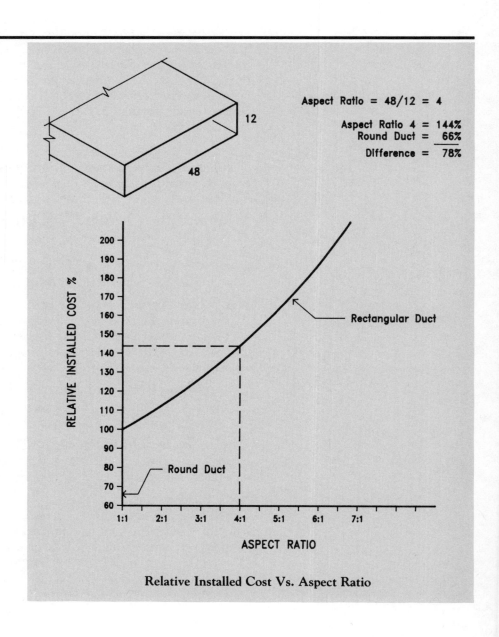

Relative Installed Cost Vs. Aspect Ratio

Figure 11.1

136

operating cost of the system. The 0.15" per 100' system also has a lower present worth, and the 2,000 fpm velocity can be tolerated with an acoustic lining.

Equal Friction Method

The equal friction method of sizing ductwork is the most common method for low and medium pressure systems. The pressure loss per foot of duct is the same for the entire system. The unit of 0.10" per 100' of duct is the most popular, but not necessarily the best basis of design.

The modified equal friction method produces a more economical installation, possibly a quieter system, and an easier system to balance. Branch ducts usually have a considerably lower pressure loss than the pressure loss of the duct system connecting to the branch duct. Rather than sizing the branch duct at the same friction loss per foot as the connecting ducts, increase the pressure loss in the branch ducts so that the pressure loss in the branch duct will be equal to the pressure loss of the connecting duct run. The branch duct sizes will be reduced to "burn up" the excess pressure rather than using only a volume damper for this purpose. The duct size in the branch should not be reduced to a point where the duct velocity will exceed the maximum recommended velocity. The example in Figure 11.4 uses the modified equal friction method.

Using Figure 11.4, the duct design procedure for the 15,000 cfm system for a bank would be as follows:

- Establish the cfm that each section of duct is to carry.
- Determine the maximum allowable main duct velocity from Figure 11.5. Note that when noise generation is the controlling factor, a maximum main duct velocity of 1,500 fpm is recommended for banks.
- Starting with the cfm at the fan, determine the area of the duct at that location, using the formula:

$$\text{area (S.F.)} = \frac{\text{cfm}}{\text{velocity (fpm)}}$$

$$\text{area} = \frac{15,000 \text{ cfm}}{1,500 \text{ fpm}} = 10 \text{ S.F.}$$

Relative Duct System Fabrication & Installation Costs for the Same Size Duct	
Duct Pressure Class	Cost Ratio
0" – 1/2"	1.0
1/2" – 1"	1.05
1" – 2"	1.15
2" – 3"	1.40
3" – 4"	1.50
4" – 6"	1.60
6" – 10"	1.80

Figure 11.2

cfm = cubic feet/min.

S.F. = square feet

fpm = feet per minute

- When the duct area is established, determine the duct dimensions to suit the available space. Assume that 24″ is the maximum possible duct height.

10 S.F. x 144 (Sq. inches per S.F.)/24″ = 60″,

Therefore, select 60″ x 24″ supply duct.

- Figure 11.6 is the chart for finding the equivalent rectangular duct size of a round duct that has the same friction loss as the rectangular duct.

Use the chart in Figure 11.6 to determine the equivalent round duct size for the 60″ x 24″ duct. The length of adjacent sides is at the top horizontal line (6-30) and at the left vertical column (6-96). The equivalent round duct is found at the intersection of the two adjacent

Life Cycle Duct System Evaluation

12,000 sq. ft. building, 20,000 cfm, system operates 3,000 hrs per yr., energy cost $0.10/kwh, installed cost of ductwork $4/lb. 12% interest, 20-year period, capital recovery factor 0.22526

Assume the duct system at 0.10″/100 ft. friction loss is 1.5 lb. per sq. ft. of building area. Duct system installed cost is 1.5 lb./sq. ft. x 12,000 sq. ft. x $4/lb. = $72,000

Duct friction in/100 ft.	Total duct s.p.	bhp	kw	Energy cost $/yr.	Inst. cost $	Main duct size	Main duct vel. fpm	Main duct perim. lin. ft.
.10	1″	4.5	3.4	1,020	72,000	72″ x 24″	1,660	16
.15	1.5″	6.8	5.1	1,530	63,000	60″ x 24″	2,000	14
.20	1.96″	8.8	6.6	1,980	56,250	52″ x 24″	2,325	12.5

bhp = cfm x s.p. / 6356 x eff

$$\left(\frac{V'}{V''}\right)^2 = \frac{P'}{P''}$$

bhp = 20,000 x 1 / 6356 x .7 = 4.5 bhp

$$\left(\frac{2000}{1660}\right)^2 = \frac{P'}{1}, P = 1.5″$$

bhp = 20,000 x 1.5 / 6356 x .7 = 6.9 bhp
20,000 x 1.96 / 6356 x .7 = 8.8 bhp

$$\left(\frac{2325}{1660}\right)^2 = \frac{P'}{1}, P = 1.96″$$

Present Worth	.10/100 ft.	.15/100 ft.	.20/100 ft.
Initial Cost	$ 72,000	$ 63,000	$ 56,250
Operating Cost			
1,020 x 1/0.22526	4,528		
1,530 x 1/0.22526		6,792	
1,980 x 1/0.22526			8,789
Present Worth	$ 76,528	$ 69,792	$ 65,039

Figure 11.3

sides. The length of the 24″ side at the top line is matched with the 60″ adjacent side at the left vertical column. The 40.4 equivalent round duct is located at the intersection of the two lines. (Over 27% more metal is required for rectangular duct than for an equivalent friction loss round duct.)

- Figure 11.7 gives the friction of air in straight round ducts for a particular cfm and velocity.

Locate 15,000 cfm at the air quantity line at the left vertical line of Figure 11.7. Follow the 15,000 cfm line horizontally until it intersects the 41″ duct diameter diagonal line. Draw a vertical line from the intersection point to the left, and read 0.065″ per 100′ on the left horizontal line. The friction loss for equal friction design for this system

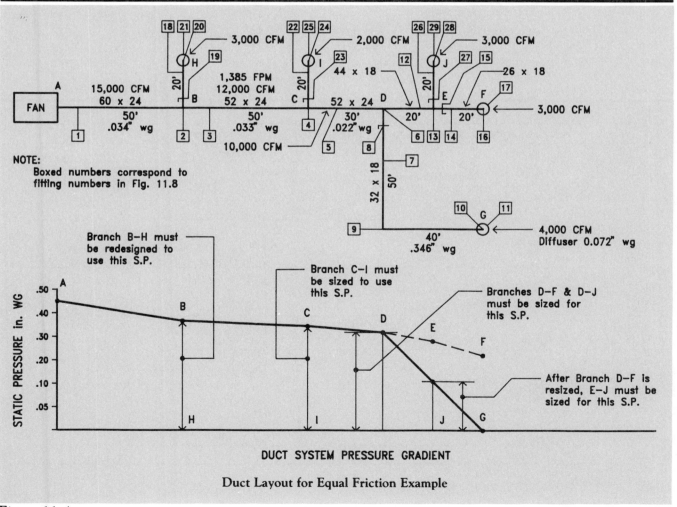

DUCT SYSTEM PRESSURE GRADIENT

Duct Layout for Equal Friction Example

Figure 11.4

is 0.065" per 100'. All of the duct in the longest run will be sized at a friction loss of 0.065" per 100'. The 1,600 fpm velocity for the 41" round duct handling is 15,000 cfm. The velocity would be different for any other duct size of equal friction loss. The velocity of any other equal friction loss rectangular duct will always be less.

The following method of sizing ducts by the equal friction method should not be used because it could lead to excessive duct velocities and the resulting high noise level.

If 15,000 cfm was plotted with 0.10" per 100' in the Figure 11.7 friction chart, the circular duct would be 39". Using Figure 11.6, the equivalent rectangular duct is found to be 56" x 24". The duct velocity in the 56" x 24" duct would be:

$$15,000 \text{ cfm}/(56" \times 24"/144) \text{ Sq. Inches/S.F.} = 1,607 \text{ fpm}$$

The 1,607 fpm duct velocity is above the 1,500 fpm maximum allowable velocity for banks.

- Determine the pressure loss in the longest run. A review of Figure 11.4 shows that the duct run from A to G would be the longest. All of the duct in this run will be sized with a 0.065" per 100' friction loss.

Using the format of Figure 11.8, the designer now determines the size, type

Application	Controlling Factor Noise Generation Main Ducts	Controlling Factor – Duct Friction			
		Main Ducts		Branch Ducts	
		Supply	Return	Supply	Return
Residences	600	1000	800	600	600
Apartments Hotel Bedrooms Hospital Bedrooms	1000	1500	1300	1200	1000
Private Offices Directors Rooms Libraries	1200	2000	1500	1600	1200
Theatres Auditoriums	800	1300	1100	1000	800
General Offices High Class Restaurants High Class Stores Banks	1500	2000	1500	1600	1200
Average Stores Cafeterias	1800	2000	1500	1600	1200
Industrial	2500	3000	1800	2200	1500

Recommended Maximum Duct Velocities for Low Velocity Systems (FPM)

(Courtesy Carrier Air Conditioning Company)

Figure 11.5

Circular Equivalents of Rectangular Duct for Equal Friction and Capacity[a]

Lgth Adj[b]	Length of One Side of Rectangular Duct (a), in.																
	4.0	4.5	5.0	5.5	6.0	6.5	7.0	7.5	8.0	9.0	10.0	11.0	12.0	13.0	14.0	15.0	16.0
3.0	3.8	4.0	4.2	4.4	4.6	4.7	4.9	5.1	5.2	5.5	5.7	6.0	6.2	6.4	6.6	6.8	7.0
3.5	4.1	4.2	4.6	4.8	5.0	5.2	5.3	5.5	5.7	6.0	6.3	6.5	6.8	7.0	7.2	7.5	7.7
4.0	4.4	4.6	4.9	5.1	5.3	5.5	5.7	5.9	6.1	6.4	6.7	7.0	7.3	7.6	7.8	8.0	8.3
4.5	4.6	4.9	5.2	5.4	5.7	5.9	6.1	6.3	6.5	6.9	7.2	7.5	7.8	8.1	8.4	8.6	8.8
5.0	4.9	5.2	5.5	5.7	6.0	6.2	6.4	6.7	6.9	7.3	7.6	8.0	8.3	8.6	8.9	9.1	9.4
5.5	5.1	5.4	5.7	6.0	6.3	6.5	6.8	7.0	7.2	7.6	8.0	8.4	8.7	9.0	9.3	9.6	9.9

Lgth Adj.[b]	Length of One Side of Rectangular Duct (a), in.																				Lgth Adj.[b]
	6	7	8	9	10	11	12	13	14	15	16	17	18	19	20	22	24	26	28	30	
6	6.6																				6
7	7.1	7.7																			7
8	7.6	8.2	8.7																		8
9	8.0	8.7	9.3	9.8																	9
10	8.4	9.1	9.8	10.4	10.9																10
11	8.8	9.5	10.2	10.9	11.5	12.0															11
12	9.1	9.9	10.7	11.3	12.0	12.6	13.1														12
13	9.5	10.3	11.1	11.8	12.4	13.1	13.7	14.2													13
14	9.8	10.8	11.4	12.2	12.9	13.5	14.2	14.7	15.3												14
15	10.1	11.0	11.8	12.6	13.3	14.0	14.6	15.3	15.8	16.4											15
16	10.4	11.3	12.2	13.0	13.7	14.4	15.1	15.7	16.4	16.9	17.5										16
17	10.7	11.6	12.5	13.4	14.1	14.9	15.6	16.2	16.8	17.4	18.0	18.6									17
18	11.0	11.9	12.9	13.7	14.5	15.3	16.0	16.7	17.3	17.9	18.5	19.1	19.7								18
19	11.2	12.2	13.2	14.1	14.9	15.7	16.4	17.1	17.8	18.4	19.0	19.6	20.2	20.8							19
20	11.5	12.6	13.5	14.4	15.2	16.0	16.8	17.5	18.2	18.9	19.5	20.1	20.7	21.3	21.9						20
22	12.0	13.0	14.1	15.0	15.9	16.8	17.6	18.3	19.1	19.8	20.4	21.1	21.7	22.3	22.9	24.0					22
24	12.4	13.5	14.6	15.6	16.5	17.4	18.3	19.1	19.9	20.6	21.3	22.0	22.7	23.3	23.9	25.1	26.2				24
26	12.8	14.0	15.1	16.2	17.1	18.1	19.0	19.8	20.6	21.4	22.1	22.9	23.5	24.2	24.9	26.1	27.3	28.4			26
28	13.2	14.5	15.6	16.7	17.7	18.7	19.6	20.5	21.3	22.1	22.9	23.7	24.4	25.1	25.8	27.1	28.3	29.5	30.6		28
30	13.6	14.9	16.1	17.2	18.3	19.3	20.2	21.1	22.0	22.9	23.7	24.4	25.2	25.9	26.6	28.0	29.3	30.5	31.7	32.8	30
32	14.0	15.3	16.5	17.7	18.8	19.8	20.8	21.8	22.7	23.5	24.4	25.2	26.0	26.7	27.5	28.9	30.2	31.5	32.7	33.9	32
34	14.4	15.7	17.0	18.2	19.3	20.4	21.4	22.4	23.3	24.2	25.1	25.9	26.7	27.5	28.3	29.7	31.0	32.4	33.7	34.9	34
36	14.7	16.1	17.4	18.6	19.8	20.9	21.9	22.9	23.9	24.8	25.7	26.6	27.4	28.2	29.0	30.5	32.0	33.3	34.6	35.9	36
38	15.0	16.5	17.8	19.0	20.2	21.4	22.4	23.5	24.5	25.4	26.4	27.2	28.1	28.9	29.8	31.3	32.8	34.2	35.6	36.8	38
40	15.3	16.8	18.2	19.5	20.7	21.8	22.9	24.0	25.0	26.0	27.0	27.9	28.8	29.6	30.5	32.1	33.6	35.1	36.4	37.8	40
42	15.6	17.1	18.5	19.9	21.1	22.3	23.4	24.5	25.6	26.6	27.6	28.5	29.4	30.3	31.2	32.8	34.4	35.9	37.3	38.7	42
44	15.9	17.5	18.9	20.3	31.5	22.7	23.9	25.0	26.1	27.1	28.1	29.1	30.0	30.9	31.8	33.5	35.1	36.7	38.1	39.5	44
46	16.2	17.8	19.3	20.6	21.9	23.2	24.4	25.5	26.6	27.7	28.7	29.7	30.6	31.6	32.5	34.2	35.9	37.4	38.9	40.4	46
48	16.5	18.1	19.6	21.0	22.3	23.6	24.8	26.0	27.1	28.2	29.2	30.2	31.2	32.2	33.1	34.9	36.6	38.2	39.7	41.2	48
50	16.8	18.4	19.9	21.4	22.7	24.0	25.2	26.4	27.6	28.7	29.8	30.8	31.8	32.8	33.7	35.5	37.2	38.9	40.5	42.0	50
52	17.1	18.7	20.2	21.7	23.1	24.4	25.7	26.9	28.0	29.2	30.3	31.3	32.3	33.3	34.3	36.2	37.9	39.6	41.2	42.8	52
54	17.3	19.0	20.6	22.0	23.5	24.8	26.1	27.3	28.5	29.7	30.8	31.8	32.9	33.9	34.9	36.8	38.6	40.3	41.9	43.5	54
56	17.6	19.3	20.9	22.4	23.8	25.2	26.5	27.7	28.9	30.1	31.2	32.3	33.4	34.4	35.4	37.4	39.2	41.0	42.7	44.3	56
58	17.8	19.5	21.2	22.7	24.2	25.5	26.9	28.2	29.4	30.6	31.7	32.8	33.9	35.0	36.0	38.0	39.8	41.6	43.3	45.0	58
60	18.1	19.8	21.5	23.0	24.5	25.9	27.3	28.6	29.8	31.0	32.2	33.3	34.4	35.5	36.5	38.5	40.4	42.3	44.0	45.7	60
62		20.1	21.7	23.3	24.8	26.3	27.6	28.9	30.2	31.5	32.6	33.8	34.9	36.0	37.1	39.1	41.0	42.9	44.7	46.4	62
64		20.3	22.0	23.6	25.1	26.6	28.0	29.3	30.6	31.9	33.1	34.3	35.4	36.5	37.6	39.6	41.6	43.5	45.3	47.1	64
66		20.6	22.3	23.9	25.5	26.9	28.4	29.7	31.0	32.3	33.5	34.7	35.9	37.0	38.1	40.2	42.2	44.1	46.0	47.7	66
68		20.8	22.6	24.2	25.8	27.3	28.7	30.1	31.4	32.7	33.9	35.2	36.3	37.5	38.6	40.7	42.8	44.7	46.6	48.4	68
70		21.1	22.8	24.5	26.1	27.6	29.1	30.4	31.8	33.1	34.4	35.6	36.8	37.9	39.1	41.2	43.3	45.3	47.2	49.0	70
72			23.1	24.8	26.4	27.9	29.4	30.8	32.2	33.5	34.8	36.0	37.2	38.4	39.5	41.7	43.8	45.8	47.8	49.6	72
74			23.3	25.1	26.7	28.2	29.7	31.2	32.5	33.9	35.2	36.4	37.7	38.8	40.0	42.2	44.4	46.4	48.4	50.3	74
76			23.6	25.3	27.0	28.5	30.0	31.5	32.9	34.3	35.6	36.8	38.1	39.3	40.5	42.7	44.9	47.0	48.9	50.9	76
78			23.8	25.6	27.3	28.8	30.4	31.8	33.3	34.6	36.0	37.2	38.5	39.7	40.9	43.2	45.4	47.5	49.5	51.4	78
80			24.1	25.8	27.5	29.1	30.7	32.2	33.6	35.0	36.3	37.6	38.9	40.2	41.4	43.7	45.9	48.0	50.1	52.0	80
82				26.1	27.8	29.4	31.0	32.5	34.0	35.4	36.7	38.0	39.3	40.6	41.8	44.1	46.4	48.5	50.6	52.6	82
84				26.4	28.1	29.7	31.3	32.8	34.3	35.7	37.1	38.4	39.7	41.0	42.2	44.6	46.9	49.0	51.1	53.2	84
86				26.6	28.3	30.0	31.6	33.1	34.6	36.1	37.4	38.8	40.1	41.4	42.6	45.0	47.3	49.6	51.7	53.7	86
88				26.9	28.6	30.3	31.9	33.4	34.9	36.4	37.8	39.2	40.5	41.8	43.1	45.5	47.8	50.0	52.2	54.3	88
90				27.1	28.9	30.6	32.2	33.8	35.3	36.7	38.2	39.5	40.9	42.2	43.5	45.9	48.3	50.5	52.7	54.8	90
92					29.1	30.8	32.5	34.1	35.6	37.1	38.5	39.9	41.3	42.6	43.9	46.4	48.7	51.0	53.2	55.3	92
96					29.6	31.4	33.0	34.7	36.2	37.7	39.2	40.6	42.0	43.3	44.7	47.2	49.6	52.0	54.2	56.4	96

(Reprinted by permission from 1989 ASHRAE Handbook Fundamentals)

Figure 11.6a

141

Circular Equivalents of Rectangular Duct for Equal Friction and Capacity[a] (Concluded)

Lgth Adj.[b]	32	34	36	38	40	42	44	46	48	50	52	56	60	64	68	72	76	80	84	88	Lgth 88 Adj.[b]
32	35.0																				32
34	36.1	37.2																			34
36	37.1	38.2	39.4																		36
38	38.1	39.3	40.4	41.5																	38
40	39.0	40.3	41.5	42.6	43.7																40
42	40.0	41.3	42.5	43.7	44.8	45.9															42
44	40.9	42.2	43.5	44.7	45.8	47.0	48.1														44
46	41.8	43.1	44.4	45.7	46.9	48.0	49.2	50.3													46
48	42.6	44.0	45.3	46.6	47.9	49.1	50.2	51.4	52.5												48
50	43.6	44.9	46.2	47.5	48.8	50.0	51.2	52.4	53.6	54.7											50
52	44.3	45.7	47.1	48.4	49.7	51.0	52.2	53.4	54.6	55.7	56.8										52
54	45.1	46.5	48.0	49.3	50.7	52.0	53.2	54.4	55.6	56.8	57.9										54
56	45.8	47.3	48.8	50.2	51.6	52.9	54.2	55.4	56.6	57.8	59.0	61.2									56
58	46.6	48.1	49.6	51.0	52.4	53.8	55.1	56.4	57.6	58.8	60.0	62.3									58
60	47.3	48.9	50.4	51.9	53.3	54.7	56.0	57.3	58.6	59.8	61.0	63.4	65.6								60
62	48.0	49.6	51.2	52.7	54.1	55.5	56.9	58.2	59.5	60.8	62.0	64.4	66.7								62
64	48.7	50.4	51.9	53.5	54.9	56.4	57.8	59.1	60.4	61.7	63.0	65.4	67.7	70.0							64
66	49.4	51.1	52.7	54.2	55.7	57.2	58.6	60.0	61.3	62.6	63.9	66.4	68.8	71.0							66
68	50.1	51.8	53.4	55.0	56.5	58.0	59.4	60.8	62.2	63.6	64.9	67.4	69.8	72.1	74.3						68
70	50.8	52.5	54.1	55.7	57.3	58.8	60.3	61.7	63.1	64.4	65.8	68.3	70.8	73.2	75.4						70
72	51.4	53.2	54.8	56.5	58.0	59.6	61.1	62.5	63.9	65.3	66.7	69.3	71.8	74.2	76.5	78.7					72
74	52.1	53.8	55.5	57.2	58.8	60.3	61.9	63.3	64.8	66.2	67.5	70.2	72.7	75.2	77.5	79.8					74
76	52.7	54.5	56.2	57.9	59.5	61.1	62.6	64.1	65.6	67.0	68.4	71.1	73.7	76.2	78.6	80.9	83.1				76
78	53.3	55.1	56.9	58.6	60.2	61.8	63.4	64.9	66.4	67.9	69.3	72.0	74.6	77.1	79.6	81.9	84.2				78
80	53.9	55.8	57.5	59.3	60.9	62.6	64.1	65.7	67.2	68.7	70.1	72.9	75.4	78.1	80.6	82.9	85.2	87.5			80
82	54.6	56.4	58.2	59.9	61.6	63.3	64.9	66.5	68.0	69.5	70.9	73.7	76.4	79.0	81.5	84.0	86.3	88.5			82
84	55.1	57.0	58.8	60.6	62.3	64.0	65.6	67.2	68.7	70.3	71.7	74.6	77.3	80.0	82.5	85.0	87.3	89.6	91.8		84
86	55.7	57.6	59.4	61.2	63.0	64.7	66.3	67.9	69.5	71.0	72.5	75.4	78.2	80.9	83.5	85.9	88.3	90.7	92.9		86
88	56.3	58.2	60.1	61.9	63.6	65.4	67.0	68.7	70.2	71.8	73.3	76.3	79.1	81.8	84.4	86.9	89.3	91.7	94.0	96.2	88
90	56.8	58.8	60.7	62.5	64.3	66.0	67.7	69.4	71.0	72.6	74.1	77.1	79.9	82.7	85.3	87.9	90.3	92.7	95.0	97.3	90
92	57.4	59.3	61.3	63.1	64.9	66.7	68.4	70.1	71.7	73.3	74.9	77.9	80.8	83.5	86.2	88.8	91.3	93.7	96.1	98.4	92
94	57.9	59.9	61.9	63.7	65.6	67.3	69.1	70.8	72.4	74.0	75.6	78.7	81.6	84.4	87.1	89.7	92.3	94.7	97.1	99.4	94
96	58.4	60.5	62.4	64.3	66.2	68.0	69.7	71.5	73.1	74.8	76.3	79.4	82.4	85.3	88.0	90.7	93.2	95.7	98.1	100.5	96

Length of One Side of Rectangular Duct (a), in.

[a] Table based on $D_e = 1.30 (ab)^{0.625}/(a + b)^{0.25}$
[b] Length of adjacent side of rectangular duct (b), in.

(Reprinted by permission from 1989 ASHRAE Handbook Fundamentals)

Figure 11.6b

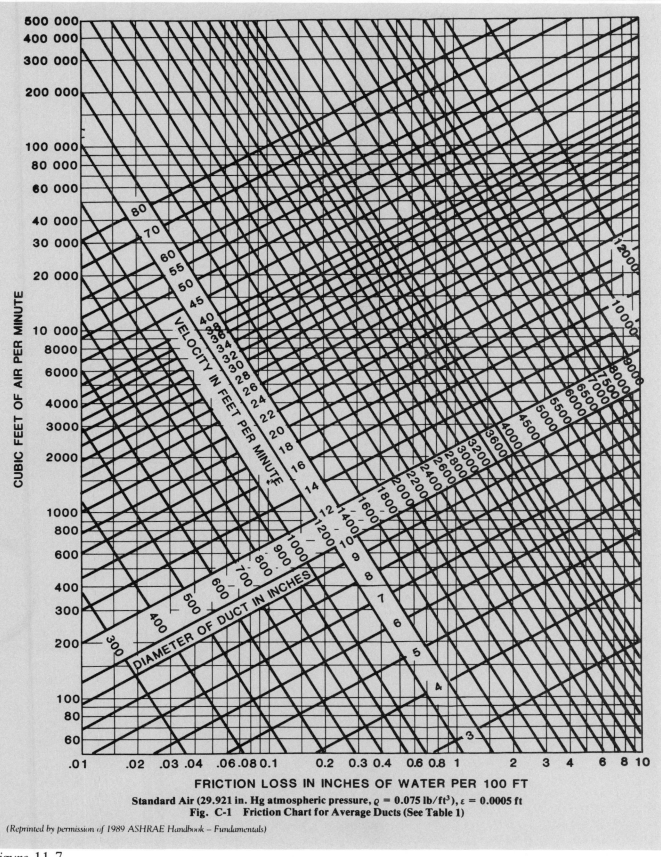

Standard Air (29.921 in. Hg atmospheric pressure, $\varrho = 0.075$ lb/ft³), $\varepsilon = 0.0005$ ft
Fig. C-1   Friction Chart for Average Ducts (See Table 1)

Figure 11.7

Duct Calculation Form

Node	No.	Fitting No.	Type of Fitting	Air-flow cfm	Duct Size	Vel fpm	Vel Pres	Duct L'th ft	Fitting Loss Coef	Duct P.D. ft/100	Tot P.D. in	Sect P.D. in
A-B	1	–	Duct	15,000	60x24	1500	.14	50	–	.065	.033	
	2	5-23	T,main	15,000	–	1500	.14	–	.01	–	.001	.034
B-C	3	–	Duct	12,000	52x24	1385	.12	50	–	.065	.033	
	4	5-23	T,main	12,000	–	1385	.12	–	0	–	0	.033
C-D	5	–	Duct	10,000	52x24	1300	.11	30	–	.065	.020	
	6	5-23	T,main	10,000	–	1300	.11	–	.02	–	.002	.022
D-G	7	–	Duct	4,000	32x18	1000	.06	90	–	.065	.059	
	6	5-27	T,br	4,000	32x18	1000	.06	–	.66	–	.040	
	8	6-2	Dpr	4,000	32x18	1000	.06	–	.04	–	.002	
	9	3-8	Elbow	4,000	32x18	1000	.06	–	.38	–	.023	
	10	3-6	Tap	4,000	24x24	1000	.06	–	1.20	–	.072	
	11	mfg	Diff	4,000	24x24	–	–	–	–	–	.150	.346
D-E	12	–	Duct	6,000	44x18	1090	.08	40	–	.065	.025	
	13	5-23	T,main	6,000	–	1090	.08	–	.02		.002	.027
E-F	14	–	Duct	3,000	26x18	923	.06	20	–	.065	.013	
	15	6-2	Dpr	3,000	26x18	923	.06	–	.04	–	.002	
	16	3-6	Tap	3,000	24x24	923	.06	–	1.20	–	.072	
	17	mfg	Diff	3,000	24x24	–	–	–	–	–	.015	.102
B-H	2	5-27	T,br	3,000	26x18	923	.06	–	.72	–	.043	
	18	–	Duct	3,000	26x18	923	.06	20	–	.065	.013	
	19	6-2	Dpr	3,000	26x18	923	.06	–	.04	–	.002	
	20	3-6	Tap	3,000	24x24	923	.06	–	1.20	–	.072	
	21	mfg	Diff	3,000	24x24	923	.06	–	–	–	.015	.146
C-I	4	5-27	T,br	2,000	24x14	860	.05	–	.75	–	.0389	
	22	–	Duct	2,000	24x14	860	.05	20	–	.065	.013	
	23	6-2	Dpr	2,000	24x14	860	.05	–	.04	–	.002	
	24	3-6	Tap	2,000	24x24	860	.05	–	1.20	–	.060	
	25	mfg	Diff	2,000	24x24	860	–	–	–	–	.015	.128
E-J	13	5-27	T,br	3,000	26x18	923	.06	–	.72	–	.043	
	26	–	Duct	3,000	26x18	923	.06	20	–	.065	.013	
	27	6-2	Dpr	3,000	26x18	923	.06	–	.04	–	.002	
	28	3-6	Tap	3,000	24x24	923	.06	–	1.20	–	.072	
	29	mfg	Diff	3,000	24x24	923	.06	–	–	–	.015	.145

System Pressure Loss = (A-B) .022 + (B-C) .033 + (C-D) .022 + (D-G) .346 = .435"

Figure 11.8

of fitting, cfm, equivalent duct size, velocity, length of duct, fitting coefficient duct pressure drop, total pressure drop, and sectional pressure drop for each duct section.

The dynamic losses of fittings are measured by the number of velocity heads of pressure loss the fitting consumes. A velocity head is the velocity pressure that corresponds to the duct velocity. Figure 11.9 gives the velocity pressure for the velocities between 300 fpm and 9,000 fpm.

Figure 11.10 provides the loss coefficients of various common duct fittings. SMACNA and ASHRAE conducted research to determine the pressure losses of the fittings.

The velocity pressure for 1,500 fpm velocity is 0.14″ wg of water. The number of velocity pressures that a fitting expends in pressure loss is called *Coefficient C*. Fitting 5-27 in Figure 11.10 is a tee, diverging, 45°F entry, rectangular main and tap. The branch coefficient C is the number of velocity pressures that are lost by the air flowing into the branch duct. Qb/Qc is the ratio of the branch cfm to the common or inlet duct cfm. In the duct layout in Figure 11.4 and the Figure 11.8 duct sizing form, the branch at B ratio is 3,000 cfm/15,000 cfm or 0.20. Vb/Vc is the velocity ratio of the branch and the common duct. For fitting B, it is 923 fpm/1,500 fpm, or 0.62. Reading Qb/Qc of 0.20 at the top of 5-27 and Vb/Vc of 0.62 at the left vertical column, the coefficient C is found to be 0.72. The velocity pressure of 1,500 fpm is 0.14″, so multiplying the velocity pressure of 0.14″ by the coefficient C of 0.72 gives a pressure loss for the branch at B of 0.10 in wg.

Note that tee fittings can produce a pressure loss in the main duct as well. Fitting 5-29 in Figure 11.10 is a tee, diverging, rectangular main and tap with damper. The loss in the main duct is found from fitting 5-31. A higher velocity in the branch duct is common in order to balance the branch duct with the following main duct loss. With a branch duct velocity of 1,600 fpm and a main duct velocity of 1,150 fpm, the splitter damper fitting produces a main duct loss of 0.27 x 0.08″ = 0.022″ s.p. If there were 50 branch connections from a main duct, the pressure loss due to the splitter damper fittings would be 50 x 0.022″ = 1.1″ s.p. The substantial 1.1″ pressure loss in the main duct is almost never addressed in fan calculations, even though it represents 1.1″/.065″/100′, or the friction loss of 1,692 feet of straight duct. Branch ducts should not be taken from a main duct with a damper connected to the main duct as a "splitter damper." Branch connections should be of the 45°F fitting 5-27 with a damper installed in the branch duct.

Sizing Branch Ducts

The pressure loss from B-G is 0.401″ wg; therefore, the pressure loss from B-H must be 0.401″ wg. The duct pressure at B is 0.401″ wg, and the pressure at the ends of both G and H is 0. The diffuser pressure loss of 0.150″ and the tap loss of 0.072″ are at point H. Their sum plus the tee branch loss of 0.10″ at B make a total loss of 0.372″ in B-H except for the 20′ of duct. 0.401″ - 0.372″ - 0.79″ wg. To determine the friction loss of the B-H duct, we must find out what loss per 100′ for a 20′ long duct will burn up the 0.079″ wg.

20′ x f/100′ = 0.079″

f = 0.395″/100′

Using Figure 11.7 with a friction loss of 0.395″/100′ and 3,000 cfm, we find a 16″ round duct, which is equivalent to a 24″ x 10″ duct. The 24″ x 10″

Velocity, fpm	Velocity Pressure, in. H_2O	Velocity, fpm	Velocity Pressure, in. H_2O	Velocity, fpm	Velocity Pressure, in. H_2O	Velocity, fpm	Velocity Pressure, in. H_2O	Velocity, fpm	Velocity Pressure, in. H_2O
300	0.01	2050	0.26	3800	0.90	5550	1.92	7300	3.32
350	0.01	2100	0.27	3850	0.92	5600	1.95	7350	3.37
400	0.01	2150	0.29	3900	0.95	5650	1.99	7400	3.41
450	0.01	2200	0.30	3950	0.97	5700	2.02	7450	3.46
500	0.02	2250	0.32	4000	1.00	5750	2.06	7500	3.51
550	0.02	2300	0.33	4050	1.02	5800	2.10	7550	3.55
600	0.02	2350	0.34	4100	1.05	5850	2.13	7600	3.60
650	0.03	2400	0.36	4150	1.07	5900	2.17	7650	3.65
700	0.03	2450	0.37	4200	1.10	5950	2.21	7700	3.70
750	0.04	2500	0.39	4250	1.13	6000	2.24	7750	3.74
800	0.04	2550	0.41	4300	1.15	6050	2.28	7800	3.79
850	0.05	2600	0.42	4350	1.18	6100	2.32	7850	3.84
900	0.05	2650	0.44	4400	1.21	6150	2.36	7900	3.89
950	0.06	2700	0.45	4450	1.23	6200	2.40	7950	3.94
1000	0.06	2750	0.47	4500	1.26	6250	2.43	8000	3.99
1050	0.07	2800	0.49	4550	1.29	6300	2.47	8050	4.04
1100	0.08	2850	0.51	4600	1.32	6350	2.51	8100	4.09
1150	0.08	2900	0.52	4650	1.35	6400	2.55	8150	4.14
1200	0.09	2950	0.54	4700	1.38	6450	2.59	8200	4.19
1250	0.10	3000	0.56	4750	1.41	6500	2.63	8250	4.24
1300	0.11	3050	0.58	4800	1.44	6550	2.67	8300	4.29
1350	0.11	3100	0.60	4850	1.47	6600	2.71	8350	4.35
1400	0.12	3150	0.62	4900	1.50	6650	2.76	8400	4.40
1450	0.13	3200	0.64	4950	1.53	6700	2.80	8450	4.45
1500	0.14	3250	0.66	5000	1.56	6750	2.84	8500	4.50
1550	0.15	3300	0.68	5050	1.59	6800	2.88	8550	4.56
1600	0.16	3350	0.70	5100	1.62	6850	2.92	8600	4.61
1650	0.17	3400	0.72	5150	1.65	6900	2.97	8650	4.66
1700	0.18	3450	0.74	5200	1.69	6950	3.01	8700	4.72
1750	0.19	3500	0.76	5250	1.72	7000	3.05	8750	4.77
1800	0.20	3550	0.79	5300	1.75	7050	3.10	8800	4.83
1850	0.21	3600	0.81	5350	1.78	7100	3.14	8850	4.88
1900	0.22	3650	0.83	5400	1.82	7150	3.19	8900	4.94
1950	0.24	3700	0.85	5450	1.85	7200	3.23	8950	4.99
2000	0.25	3750	0.88	5500	1.89	7250	3.28	9000	5.05

Velocities vs. Velocity Pressures

(Reprinted by permission from 1985 ASHRAE Handbook – Fundamentals)

Figure 11.9

146

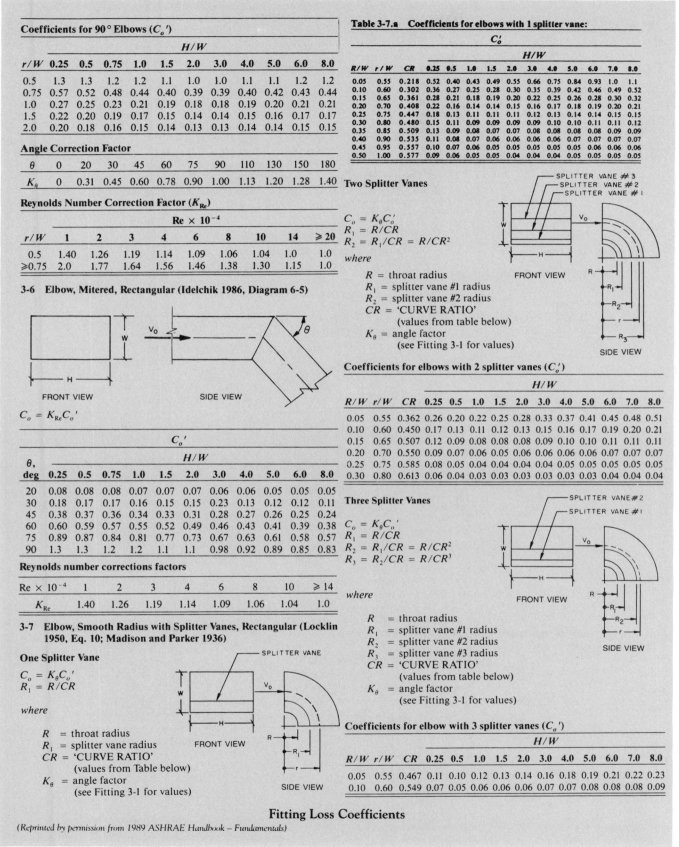

Coefficients for 90° Elbows (C_o')

	H/W										
r/W	0.25	0.5	0.75	1.0	1.5	2.0	3.0	4.0	5.0	6.0	8.0
0.5	1.3	1.3	1.2	1.2	1.1	1.0	1.0	1.1	1.1	1.2	1.2
0.75	0.57	0.52	0.48	0.44	0.40	0.39	0.39	0.40	0.42	0.43	0.44
1.0	0.27	0.25	0.23	0.21	0.19	0.18	0.18	0.19	0.20	0.21	0.21
1.5	0.22	0.20	0.19	0.17	0.15	0.14	0.14	0.15	0.16	0.17	0.17
2.0	0.20	0.18	0.16	0.15	0.14	0.13	0.13	0.14	0.14	0.15	0.15

Angle Correction Factor

θ	0	20	30	45	60	75	90	110	130	150	180
K_θ	0	0.31	0.45	0.60	0.78	0.90	1.00	1.13	1.20	1.28	1.40

Reynolds Number Correction Factor (K_{Re})

	Re × 10⁻⁴								
r/W	1	2	3	4	6	8	10	14	≥20
0.5	1.40	1.26	1.19	1.14	1.09	1.06	1.04	1.0	1.0
≥0.75	2.0	1.77	1.64	1.56	1.46	1.38	1.30	1.15	1.0

3-6   Elbow, Mitered, Rectangular (Idelchik 1986, Diagram 6-5)

FRONT VIEW   SIDE VIEW

$C_o = K_{Re} C_o'$

	C_o'										
	H/W										
θ, deg	0.25	0.5	0.75	1.0	1.5	2.0	3.0	4.0	5.0	6.0	8.0
20	0.08	0.08	0.08	0.07	0.07	0.07	0.06	0.06	0.05	0.05	0.05
30	0.18	0.17	0.17	0.16	0.15	0.15	0.23	0.13	0.12	0.12	0.11
45	0.38	0.37	0.36	0.34	0.33	0.31	0.28	0.27	0.26	0.25	0.24
60	0.60	0.59	0.57	0.55	0.52	0.49	0.46	0.43	0.41	0.39	0.38
75	0.89	0.87	0.84	0.81	0.77	0.73	0.67	0.63	0.61	0.58	0.57
90	1.3	1.3	1.2	1.2	1.1	1.1	0.98	0.92	0.89	0.85	0.83

Reynolds number corrections factors

Re × 10⁻⁴	1	2	3	4	6	8	10	≥14
K_{Re}	1.40	1.26	1.19	1.14	1.09	1.06	1.04	1.0

3-7   Elbow, Smooth Radius with Splitter Vanes, Rectangular (Locklin 1950, Eq. 10; Madison and Parker 1936)

One Splitter Vane

$C_o = K_\theta C_o'$
$R_1 = R/CR$

where

R = throat radius
R_1 = splitter vane radius
CR = 'CURVE RATIO' (values from Table below)
K_θ = angle factor (see Fitting 3-1 for values)

SPLITTER VANE

FRONT VIEW   SIDE VIEW

Table 3-7.a   Coefficients for elbows with 1 splitter vane:

			C_o'										
			H/W										
R/W	r/W	CR	0.25	0.5	1.0	1.5	2.0	3.0	4.0	5.0	6.0	7.0	8.0
0.05	0.55	0.218	0.52	0.40	0.43	0.49	0.55	0.66	0.75	0.84	0.93	1.0	1.1
0.10	0.60	0.302	0.36	0.27	0.25	0.28	0.30	0.35	0.39	0.42	0.46	0.49	0.52
0.15	0.65	0.361	0.28	0.21	0.18	0.19	0.20	0.22	0.25	0.26	0.28	0.30	0.32
0.20	0.70	0.408	0.22	0.16	0.14	0.14	0.15	0.16	0.17	0.18	0.19	0.20	0.21
0.25	0.75	0.447	0.18	0.13	0.11	0.11	0.11	0.12	0.13	0.14	0.14	0.15	0.15
0.30	0.80	0.480	0.15	0.11	0.09	0.09	0.09	0.09	0.10	0.10	0.11	0.11	0.12
0.35	0.85	0.509	0.13	0.09	0.08	0.07	0.07	0.08	0.08	0.08	0.08	0.09	0.09
0.40	0.90	0.535	0.11	0.08	0.07	0.06	0.06	0.06	0.06	0.07	0.07	0.07	0.07
0.45	0.95	0.557	0.10	0.07	0.06	0.05	0.05	0.05	0.05	0.05	0.06	0.06	0.06
0.50	1.00	0.577	0.09	0.06	0.05	0.05	0.04	0.04	0.04	0.05	0.05	0.05	0.05

Two Splitter Vanes

$C_o = K_\theta C_o'$
$R_1 = R/CR$
$R_2 = R_1/CR = R/CR^2$

where

R = throat radius
R_1 = splitter vane #1 radius
R_2 = splitter vane #2 radius
CR = 'CURVE RATIO' (values from table below)
K_θ = angle factor (see Fitting 3-1 for values)

SPLITTER VANE #3
SPLITTER VANE #2
SPLITTER VANE #1

FRONT VIEW   SIDE VIEW

Coefficients for elbows with 2 splitter vanes (C_o')

			H/W										
R/W	r/W	CR	0.25	0.5	1.0	1.5	2.0	3.0	4.0	5.0	6.0	7.0	8.0
0.05	0.55	0.362	0.26	0.20	0.22	0.25	0.28	0.33	0.37	0.41	0.45	0.48	0.51
0.10	0.60	0.450	0.17	0.13	0.11	0.12	0.13	0.15	0.16	0.17	0.19	0.20	0.21
0.15	0.65	0.507	0.12	0.09	0.08	0.08	0.08	0.09	0.10	0.10	0.11	0.11	0.11
0.20	0.70	0.550	0.09	0.07	0.06	0.05	0.06	0.06	0.06	0.06	0.07	0.07	0.07
0.25	0.75	0.585	0.08	0.05	0.04	0.04	0.04	0.04	0.05	0.05	0.05	0.05	0.05
0.30	0.80	0.613	0.06	0.04	0.03	0.03	0.03	0.03	0.03	0.03	0.04	0.04	0.04

Three Splitter Vanes

$C_o = K_\theta C_o'$
$R_1 = R/CR$
$R_2 = R_1/CR = R/CR^2$
$R_3 = R_2/CR = R/CR^3$

where

R = throat radius
R_1 = splitter vane #1 radius
R_2 = splitter vane #2 radius
R_3 = splitter vane #3 radius
CR = 'CURVE RATIO' (values from table below)
K_θ = angle factor (see Fitting 3-1 for values)

SPLITTER VANE #2
SPLITTER VANE #1

FRONT VIEW   SIDE VIEW

Coefficients for elbow with 3 splitter vanes (C_o')

			H/W										
R/W	r/W	CR	0.25	0.5	1.0	1.5	2.0	3.0	4.0	5.0	6.0	7.0	8.0
0.05	0.55	0.467	0.11	0.10	0.12	0.13	0.14	0.16	0.18	0.19	0.21	0.22	0.23
0.10	0.60	0.549	0.07	0.05	0.06	0.06	0.06	0.07	0.07	0.08	0.08	0.08	0.09

Fitting Loss Coefficients

(Reprinted by permission from 1989 ASHRAE Handbook – Fundamentals)

Figure 11.10

3-8 Elbow, Mitered, with Single-Thickness Vanes, Rectangular (Rozell 1974)

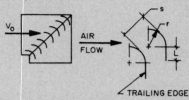

Design No.	Dimensions, in.			
	r	s	L	C_o
1[a]	2.0	1.5	0.75	0.12
2	4.5	2.25	0	0.15
3	4.5	3.25	1.60	0.18

[a]When extension of trailing edge is not provided for this vane, losses are approximately unchanged for single elbows, but increase considerably for elbows in series.

3-9 Elbow, Mitered, with Double-Thickness Vanes, Rectangular (Rozell 1974)

EMBOSSED VANE RUNNER

PUSH-ON VANE RUNNER

Design No.	Dimensions, in.		C_o				Remarks
			Velocity V_o, fpm				
	r	s	1000	2000	3000	4000	
1	2.0	1.5	0.27	0.22	0.19	0.17	Embossed Vane Runner
2	2.0	1.5	0.33	0.29	0.26	0.23	Push-On Vane Runner
3	2.0	2.13	0.38	0.31	0.27	0.24	Embossed Vane Runner
4	4.5	3.25	0.26	0.21	0.18	0.16	Embossed Vane Runner

3-10 Elbow, Variable Inlet/Outlet Areas, Rectangular (Idelchik 1986, Diagram 6-4)

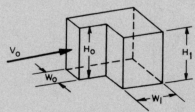

$$C_o = K_{Re} C_o'$$

C_o'						
	W_1/W_o					
H_o/W_o	0.6	0.8	1.2	1.4	1.6	2.0
0.25	1.8	1.4	1.1	1.1	1.1	1.1
1.0	1.7	1.4	1.0	0.95	0.90	0.84
4.0	1.5	1.1	0.81	0.76	0.72	0.66
∞	1.5	1.0	0.69	0.63	0.60	0.55

Reynolds number correction factor

Re × 10⁻⁴	1	2	3	4	6	8	10	≥ 14
K_{Re}	1.40	1.26	1.19	1.14	1.09	1.06	1.04	1.0

3-11 Elbows, 90°, Z-Shaped, Rectangular (Idelchik 1986, Diagram 6-11)

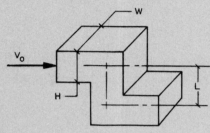

$$C = K K_{Re} C_o'$$

Coefficients for $W/H = 1.0$

L/H	0	0.4	0.6	0.8	1.0	1.2	1.4	1.6	1.8	2.0
C_o'	0	0.62	0.90	1.6	2.6	3.6	4.0	4.2	4.2	4.2
L/H	2.4	2.8	3.2	4.0	5.0	6.0	7.0	9.0	10.0	∞
C_o'	3.7	3.3	3.2	3.1	2.9	2.9	2.8	2.7	2.5	2.3

For W/H values other than 1.0, apply the following factor

W/H	0.25	0.50	0.75	1.0	1.5	2.0	3.0	4.0	6.0	8.0
K	1.10	1.07	1.04	1.0	0.95	0.90	9.83	0.78	0.72	0.70

Reynolds Number Correction Factor

Re × 10⁻⁴	1	2	3	4	6	8	10	≥14
K_{Re}	1.40	1.26	1.19	1.14	1.09	1.06	1.04	1.0

3-12 Combined 90° Elbows Lying in Different Planes, Rectangular (Idelchik 1986, Diagram 6-11)

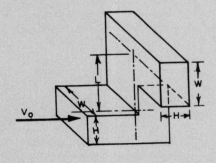

$$C_o = K K_{Re} C_o'$$

Fitting Loss Coefficients

(Reprinted by permission from 1989 ASHRAE Handbook – Fundamentals)

Figure 11.10 (cont.)

Coefficients for Square Ducts

L/W	0	0.4	0.6	0.8	1.0	1.2	1.4	1.6	1.8	2.0
C_o'	1.2	2.4	2.9	3.3	3.4	3.4	3.4	3.3	3.2	3.1
L/W	2.4	2.8	3.2	4.0	5.0	6.0	7.0	9.0	10.0	∞
C_o'	3.2	3.2	3.2	3.0	2.9	2.8	2.7	2.5	2.4	2.3

Apply the following factor for other than $H/W = 1.0$

H/W	0.25	0.50	0.75	1.0	1.5	2.0	3.0	4.0	6.0	8.0
K	1.10	1.07	1.04	1.0	0.95	0.90	0.83	0.78	0.72	0.70

Reynolds Number Correction Factor

Re × 10⁻⁴	1	2	3	4	6	8	10	≥14
K_{Re}	1.40	1.26	1.19	1.14	1.09	1.06	1.04	1.0

3-13 Offset, S-Shaped (Gooseneck), Rectangular and Round (Idelchik 1986, Diagram 6-16)

	K							
	L/D							
θ, deg	0	1	2	3	4	6	8	10
15	0.20	0.42	0.60	0.78	0.94	1.16	1.20	1.15
30	0.40	0.65	0.88	1.16	1.20	1.18	1.12	1.06
45	0.60	1.06	0.20	1.23	1.20	1.08	1.03	1.08
60	1.05	1.38	1.37	1.28	1.15	1.06	1.16	1.30
75	1.50	1.58	1.46	1.30	1.27	1.30	1.37	1.47
90	1.70	1.67	1.40	1.37	1.38	1.47	1.55	1.63

	K						
	L/D						
θ, deg	12	14	16	18	20	25	≥ 40
15	1.08	1.05	1.02	1.00	1.10	1.25	2.0
30	1.06	1.15	1,28	1.40	1.50	1.70	2.0
45	1.17	1.30	1.42	1.55	1.65	1.80	2.0
60	1.42	1.54	1.66	1.76	1.85	1.95	2.0
75	1.57	1.68	1.75	1.80	1.88	1.97	2.0
90	1.70	1.76	1.82	1.88	1.92	1.98	2.0

3-14 Offset, S-Shaped in Two Planes 90° Apart, Rectangular and Round (Idelchik 1986, Diagram 6-16)

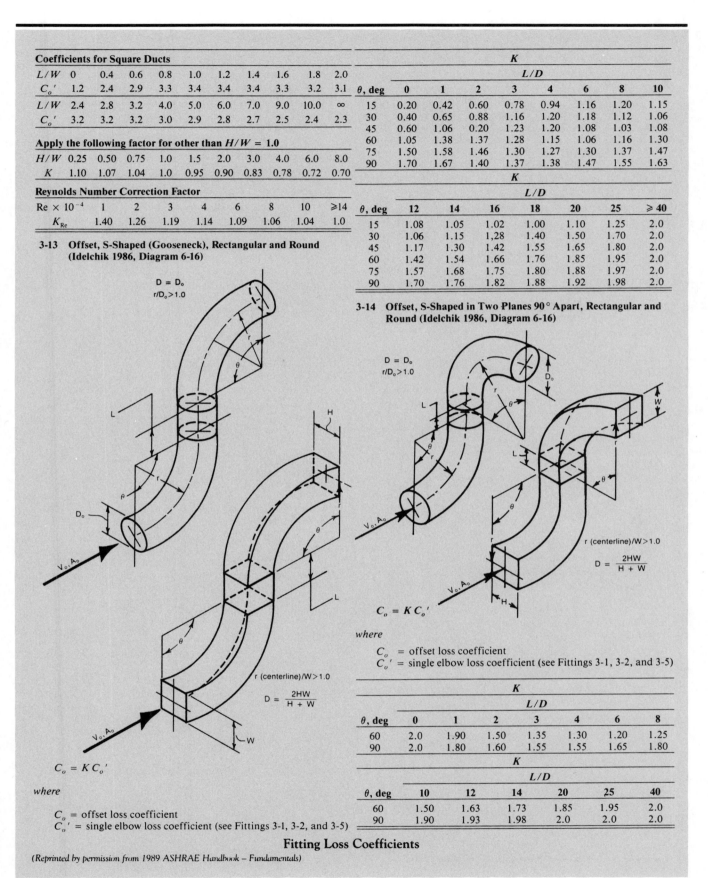

$$C_o = K C_o'$$

where

C_o = offset loss coefficient
C_o' = single elbow loss coefficient (see Fittings 3-1, 3-2, and 3-5)

$$C_o = K C_o'$$

where

C_o = offset loss coefficient
C_o' = single elbow loss coefficient (see Fittings 3-1, 3-2, and 3-5)

	K						
	L/D						
θ, deg	0	1	2	3	4	6	8
60	2.0	1.90	1.50	1.35	1.30	1.20	1.25
90	2.0	1.80	1.60	1.55	1.55	1.65	1.80

	K					
	L/D					
θ, deg	10	12	14	20	25	40
60	1.50	1.63	1.73	1.85	1.95	2.0
90	1.90	1.93	1.98	2.0	2.0	2.0

Fitting Loss Coefficients

(*Reprinted by permission from 1989 ASHRAE Handbook – Fundamentals*)

Figure 11.10 (cont.)

Branch, $C_{c,b}$

$\frac{A_s}{A_c}$	$\frac{A_b}{A_c}$	\multicolumn{10}{c}{Q_b/Q_s}									
		0.2	0.4	0.6	0.8	1.0	1.2	1.4	1.6	1.8	2.0
0.6	0.2	−.55	1.3	3.1	4.7	6.1	7.4	8.6	9.6	11	12
	0.3	−1.1	0	0.88	1.6	2.3	2.8	3.3	3.7	4.1	4.5
	0.4	−1.2	−.48	0.10	0.54	0.89	1.2	1.4	1.6	1.8	2.0
	0.5	−1.3	−.62	−.14	0.21	0.47	0.68	0.85	0.99	1.1	1.2
	0.6	−1.3	−.69	−.26	0.04	0.26	0.42	0.57	0.66	0.75	0.82
0.8	0.2	0.06	1.8	3.5	5.1	6.5	7.8	8.9	10	11	12
	0.3	−.52	0.35	1.1	1.7	2.3	2.8	3.2	3.6	3.9	4.2
	0.4	−.67	−.05	0.43	0.80	1.1	1.4	1.6	1.8	1.9	2.1
	0.6	−.75	−.27	0.05	0.28	0.45	0.58	0.68	0.76	0.83	0.88
	0.7	−.77	−.31	−.02	0.18	0.32	0.43	0.50	0.56	0.61	0.65
	0.8	−.78	−.34	−.07	0.12	0.24	0.33	0.39	0.44	0.47	0.50
1.0	0.2	0.40	2.1	3.7	5.2	6.6	7.8	9.0	11	11	12
	0.3	−.21	0.54	1.2	1.8	2.3	2.7	3.1	3.7	3.7	4.0
	0.4	−.33	0.21	0.62	0.96	1.2	1.5	1.7	2.0	2.0	2.1
	0.5	−.38	0.05	0.37	0.60	0.79	0.93	1.1	1.2	1.2	1.3
	0.6	−.41	−.02	0.23	0.42	0.55	0.66	0.73	0.80	0.85	0.89
	0.8	−.44	−.10	0.11	0.24	0.33	0.39	0.43	0.46	0.47	0.48
	1.0	−.46	−.14	0.05	0.16	0.23	0.27	0.29	0.30	0.30	0.29

Main, $C_{c,s}$

$\frac{A_s}{A_c}$	$\frac{A_b}{A_c}$	\multicolumn{10}{c}{Q_b/Q_s}									
		0.2	0.4	0.6	0.8	1.0	1.2	1.4	1.6	1.8	2.0
0.3	0.2	5.3	−.01	2.0	1.1	0.34	−.20	−.61	−.93	−1.2	−1.4
	0.3	5.4	3.7	2.5	1.6	1.0	0.53	0.16	−.14	−.38	−.58
0.4	0.2	1.9	1.1	0.46	−.07	−.49	−.83	−1.1	−1.3	−1.5	−1.7
	0.3	2.0	1.4	0.81	0.42	0.08	−.20	−.43	−.62	−.78	−.92
	0.4	2.0	1.5	1.0	0.68	0.39	0.16	−.04	−.21	−.35	−.47
0.5	0.2	0.77	0.34	−.09	−.48	−.81	−1.1	−1.3	−1.5	−1.7	−1.8
	0.3	0.85	0.56	0.25	−.03	−.27	−.48	−.67	−.82	−.96	−1.1
	0.4	0.88	0.66	0.43	0.21	0.02	−.15	−.30	−.42	−.54	−.64
	0.5	0.91	0.73	0.54	0.36	0.21	0.06	−.06	−.17	−.26	−.35
0.6	0.2	0.30	0	−.34	−.67	−.96	−1.2	−1.4	−1.6	−1.8	−1.9
	0.3	0.37	0.21	−.02	−.24	−.44	−.63	−.79	−.93	−1.1	−1.2
	0.4	0.40	0.31	0.16	−.1	−.16	−.30	−.43	−.54	−.64	−.73
	0.5	0.43	0.37	0.26	0.14	0.02	−.09	−.20	−.29	−.37	−.45
	0.6	0.44	0.41	0.33	0.24	0.14	0.05	−.03	−.11	−.18	−.25
0.8	0.2	−.06	−.27	−.57	−.86	−1.1	−1.4	−1.6	−1.7	−1.9	−2.0
	0.3	0	−.08	−.25	−.43	−.62	−.78	−.93	−1.1	−1.2	−1.3
	0.4	0.04	0.02	−.08	−.21	−.34	−.46	−.57	−.67	−.77	−.85
	0.5	0.06	0.08	0.02	−.06	−.16	−.25	−.34	−.42	−.50	−.57
	0.6	0.07	0.12	0.09	0.03	−.04	−.11	−.18	−.25	−.31	−.37
	0.7	0.08	0.15	0.14	0.10	0.05	−.01	−.07	−.12	−.17	−.22
	0.8	0.09	0.17	0.18	0.16	0.11	0.07	0.02	−.02	−.07	−.11
1.0	0.2	−.19	−.39	−.67	−.96	−1.2	−1.5	−1.6	−1.8	−2.0	−2.1
	0.3	−.12	−.19	−.35	−.54	−.71	−.87	−1.0	−1.2	−1.3	−1.4
	0.4	−.09	−.10	−.19	−.31	−.43	−.55	−.66	−.77	−.86	−.94
	0.5	−.07	−.04	−.09	−.17	−.26	−.35	−.44	−.52	−.59	−.66
	0.6	−.06	0	−.02	−.07	−.14	−.21	−.28	−.34	−.40	−.46
	0.8	−.04	0.06	0.07	0.05	0.02	−.03	−.07	−.12	−.16	−.20
	1.0	−.3	0.09	0.13	0.13	0.11	0.08	0.06	0.03	−.01	−.03

5-6 Tee, Converging, Rectangular (Idelchik 1986, Diagram 7-11)

$r/w_b = 1$

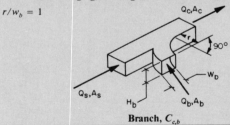

Branch, $C_{c,b}$

$\frac{A_b}{A_s}$	$\frac{A_b}{A_c}$	\multicolumn{9}{c}{Q_b/Q_c}								
		0.1	0.2	0.3	0.4	0.5	0.6	0.7	0.8	0.9
0.33	0.25	−1.2	−.40	0.40	1.6	3.0	4.8	6.8	8.9	11
0.5	0.5	−.50	−.20	0	0.25	0.45	0.70	1.0	1.5	2.0
0.67	0.5	−1.0	−.60	−.20	0.10	0.30	0.60	1.0	1.5	2.0
1.0	0.5	−2.2	−1.5	−.95	−.50	0	0.40	0.80	1.3	1.9
1.0	1.0	−.60	−.30	−.10	−.04	0.13	0.21	0.29	0.36	0.42
1.33	1.0	−1.2	−.80	−.40	−.20	0	0.16	0.24	0.32	0.38
2.0	1.0	−2.1	−1.4	−.90	−.50	−.20	0	0.20	0.25	0.30

Main, $C_{c,s}$

$\frac{A_s}{A_c}$	$\frac{A_b}{A_c}$	\multicolumn{9}{c}{Q_b/Q_c}								
		0.1	0.2	0.3	0.4	0.5	0.6	0.7	0.8	0.9
0.33	0.25	0.30	0.30	0.20	−.10	−.45	−.92	−1.5	−2.0	−2.6
0.5	0.5	0.17	0.16	0.10	0	−0.08	−.18	−.27	−.37	−.46
0.67	0.5	0.27	0.35	0.32	0.25	0.12	−.03	−.23	−.42	−.58
1.0	0.5	1.2	1.1	0.90	0.65	0.35	0	−.40	−.80	−1.3
1.0	1.0	0.18	0.24	0.27	0.26	0.23	0.18	0.10	0	−.12
1.33	1.0	0.75	0.36	0.38	0.35	0.27	0.18	0.05	−.08	−.22
2.0	1.0	0.80	0.87	0.80	0.68	0.55	0.40	0.25	0.08	−.10

5-7 Tee, Converging, Round Tap to Rectangular Main (SMACNA 1981, Table 6-9C)

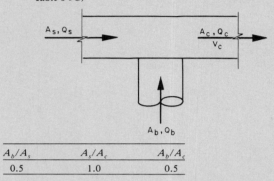

A_b/A_s	A_s/A_c	A_b/A_c
0.5	1.0	0.5

Branch, $C_{c,b}$

V_c	\multicolumn{10}{c}{Q_b/Q_c}									
(fpm)	0.1	0.2	0.3	0.4	0.5	0.6	0.7	0.8	0.9	1.0
<1200	−.63	−.55	0.13	0.23	0.78	1.30	1.93	3.10	4.88	5.60
>1200	−.49	−.21	0.23	0.60	1.27	2.06	2.75	3.70	4.93	5.95

For main coefficient ($C_{c,s}$), see Fitting 5-3.

5-8 Tee, Converging, Rectangular Main and Tap (SMACNA 1981, Table 6-9D)

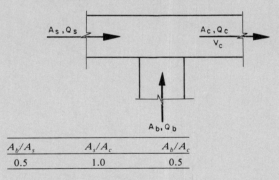

A_b/A_s	A_s/A_c	A_b/A_c
0.5	1.0	0.5

Branch, $C_{c,b}$

V_c	\multicolumn{10}{c}{Q_b/Q_c}									
(fpm)	0.1	0.2	0.3	0.4	0.5	0.6	0.7	0.8	0.9	1.0
<1200	−.75	−.53	−.03	0.33	1.03	1.10	2.15	2.93	4.18	4.78
>1200	−.69	−.21	0.23	0.67	1.17	1.66	2.67	3.36	3.93	5.13

For main coefficient ($C_{c,s}$), see Fitting 5-3.

Fitting Loss Coefficients

(Reprinted by permission from 1989 ASHRAE Handbook – Fundamentals)

Figure 11.10 (cont.)

5-9 Converging, Rectangular Main and Tap (45° Entry) (SMACNA 1981, Table 6-9F)

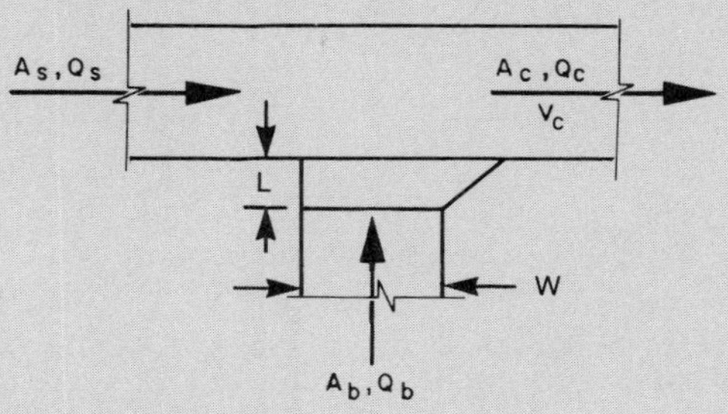

$L = 0.25\ W$, 3 in. min.

A_b/A_s	A_s/A_c	A_b/A_c
0.5	1.0	0.5

Branch, $C_{c,b}$

V_c	Q_b/Q_c									
(fpm)	**0.1**	**0.2**	**0.3**	**0.4**	**0.5**	**0.6**	**0.7**	**0.8**	**0.9**	**1.0**
<1200	−.83	−.68	−.30	0.28	0.55	1.03	1.50	1.93	2.50	3.03
>1200	−.72	−.52	−.23	0.34	0.76	1.14	1.83	2.01	2.90	3.63

For main coefficient ($C_{c,s}$), see Fitting 5-3.

5-10 Tee, Diverging, Round, Conical Branch (Jones 1969, Fig. 12)

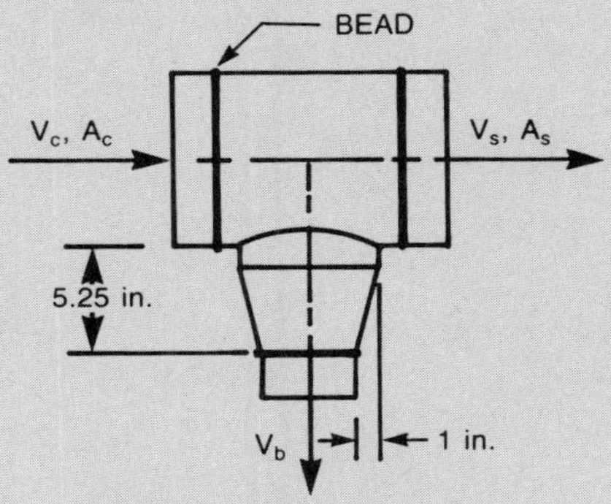

$A_c = A_s$

Branch

V_b/V_c	0	0.2	0.4	0.6	0.8	1.0	1.2	1.4	1.6	1.8	2.0
$C_{c,b}$	1.0	0.85	0.74	0.62	0.52	0.42	0.36	0.32	0.32	0.37	0.52

For main loss coefficient ($C_{c,s}$), see Fitting 5-23.

5-11 Wye, 45°, Diverging, Round, Conical Branch (Jones 1969, Fig. 14)

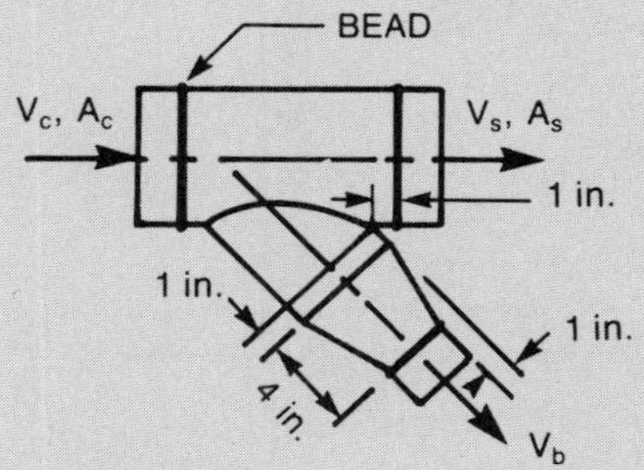

$A_c = A_s$

Branch

V_b/V_c	0	0.2	0.4	0.6	0.8	1.0	1.2	1.4	1.6	1.8	2.0
$C_{c,b}$	1.0	0.84	0.61	0.41	0.27	0.17	0.12	0.12	0.14	0.18	0.27

For main loss coefficient ($C_{c,s}$), see Fitting 5-23.

5-12 Tee, Diverging, Round, with 90° Elbow, Branch 90° to Main (Jones 1969, Fig. 17)

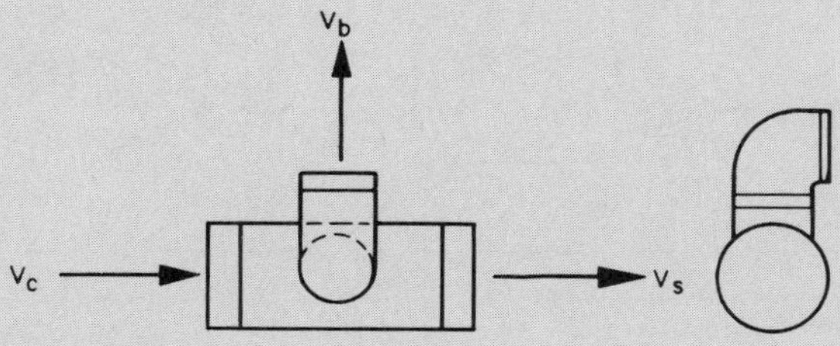

$A_c = A_s$

Branch

V_b/V_c	0	0.2	0.4	0.6	0.8	1.0	1.2	1.4	1.6	1.8	2.0
$C_{c,b}$	1.0	1.03	0.08	1.18	1.33	1.56	1.86	2.2	2.6	3.0	3.4

For main loss coefficient ($C_{c,s}$), see Fitting 5-23.

5-13 Tee, Diverging, Round, with 45° Elbow, Branch 90° to Main (Jones 1969, Fig. 18)

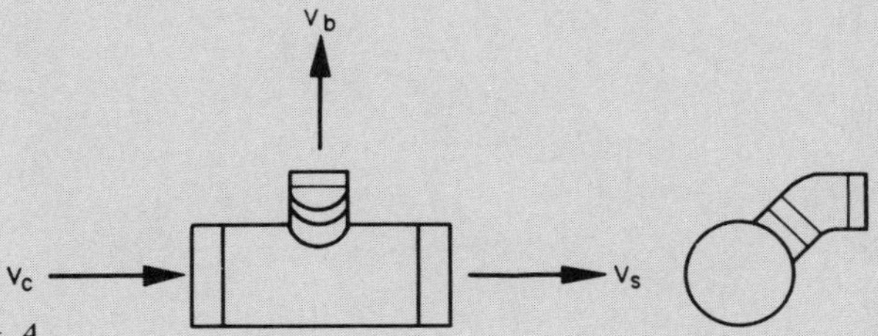

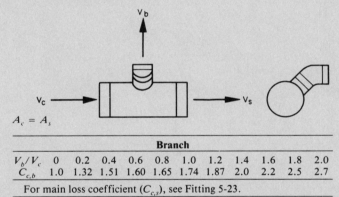

$A_c = A_s$

Branch

V_b/V_c	0	0.2	0.4	0.6	0.8	1.0	1.2	1.4	1.6	1.8	2.0
$C_{c,b}$	1.0	1.32	1.51	1.60	1.65	1.74	1.87	2.0	2.2	2.5	2.7

For main loss coefficient ($C_{c,s}$), see Fitting 5-23.

5-14 Tee, Diverging, Round (Conical Branch), with 45° Elbow, Branch 90° to Main (Jones 1969, Fig. 19)

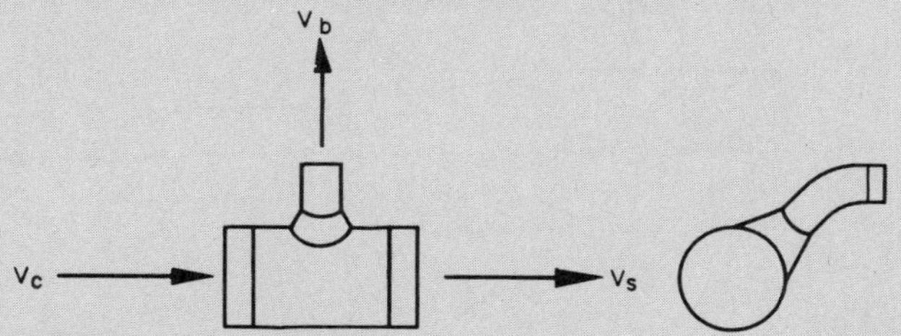

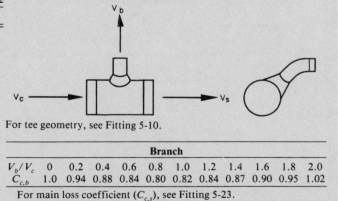

For tee geometry, see Fitting 5-10.

Branch

V_b/V_c	0	0.2	0.4	0.6	0.8	1.0	1.2	1.4	1.6	1.8	2.0
$C_{c,b}$	1.0	0.94	0.88	0.84	0.80	0.82	0.84	0.87	0.90	0.95	1.02

For main loss coefficient ($C_{c,s}$), see Fitting 5-23.

Fitting Loss Coefficients

(Reprinted by permission from 1989 ASHRAE Handbook – Fundamentals)

Figure 11.10 (cont.)

5-15 Wye, 45°, Round, with 60° Elbow, Branch 90° to Main (Jones 1969, Fig. 3)

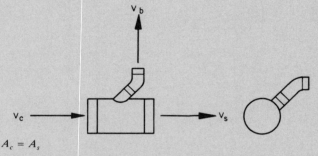

$A_c = A_s$

				Branch						
V_b/V_c	0	0.2	0.4	0.6	0.8	1.0	1.2	1.4	1.6	1.8	2.0
$C_{c,b}$	1.0	0.88	0.77	0.68	0.65	0.69	0.73	0.88	1.14	1.54	2.2

For main loss coefficient ($C_{c,s}$), see Fitting 5-23.

5-16 Wye, 45°, Diverging, Round (Conical Branch), with 60° Elbow, Branch 90° to Main (Jones 1969, Fig. 20)

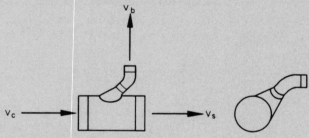

For wye geometry, see Fitting 5-11.

$A_c = A_s$

				Branch						
V_b/V_c	0	0.2	0.4	0.6	0.8	1.0	1.2	1.4	1.6	1.8	2.0
$C_{c,b}$	1.0	0.82	0.63	0.52	0.45	0.42	0.41	0.40	0.41	0.45	0.56

For main loss coefficient ($C_{c,s}$), see Fitting 5-23.

5-17 Wye, 48°, Diverging, Conical Main and Branch, with 45° Elbow, Branch 90° to Main (Idelchik 1986, Diagram 7-19)

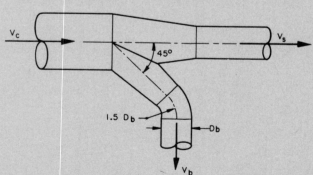

Branch

V_b/V_c	0.2	0.4	0.6	0.7	0.8	0.9	1.0	1.1	1.2
$C_{c,b}$	0.76	0.60	0.52	0.50	0.51	0.52	0.56	0.61	0.68

V_b/V_c	1.4	1.6	1.8	2.0	2.2	2.4	2.6	2.8	3.0
$C_{c,b}$	0.86	1.1	1.4	1.8	2.2	2.6	3.1	3.7	4.2

Main

V_s/V_c	0.2	0.4	0.6	0.8	1.0	1.2	1.4	1.6	1.8	2.0
$C_{c,s}$	0.14	0.06	0.05	0.09	0.18	0.30	0.46	0.64	0.84	1.0

5-18 Tee, Diverging, Round, with 60° Elbow, Branch 45° to Main (Jones 1969, Fig. 22)

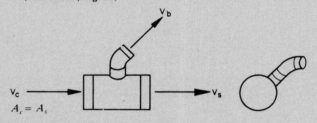

$A_c = A_s$

				Branch						
V_b/V_c	0	0.2	0.4	0.6	0.8	1.0	1.2	1.4	1.6	1.8	2.0
$C_{c,b}$	1.0	1.06	1.15	1.29	1.45	1.65	1.89	2.2	2.5	2.9	3.3

For main loss coefficient ($C_{c,s}$), see Fitting 5-23.

5-19 Tee, Diverging, Round (Conical Branch), with 60° Elbow, Branch 45° to Main (Jones 1969, Fig. 23)

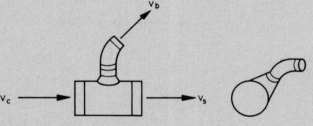

For tee geometry, see Fitting 5-10.

$A_c = A_s$

				Branch						
V_b/V_c	0	0.2	0.4	0.6	0.8	1.0	1.2	1.4	1.6	1.8	2.0
$C_{c,b}$	1.0	0.95	0.90	0.86	0.81	0.79	0.79	0.81	0.86	0.96	1.10

For main loss coefficient ($C_{c,s}$), see Fitting 5-23.

5-20 Wye, 45°, Diverging, Round, with 30° Elbow, Branch 45° to Main (Jones 1969, Fig. 2)

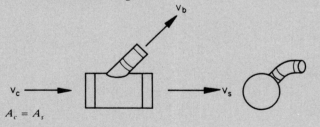

$A_c = A_s$

Fitting Loss Coefficients

(Reprinted by permission from 1989 ASHRAE Handbook – Fundamentals)

Figure 11.10 (cont.)

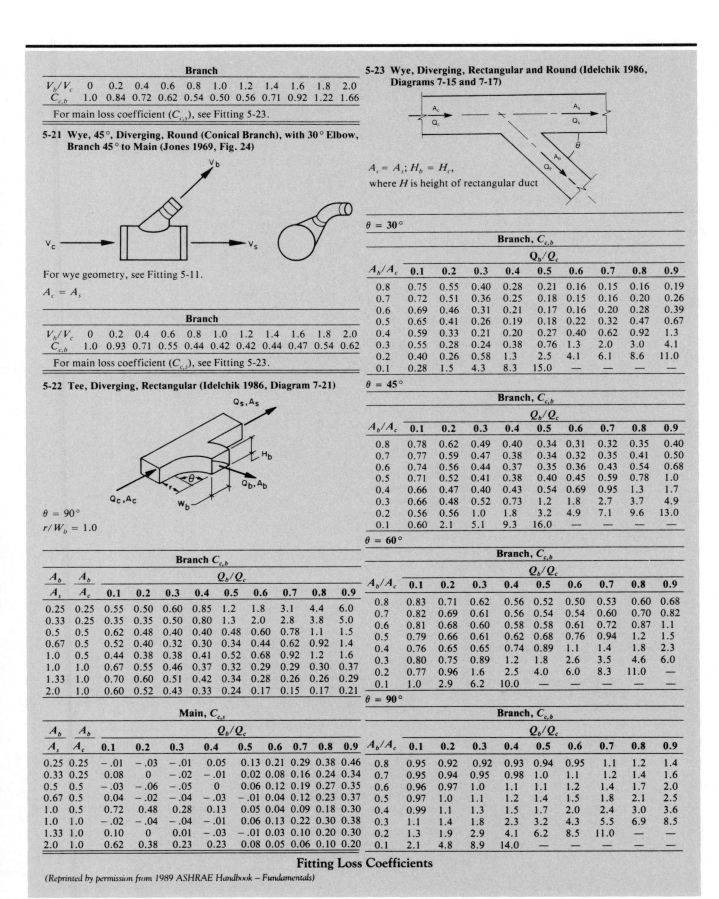

Branch

V_b/V_c	0	0.2	0.4	0.6	0.8	1.0	1.2	1.4	1.6	1.8	2.0
$C_{c,b}$	1.0	0.84	0.72	0.62	0.54	0.50	0.56	0.71	0.92	1.22	1.66

For main loss coefficient ($C_{c,s}$), see Fitting 5-23.

5-21 Wye, 45°, Diverging, Round (Conical Branch), with 30° Elbow, Branch 45° to Main (Jones 1969, Fig. 24)

For wye geometry, see Fitting 5-11.

$A_c = A_s$

Branch

V_b/V_c	0	0.2	0.4	0.6	0.8	1.0	1.2	1.4	1.6	1.8	2.0
$C_{c,b}$	1.0	0.93	0.71	0.55	0.44	0.42	0.42	0.44	0.47	0.54	0.62

For main loss coefficient ($C_{c,s}$), see Fitting 5-23.

5-22 Tee, Diverging, Rectangular (Idelchik 1986, Diagram 7-21)

$\theta = 90°$
$r/W_b = 1.0$

Branch $C_{c,b}$

$\dfrac{A_b}{A_s}$	$\dfrac{A_b}{A_c}$	0.1	0.2	0.3	0.4	0.5	0.6	0.7	0.8	0.9
0.25	0.25	0.55	0.50	0.60	0.85	1.2	1.8	3.1	4.4	6.0
0.33	0.25	0.35	0.35	0.50	0.80	1.3	2.0	2.8	3.8	5.0
0.5	0.5	0.62	0.48	0.40	0.40	0.48	0.60	0.78	1.1	1.5
0.67	0.5	0.52	0.40	0.32	0.30	0.34	0.44	0.62	0.92	1.4
1.0	0.5	0.44	0.38	0.38	0.41	0.52	0.68	0.92	1.2	1.6
1.0	1.0	0.67	0.55	0.46	0.37	0.32	0.29	0.29	0.30	0.37
1.33	1.0	0.70	0.60	0.51	0.42	0.34	0.28	0.26	0.26	0.29
2.0	1.0	0.60	0.52	0.43	0.33	0.24	0.17	0.15	0.17	0.21

Main, $C_{c,s}$

$\dfrac{A_b}{A_s}$	$\dfrac{A_b}{A_c}$	0.1	0.2	0.3	0.4	0.5	0.6	0.7	0.8	0.9
0.25	0.25	-.01	-.03	-.01	0.05	0.13	0.21	0.29	0.38	0.46
0.33	0.25	0.08	0	-.02	-.01	0.02	0.08	0.16	0.24	0.34
0.5	0.5	-.03	-.06	-.05	0	0.06	0.12	0.19	0.27	0.35
0.67	0.5	0.04	-.02	-.04	-.03	-.01	0.04	0.12	0.23	0.37
1.0	0.5	0.72	0.48	0.28	0.13	0.05	0.04	0.04	0.09	0.18	0.30
1.0	1.0	-.02	-.04	-.04	-.01	0.06	0.13	0.22	0.30	0.38
1.33	1.0	0.10	0	0.01	-.03	-.01	0.03	0.10	0.20	0.30
2.0	1.0	0.62	0.38	0.23	0.23	0.08	0.05	0.06	0.10	0.20

5-23 Wye, Diverging, Rectangular and Round (Idelchik 1986, Diagrams 7-15 and 7-17)

$A_c = A_s$; $H_b = H_c$,
where H is height of rectangular duct

$\theta = 30°$

Branch, $C_{c,b}$

A_b/A_c	0.1	0.2	0.3	0.4	0.5	0.6	0.7	0.8	0.9
0.8	0.75	0.55	0.40	0.28	0.21	0.16	0.15	0.16	0.19
0.7	0.72	0.51	0.36	0.25	0.18	0.15	0.16	0.20	0.26
0.6	0.69	0.46	0.31	0.21	0.17	0.16	0.20	0.28	0.39
0.5	0.65	0.41	0.26	0.19	0.18	0.22	0.32	0.47	0.67
0.4	0.59	0.33	0.21	0.20	0.27	0.40	0.62	0.92	1.3
0.3	0.55	0.28	0.24	0.38	0.76	1.3	2.0	3.0	4.1
0.2	0.40	0.26	0.58	1.3	2.5	4.1	6.1	8.6	11.0
0.1	0.28	1.5	4.3	8.3	15.0	—	—	—	—

$\theta = 45°$

Branch, $C_{c,b}$

A_b/A_c	0.1	0.2	0.3	0.4	0.5	0.6	0.7	0.8	0.9
0.8	0.78	0.62	0.49	0.40	0.34	0.31	0.32	0.35	0.40
0.7	0.77	0.59	0.47	0.38	0.34	0.32	0.35	0.41	0.50
0.6	0.74	0.56	0.44	0.37	0.35	0.36	0.43	0.54	0.68
0.5	0.71	0.52	0.41	0.38	0.40	0.45	0.59	0.78	1.0
0.4	0.66	0.47	0.40	0.43	0.54	0.69	0.95	1.3	1.7
0.3	0.66	0.48	0.52	0.73	1.2	1.8	2.7	3.7	4.9
0.2	0.56	0.56	1.0	1.8	3.2	4.9	7.1	9.6	13.0
0.1	0.60	2.1	5.1	9.3	16.0	—	—	—	—

$\theta = 60°$

Branch, $C_{c,b}$

A_b/A_c	0.1	0.2	0.3	0.4	0.5	0.6	0.7	0.8	0.9
0.8	0.83	0.71	0.62	0.56	0.52	0.50	0.53	0.60	0.68
0.7	0.82	0.69	0.61	0.56	0.54	0.54	0.60	0.70	0.82
0.6	0.81	0.68	0.60	0.58	0.58	0.61	0.72	0.87	1.1
0.5	0.79	0.66	0.61	0.62	0.68	0.76	0.94	1.2	1.5
0.4	0.76	0.65	0.65	0.74	0.89	1.1	1.4	1.8	2.3
0.3	0.80	0.75	0.89	1.2	1.8	2.6	3.5	4.6	6.0
0.2	0.77	0.96	1.6	2.5	4.0	6.0	8.3	11.0	—
0.1	1.0	2.9	6.2	10.0	—	—	—	—	—

$\theta = 90°$

Branch, $C_{c,b}$

A_b/A_c	0.1	0.2	0.3	0.4	0.5	0.6	0.7	0.8	0.9
0.8	0.95	0.92	0.92	0.93	0.94	0.95	1.1	1.2	1.4
0.7	0.95	0.94	0.95	0.98	1.0	1.1	1.2	1.4	1.6
0.6	0.96	0.97	1.0	1.1	1.1	1.2	1.4	1.7	2.0
0.5	0.97	1.0	1.1	1.2	1.4	1.5	1.8	2.1	2.5
0.4	0.99	1.1	1.3	1.5	1.7	2.0	2.4	3.0	3.6
0.3	1.1	1.4	1.8	2.3	3.2	4.3	5.5	6.9	8.5
0.2	1.3	1.9	2.9	4.1	6.2	8.5	11.0	—	—
0.1	2.1	4.8	8.9	14.0	—	—	—	—	—

Fitting Loss Coefficients

(*Reprinted by permission from 1989 ASHRAE Handbook – Fundamentals*)

Figure 11.10 (cont.)

	Main								
V_s/V_c	0	0.1	0.2	0.3	0.4	0.5	0.6	0.8	1.0
$C_{c,s}$	0.40	0.32	0.26	0.20	0.14	0.10	0.06	0.02	0

5-24 Diverging Wye, Rectangular (Idelchik 1986, Diagrams 7-16 and 7-17)

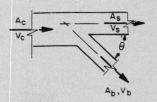

$\theta = 15°$ to $90°$ and $A_c = A_s + A_b$

	Branch, $C_{c,b}$												
θ, deg	V_b/V_c												
	0.1	0.2	0.3	0.4	0.5	0.6	0.8	1.0	1.2	1.4	1.6	1.8	2.0
15	0.81	0.65	0.51	0.38	0.28	0.20	0.11	0.06	0.14	0.30	0.51	0.76	1.0
30	0.84	0.69	0.56	0.44	0.34	0.26	0.19	0.15	0.15	0.30	0.51	0.76	1.0
45	0.87	0.74	0.63	0.54	0.45	0.38	0.29	0.24	0.23	0.30	0.51	0.76	1.0
60	0.90	0.82	0.79	0.66	0.59	0.53	0.43	0.36	0.33	0.39	0.51	0.76	1.0
90	1.0	1.0	1.0	1.0	1.0	1.0	1.0	1.0	1.0	1.0	1.0	1.0	1.0

	Main, $C_{c,s}$					
θ, deg	15–60	90				
$\dfrac{V_s}{V_c}$	0–1.0	A_s/A_c				
		0–0.4	0.5	0.6	0.7	≥0.8
0	1.0	1.0	1.0	1.0	1.0	1.0
0.1	0.81	0.81	0.81	0.81	0.81	0.81
0.2	0.64	0.64	0.64	0.64	0.64	0.64
0.3	0.50	0.50	0.52	0.52	0.50	0.50
0.4	0.36	0.36	0.40	0.38	0.37	0.36
0.5	0.25	0.25	0.30	0.28	0.27	0.25
0.6	0.16	0.16	0.23	0.20	0.18	0.16
0.8	0.04	0.04	0.17	0.10	0.07	0.04
1.0	0	0	0.20	0.10	0.05	0
1.2	0.07	0.07	0.36	0.21	0.14	0.07
1.4	0.39	0.39	0.79	0.59	0.39	—
1.6	0.90	0.90	1.4	1.2	—	—
1.8	1.8	1.8	2.4	—	—	—
2.0	3.2	3.2	4.0	—	—	—

5-25 Tee, Diverging, Rectangular Main to Round Tap (SMACNA 1981, Table 6-10T)

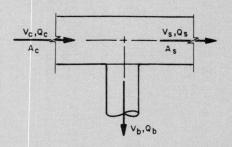

	Branch, $C_{c,b}$								
V_b/V_c	Q_b/Q_c								
	0.1	0.2	0.3	0.4	0.5	0.6	0.7	0.8	0.9
0.2	1.00								
0.4	1.01	1.07							
0.6	1.14	1.10	1.08						
0.8	1.18	1.31	1.12	1.13					
1.0	1.30	1.38	1.20	1.23	1.26				
1.2	1.46	1.58	1.45	1.31	1.39	1.48			
1.4	1.70	1.82	1.65	1.51	1.56	1.64	1.71		
1.6	1.93	2.06	2.00	1.85	1.70	1.76	1.80	1.88	
1.8	2.06	2.17	2.10	2.13	2.06	1.98	1.99	2.00	2.07

For main coefficient ($C_{c,s}$), see Fitting 5-23.

5-26 Tee, Diverging, Rectangular Main to Round Tap (Conical) (Inoue 1980, Korst 1950)

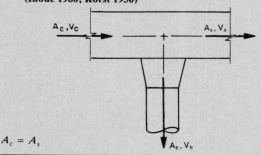

$A_c = A_s$

	Branch					
V_b/V_c	0.40	0.50	0.75	1.0	1.3	1.5
$C_{c,b}$	0.80	0.83	0.90	1.0	1.1	1.4

For main coefficient ($C_{c,s}$), see Fitting 5-23.

5-27 Tee, Diverging, Rectangular Main, and Tap (45° Entry) (SMACNA 1981, Table 6-10N)

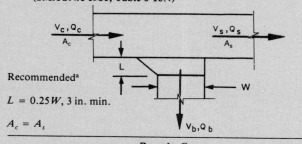

Recommended[a]

$L = 0.25W$, 3 in. min.

$A_c = A_s$

	Branch, $C_{c,b}$								
V_b/V_c	Q_b/Q_c								
	0.1	0.2	0.3	0.4	0.5	0.6	0.7	0.8	0.9
0.2	0.91								
0.4	0.81	0.79							
0.6	0.77	0.72	0.70						
0.8	0.78	0.73	0.69	0.66					
1.0	0.78	0.98	0.85	0.79	0.74				
1.2	0.90	1.11	1.16	1.23	1.03	0.86			
1.4	1.19	1.22	1.26	1.29	1.54	0.25	0.92		
1.6	1.35	1.42	1.55	1.59	1.63	1.50	1.31	1.09	
1.8	1.44	1.50	1.75	1.74	1.72	2.24	1.63	1.40	1.17

For main coefficient ($C_{c,s}$), see Fitting 5-23.
[a] For performance study, see SMACNA (1987).

Fitting Loss Coefficients

(Reprinted by permission from 1989 ASHRAE Handbook – Fundamentals)

Figure 11.10 (cont.)

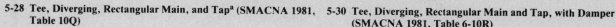

5-28 Tee, Diverging, Rectangular Main, and Tap[a] (SMACNA 1981, Table 10Q)

$A_c = A_s$

	Branch, $C_{c,b}$								
	Q_b/Q_c								
V_b/V_c	0.1	0.2	0.3	0.4	0.5	0.6	0.7	0.8	0.9
0.2	1.03								
0.4	1.04	1.01							
0.6	1.11	1.03	0.05						
0.8	1.16	1.21	1.17	1.12					
1.0	1.38	1.40	1.30	1.36	1.27				
1.2	1.52	1.61	1.68	1.91	1.47	1.66			
1.4	1.79	2.01	1.90	2.31	2.28	2.20	1.95		
1.6	2.07	2.28	2.13	2.71	2.99	2.81	2.09	2.20	
1.8	2.32	2.54	2.64	3.09	3.72	2.48	2.21	2.29	2.57

For main coefficient ($C_{c,s}$), see Fitting 5-23.
[a] For performance study, see SMACNA (1987).

5-30 Tee, Diverging, Rectangular Main and Tap, with Damper (SMACNA 1981, Table 6-10R)

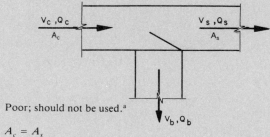

Poor; should not be used.[a]

$A_c = A_s$

	Branch, $C_{c,b}$								
	Q_b/Q_c								
V_b/V_c	0.1	0.2	0.3	0.4	0.5	0.6	0.7	0.8	0.9
0.2	0.58								
0.4	0.67	0.64							
0.6	0.78	0.76	0.75						
0.8	0.88	0.98	0.81	1.01					
1.0	1.12	1.05	1.08	1.18	1.29				
1.2	1.49	1.48	1.40	1.51	1.70	1.91			
1.4	2.10	2.21	2.25	2.29	2.32	2.48	2.53		
1.6	2.72	3.30	2.84	3.09	3.30	3.19	3.29	3.16	
1.8	3.42	4.58	3.65	3.92	4.20	4.15	4.14	4.10	4.05

For main coefficient ($C_{c,s}$), see Fitting 5-31.
[a] For performance study, see SMACNA (1987).

5-29 Tee, Diverging, Rectangular Main and Tap (45° Entry), with Damper (SMACNA 1981, Table 6-10P)

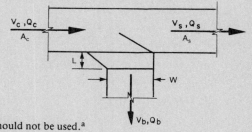

Poor; should not be used.[a]

$L = 0.25W$, 3 in. min.

$A_c = A_s$

	Branch, $C_{c,b}$								
	Q_b/Q_c								
V_b/V_c	0.1	0.2	0.3	0.4	0.5	0.6	0.7	0.8	0.9
0.2	0.61								
0.4	0.46	0.61							
0.6	0.43	0.50	0.54						
0.8	0.39	0.43	0.62	0.53					
1.0	0.34	0.57	0.77	0.73	0.68				
1.2	0.37	0.64	0.85	0.98	1.07	0.83			
1.4	0.57	0.71	1.04	1.16	1.54	0.36	1.18		
1.6	0.89	1.08	1.28	1.30	1.69	2.09	1.81	1.47	
1.8	1.33	1.34	2.04	1.78	1.90	2.40	2.77	2.23	1.92

For main coefficient ($C_{c,s}$), see Fitting 5-31.
[a] For performance study, see SMACNA (1987).

5-31 Tee, Diverging, Rectangular, with Extractor (SMACNA 1981, Table 6-10S)

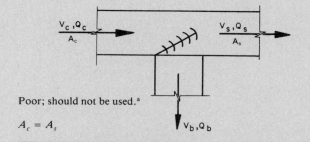

Poor; should not be used.[a]

$A_c = A_s$

	Branch, $C_{c,b}$								
	Q_b/Q_c								
V_b/V_c	0.1	0.2	0.3	0.4	0.5	0.6	0.7	0.8	0.9
0.2	0.60								
0.4	0.62	0.69							
0.6	0.74	0.80	0.82						
0.8	0.99	1.10	0.95	0.90					
1.0	1.48	1.12	1.41	1.24	1.21				
1.2	1.91	1.33	1.43	1.52	1.55	1.64			
1.4	2.47	1.67	1.70	2.04	1.86	1.98	2.47		
1.6	3.17	2.40	2.33	2.53	2.31	2.51	3.13	3.25	
1.8	3.85	3.37	2.89	3.23	3.09	3.03	3.30	3.74	4.11

	Main								
V_b/V_c	0.2	0.4	0.6	0.8	1.0	1.2	1.4	1.6	1.8
$C_{c,s}$	0.03	0.04	0.07	0.12	0.13	0.14	0.27	0.30	0.25

[a] For performance study, see SMACNA (1987).

Fitting Loss Coefficients

(Reprinted by permission from 1989 ASHRAE Handbook – Fundamentals)

Figure 11.10 (cont.)

5-32 Symmetrical Wye, Dovetail, Rectangular (Idelchik 1986, Diagram 7-24)

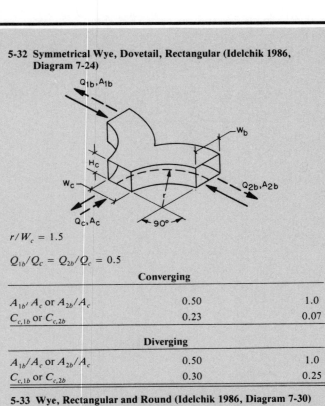

$r/W_c = 1.5$

$Q_{1b}/Q_c = Q_{2b}/Q_c = 0.5$

Converging

A_{1b}/A_c or A_{2b}/A_c	0.50	1.0
$C_{c,1b}$ or $C_{c,2b}$	0.23	0.07

Diverging

A_{1b}/A_c or A_{2b}/A_c	0.50	1.0
$C_{c,1b}$ or $C_{c,2b}$	0.30	0.25

5-33 Wye, Rectangular and Round (Idelchik 1986, Diagram 7-30)

$A_{1b} = A_{2b}$

$A_c = A_{1b} + A_{2b}$

Converging $\quad C_{c,1b}$ or $C_{c,2b}$

θ, deg	Q_{1b}/Q_c or Q_{2b}/Q_c										
	0	0.1	0.2	0.3	0.4	0.5	0.6	0.7	0.8	0.9	1.0
15	-2.6	-1.9	-1.3	-.77	-.30	0.10	0.41	0.67	0.85	0.97	1.0
30	-2.1	-1.5	-1.0	-.53	-.10	0.28	0.69	0.91	1.1	1.4	1.6
45	-1.3	-.93	-.55	-.16	0.20	0.56	0.92	1.3	1.6	2.0	2.3

Diverging $\quad C_{c,1b}$ or $C_{c,2b}$

θ, deg	V_{1b}/V_c or V_{2b}/V_c												
	0.1	0.2	0.3	0.4	0.5	0.6	0.8	1.0	1.2	1.4	1.6	1.8	2.0
15	0.81	0.65	0.51	0.38	0.28	0.20	0.11	0.06	0.14	0.30	0.51	0.76	1.0
30	0.84	0.69	0.56	0.44	0.34	0.26	0.19	0.15	0.15	0.30	0.51	0.76	1.0
45	0.87	0.74	0.63	0.54	0.45	0.38	0.29	0.24	0.23	0.30	0.51	0.76	1.0
60	0.90	0.82	0.79	0.66	0.59	0.53	0.43	0.36	0.33	0.39	0.51	0.76	1.0
90	1.0	1.0	1.0	1.0	1.0	1.0	1.0	1.0	1.0	1.0	1.0	1.0	1.0

5-34 Wye (Double), 45°, Rectangular and Round (Idelchik 1986, Diagram 7-27)

$A_{1b} = A_{2b}$

$A_s = A_c$

→ Converging

⇠ Diverging

Converging Flow

Branch, $C_{c,b}$

$\dfrac{Q_{2b}}{Q_{1b}}$	Q_{1b}/Q_c						
	0	0.1	0.2	0.3	0.4	0.5	0.6
$A_{1b}/A_c = 0.2$							
0.5	-1.0	-.36	0.59	1.8	3.2	4.9	6.8
1.0	-1.0	-.24	0.63	1.7	2.6	3.7	—
2.0	-1.0	-.19	0.21	0.04	—	—	—
$A_{1b}/A_c = 0.4$							
0.5	-1.0	-.48	-.02	0.58	0.92	1.3	16
1.0	-1.0	-.36	0.17	0.55	0.72	0.78	—
2.0	-1.0	-.18	0.16	-.06	—	—	—
$A_{1b}/A_c = 0.6$							
0.5	-1.0	-.50	-.07	0.31	0.60	0.82	0.92
1.0	-1.0	-.37	0.12	0.55	0.60	0.52	—
2.0	-1.0	-.18	0.26	0.16	—	—	—
$A_{1b}/A_c = 1.0$							
0.5	-1.0	-.51	-.09	0.25	0.50	0.65	0.64
1.0	-1.0	-.37	0.13	0.46	0.61	0.54	—
2.0	-1.0	-.15	0.38	0.42	—	—	—

Main, $C_{c,s}$

$\dfrac{A_{2b}}{A_{1b}}$	Q_s/Q_c										
	0	0.1	0.2	0.3	0.4	0.5	0.6	0.7	0.8	0.9	1.0
$A_{1b}/A_c = 0.2$											
0.5 & 2.0	-2.9	-1.9	-1.3	-.80	-.56	-.23	-.01	0.16	0.22	0.15	0
1.0	-2.5	-1.9	-1.3	-.80	-.42	-.12	0.08	0.20	0.22	0.15	0
$A_{1b}/A_c = 0.4$											
0.5 & 2.0	-.98	-.61	-.30	-.05	0.14	0.26	0.33	0.34	0.28	0.17	0
1.0	-.77	-.44	-.16	0.05	0.21	0.31	0.36	0.35	0.29	0.17	0

Fitting Loss Coefficients

(*Reprinted by permission from 1989 ASHRAE Handbook – Fundamentals*)

Figure 11.10 (cont.)

$A_{1b}/A_c = 0.6$											
0.5 & 2.0	−.32	0.08	0.11	0.27	0.37	0.43	0.44	0.40	0.31	0.18	0
1.0	−.18	−.04	0.21	0.34	0.42	0.46	0.46	0.41	0.31	0.18	0
$A_{1b}/A_c = 1.0$											
0.5 & 2.0	0.11	0.36	0.46	0.53	0.57	0.56	0.52	0.44	0.33	0.18	0
1.0	0.29	0.42	0.51	0.57	0.58	0.58	0.54	0.45	0.33	0.18	0

Diverging Flow: Use Fitting 5-23.

5-35 Cross, 90°, Rectangular and Round (Idelchik 1986, Diagram 7-29)

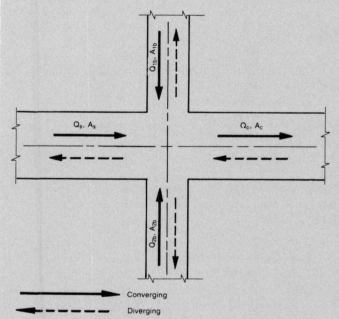

Converging
Diverging

$A_{1b} = A_{2b}$

$A_s = A_c$

Converging Flow

Branch, $C_{c,b}$							
$\frac{Q_{2b}}{Q_{1b}}$	Q_{1b}/Q_c or Q_{2b}/Q_c						
	0	0.1	0.2	0.3	0.4	0.5	0.6
$A_{1b}/A_c = 0.2$							
0.5	−.85	−.10	1.1	2.7	4.8	7.3	10
1.0	−.85	−.05	1.4	3.1	5.1	7.4	—
2.0	−.85	−.31	1.8	3.4	—	—	—
$A_{1b}/A_c = 0.4$							
0.5	−.85	−.29	0.34	1.0	1.8	2.6	3.4
1.0	−.85	−.14	0.60	1.3	2.1	2.7	—
2.0	−.85	0.12	1.0	1.7	—	—	—
$A_{1b}/A_c = 0.6$							
0.5	−.85	−.32	0.20	0.72	1.2	1.7	2.1
1.0	−.85	−.18	0.46	1.0	1.5	1.9	—
2.0	−.85	0.09	0.88	1.4	—	—	—

$A_{1b}/A_c = 0.8$						
0.5	−.85	−.33	0.13	0.61	1.0	1.4	1.7
1.0	−.85	−.18	0.41	0.91	1.3	1.5	—
2.0	−.85	0.08	0.83	1.3	—	—	—
$A_{1b}/A_c = 1.0$							
0.5	−.85	−.34	0.13	0.56	0.93	1.3	1.5
1.0	−.85	−.19	0.39	0.86	1.2	1.4	—
2.0	−.85	0.07	0.81	1.2	—	—	—

	Main					
Q_s/Q_c	0	0.1	0.2	0.3	0.4	0.5
$C_{c,s}$	1.2	1.2	1.2	1.1	1.1	0.96
Q_s/Q_c	0.6	0.7	0.8	0.9	1.0	
$C_{c,s}$	0.85	0.72	0.56	0.39	0.20	

Diverging Flow: Use Fitting 5-23.

OBSTRUCTIONS

6-1 Damper, Butterfly, Round (Idelchik 1986, Diagram 9-16; Zolotov 1967)

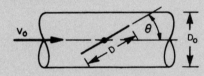

	C_o										
$\frac{D}{D_o}$	θ, degrees										
	0	10	20	30	40	50	60	70	75	80	85
0.5	0.19	0.27	0.37	0.49	0.61	0.74	0.86	0.96	0.99	1.0	1.0
0.6	0.19	0.32	0.48	0.69	0.94	1.2	1.5	1.7	1.8	1.9	1.9
0.7	0.19	0.37	0.64	1.0	1.5	2.1	2.8	3.5	3.7	3.9	4.1
0.8	0.19	0.45	0.87	1.6	2.6	4.1	6.1	8.4	9.4	10	10
0.9	0.19	0.54	1.2	2.5	5.0	9.6	17	30	38	45	50
1.0	0.19	0.67	1.8	4.4	11	32	113	—	—	—	—

6-2 Damper, Butterfly, Rectangular (Idelchik 1986, Diagram 9-17; Zolotov 1967)

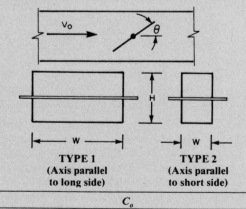

TYPE 1
(Axis parallel to long side)

TYPE 2
(Axis parallel to short side)

		C_o								
		θ, degrees								
Type	H/W	0	10	20	30	40	50	60	65	70
1	< 0.25	0.04	0.30	1.1	3.0	8.0	23	60	100	190
1	0.25–1.0	0.08	0.33	1.2	3.3	9.0	26	70	128	210
2	> 1.0	0.13	0.35	1.3	3.6	10	29	80	155	230

Fitting Loss Coefficients

(Reprinted by permission from 1989 ASHRAE Handbook – Fundamentals)

Figure 11.10 (cont.)

duct carrying 3,000 cfm has a velocity of 1,800 fpm. This is above the 1,500 fpm maximum duct velocity for banks. To find the minimum duct size for 3,000 cfm at 1,500 fpm:

area = cfm/fpm, 3,000/1,500 = 2 S.F. x 144/24" = 12"

A 24" x 12" duct should be used in lieu of the 24" x 10" duct, to use up as much pressure as possible without generating noise. A balancing damper in the branch duct will fine tune the balancing.

The reduced branch duct size saves sheet metal.

The 3,000 cfm branch duct would be 24" x 18" if it were sized as equal friction using 0.65 in wg/100:

$$\frac{(24" + 18") \times 2 \text{ (sides)}}{12"/\text{ft.}} \times 20' = 140 \text{ S.F.}$$

$$\frac{(24" + 12") \times 2}{12"/\text{ft.} \times 20'} = 120 \text{ S.F.}$$

A savings of 20 S.F. or 15% of the sheet metal can be realized while providing a better job.

Duct Leakage

Duct leakage is responsible for one of the major air distribution problems. Unsealed ductwork can cause 40% of the air that enters a supply duct to leave the duct before reaching its intended destination. The rate of duct air leakage increases as the pressure in the duct increases. Duct sealing should be specified and practiced, but it is always prudent to keep duct pressures as low as practical. Methods for avoiding and correcting duct leakage are addressed in Chapter 19.

Acoustic Lining and Fibrous Glass Duct

Acoustic lining is insulating material applied to the interior duct surface to attenuate noise. ASHRAE and SMACNA have found that fibrous glass liners that are air side spray-coated and mechanically fastened propagate a duct friction loss that is twice that of sheet metal, "based on good workmanship." In actual practice, unfinished edges, gaps between the liner sections, and physical abuse increase these values. Acoustically-lined duct friction factors must be addressed in determining system friction losses. A friction loss of 2-1/2 times the sheet metal friction loss is appropriate.

Rigid fibrous glass air duct increases the duct friction by 1.4 times. Fiberglass air duct usually offers an installed cost that is about 25% less than the installed cost of insulated sheet metal duct.

The maximum velocity of fiberglass air duct is 2,400 fpm and the 2" maximum static pressure must not be exceeded. Although fiberglass air duct results in a quieter duct system, the duct static pressure must be kept low, to avoid excessive leakage.

Flexible duct "when fully extended" has double the friction loss of sheet metal duct. Because flexible duct is rarely installed fully extended, a friction loss of three to five times greater than that of sheet metal duct should be used.

Round Duct

Round ductwork is the most efficient configuration for ductwork design. In general, round ductwork is less expensive to install than rectangular duct. Local labor regulations and shop practices can be a deciding factor for the use of round ductwork. As shown previously in this chapter, round duct can use 27% less metal than rectangular duct. Higher duct velocities can be tolerated in round duct than in rectangular duct, because the rumbling pulsations found in rectangular duct systems are avoided. The use of round duct mains is, however, handicapped by the available duct space above ceilings. The 42″ round duct for the 15,000 cfm systems requires 18″ more space than the 60″ x 24″ rectangular duct. If it is not economical to install multiple round main ducts, a rectangular main duct and round branch ducts can be installed.

Rooftop Ductwork

Rooftop ductwork is addressed in Chapter 19, in Part II of this book.

Fume Hood Ductwork

Ductwork used to exhaust laboratory fume hoods must be sealed airtight to eliminate the possibility of contaminated air escaping into the occupied space. Laboratory exhaust negative duct pressures could be 3″-4″ wg, especially if HEPA (high efficiency particulate air) filters are involved. The leakage rate at these pressures could be devastating to system performance. Welded duct joints should ensure a tight system, but dramatically increase the cost of the ductwork installed. Ductwork installed outdoors should be stainless steel with welded joints, to enhance performance and minimize maintenance.

Many local codes require stainless steel duct systems for laboratory fume hood exhaust, on the basis that there is no assurance that although noxious chemicals are not planned to be used at the time of design, chemicals may be used in the laboratory that may not be compatible with galvanized ducts.

Fume hood exhaust systems are under negative pressure, and duct construction must follow the SMACNA "Accepted Industry Practice for Industrial Duct Construction." The SMACNA duct construction standard permits 24 gauge for 18″ duct. SMACNA's "Industrial Duct Construction" guide requires 20 gauge. Ducts under negative pressure have different construction standards than ducts under positive pressure. Round ducts have different construction standards than rectangular ducts. It is important to use the appropriate design guides for each application.

Round ducts should be used as much as possible for fume hood exhaust systems. In addition to the reasons previously mentioned, round ductwork is easier to seal airtight and much easier to weld than rectangular duct.

Fume hood exhaust stacks must not terminate with a rain cap. Rain caps not only add a large pressure loss to the system, but cause the contaminated exhaust air to be deflected down toward the roof where it can come into contact with people and possibly be recirculated into the building. Exhaust stacks should discharge vertically at a 3,000 fpm velocity to disperse the plume into the atmosphere. As mentioned earlier, the exterior ductwork should be stainless steel with welded joints. Figure 11.11 shows a stackhead design.

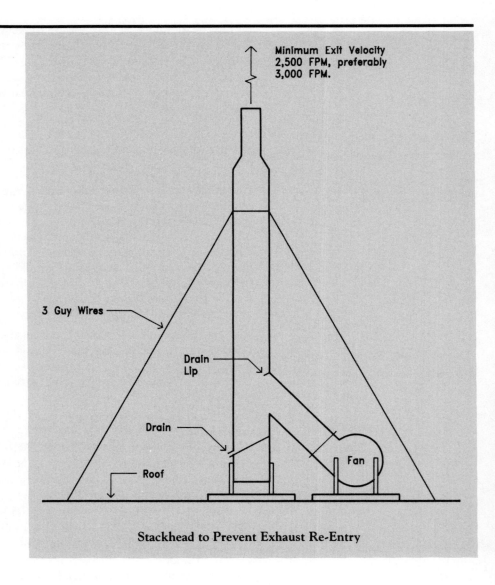

Figure 11.11

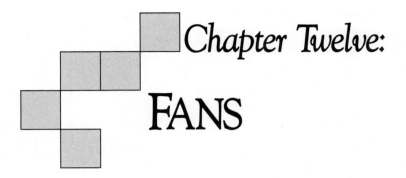

Chapter Twelve:

FANS

Although there are many types of fans, there are just two basic fan categories: centrifugal and axial. In a centrifugal fan, airflow enters the rotor and makes a 90°F turn in all directions. The air is captured by a scroll-shaped housing and directed through the fan discharge. Axial fans have propeller-type rotors and produce an airflow in a straight path through the fan. Figure 12.1 shows the various types of fans.

Centrifugal Fans   Backward-Curved Fans

Backward-curved fans have flat wheel blades that lean away from the direction that the fan is turning. The fan has a higher running speed and efficiency for a given amount of air than the forward-curved or radial-curved fans. The horsepower characteristic curve reaches a maximum near the peak efficiency and becomes lower toward the free delivery end of the fan curve.

Motors selected above the maximum horsepower curve will give a non-overloading benefit to the selection. The backward-curved fan is used for general HVAC systems. Although it is used for low, medium and high pressure applications, it is especially effective when large air volumes and high static pressures are encountered. The fan is used in industrial applications where airfoil fans would not be acceptable because of corrosion or erosion to the airfoil blades.

Airfoil Fans
Airfoil fans are backward-curved fans with deep airfoil blades. These are the most efficient of all centrifugal fans and turn at a higher speed than the other centrifugal fans.

Forward-Curved Fans
Forward-curved fans use small blades that are curved forward in the direction of the wheel's rotation. Forward-curved fans run at speed of backward-curved fans. Because of the low speed and smaller wheel, the construction is lighter and the cost is lower than that of the backward-curved fan. The pressure curve is less steep than the backward-curved fan, which makes it more effective for most VAV systems. The pressure curve also has a dip, and operation to the left of the peak should be avoided. The bhp curve rises continuously, which can cause overloading to occur. Forward-curved fans cannot develop the

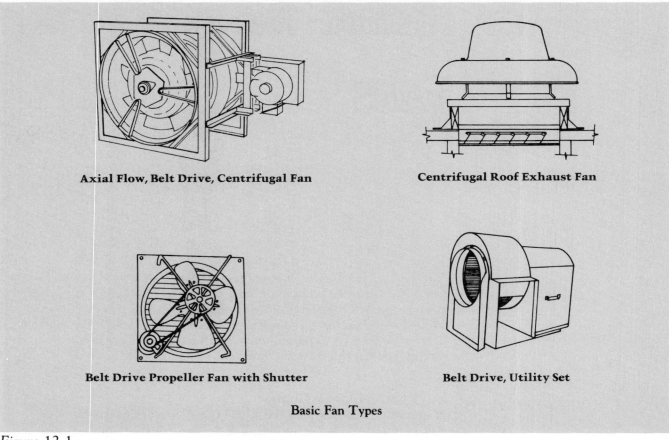

Axial Flow, Belt Drive, Centrifugal Fan

Centrifugal Roof Exhaust Fan

Belt Drive Propeller Fan with Shutter

Belt Drive, Utility Set

Basic Fan Types

Figure 12.1

pressures of backward-curved fans and their efficiency is lower. These fans are used in low pressure HVAC applications, including packaged air-conditioning units and window air conditioners, and residential furnaces.

Axial Fans

Vaneaxial Fans

The vaneaxial fan is the most efficient of the axial fans. It uses straightening vanes to improve both efficiency and pressure capability. The most effective vaneaxial fans use airfoil blades, which can be fixed or adjustable. Vaneaxial fans used in VAV applications employ pitch-in-motion blades to automatically control the blade pitch for the desired air volume. The vaneaxial performance curve has a dip similar to that of the forward-curved fan. Vaneaxial fans are used in general HVAC systems in low, medium, and high pressure applications. They are an especially attractive option in cases where straight-through airflow applications apply. The fact that the vaneaxial fan is conveniently configured for a direct drive arrangement eliminates the major fan maintenance activity, fan belts. It should be noted that vaneaxial fans tend to be noisier than centrifugal fans, and inlet and outlet sound traps may be advisable.

Propeller Fans

Propeller fan wheels are usually two or more single thickness blades in a ring enclosure. The efficiencies are low, and the pressure capability extremely low. They are best applied to moving large volumes of air at very low pressure, such as circulating air from ceiling fans, desk and floor fans. Propeller fans are also used to exhaust large volumes of air direction outdoors (via wall-mounted fans).

Tube Axial Fans

Tube axial fans are used where space limitations prohibit the larger housing required for a centrifugal fan. These high efficiency fans can also be used as roof ventilators. Fans operating in parallel and in series are discussed in Chapter 20.

System Effect Factor

One of the most frequent reasons for poor air distribution system performance is failure to consider the system effect factor in selecting the fan. It should be noted that fans are rated based on tests that are performed according to the standard testing procedure established by the Air Movement and Control Association, Inc. (AMCA). The only place one will find fans connected as they were in the test lab, is in the test lab. It is almost impossible to find a fan installed in the real world, exactly as it was in the testing lab.

The published data of a fan's performance is based on lab tests. In order for us to know what a fan will produce when it is not connected as it was in the lab, AMCA has established the system effect factor. The loss of fan performance due to fan inlet and outlet connections that will reduce the fan output, can be calculated from published rating tables. By knowing in advance what the deficiency is, we can select a fan that will compensate for the system effect factor. It is not uncommon to find that a fan is delivering half of the cfm that was intended, because the system effect factor has been overlooked. (The system effect factor is covered extensively in Chapter 20.)

Fans and Air Density

Density is the ratio of a specimen's mass to its volume. It is the mass of a unit volume of a substance. The substance that we are concerned with here is air. The term "standard air" applies to air that is 0.075 pounds per cubic foot at 68°F and at sea level. When a cubic foot of air is heated, its density changes. Its mass is reduced. Air heated to 250°F weighs only 0.055 pounds per cubic foot. The density (and mass) of air increases when it is cooled. Air at 0°F weighs 0.087 pounds per cubic foot.

Density cannot be ignored when selecting fans and air distribution systems. A duct system's resistance is directly related to the density of the air that flows through it. Fans are rated for 0.075 pounds per cubic foot of air. When air at any other density is moved through the fan, the fan's performance is affected.

Overlooking air density will produce enormous operating problems with induced draft fan systems for boilers, engine tail pipe exhaust systems, kitchen exhaust systems, foundry exhaust systems, etc. Chapter 27 explores air density in detail.

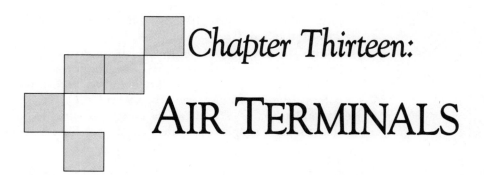

Chapter Thirteen:

AIR TERMINALS

The last thing that air contacts before it enters the space it is to condition is the supply (register or diffuser). The effectiveness of that air as it travels through the room depends on the proper selection of the register or diffuser.

Supply Air Outlets

Comfort air distribution conditions can be obtained with room average air velocities of 50 fpm in the occupied zone, within 6′ of the floor. Without adequate air velocity, stagnant air will result in high temperatures and discomfort. The velocity of the supply air at the air outlet to the room induces the surrounding room air into the supply airflow. This process of air entrainment is essential to good air distribution because it is responsible for eliminating stagnant air areas.

When an air outlet is located so that it discharges air against the ceiling, air is entrained into the lower portion of the air stream. If air is discharged from a grille or diffuser mounted on an exposed duct into the free open space, the throw (distance the supply air is blown from the terminal unit) will be shortened by one-third.

Another consideration is smudging. The deposit of dirt particles (held in suspension in the surrounding room air which is induced into the airstream) leaves marked areas on the ceiling. Anti-smudge rings will help reduce the deposit, but the best solution is to select outlets with lower velocities.

Selection Guidelines

The following procedure should be followed for selecting supply air outlets:
- Determine how much air is to be supplied to each room (based on the heating and cooling calculations).
- Select the type and quantity of outlets for each room by determining each of the following:
  — The cfm of each outlet.
  — The throw from each outlet. The throw is the horizontal distance to a point where either 150, 100, or 50 fpm velocity is reached.
  — Building structural characteristics. These will include any beams or soffits that may interfere with the airflow from the air outlet. For example, suspended light fixtures could impede airflow.
  — Architectural aesthetic requirements. Generally, architects prefer to have air outlets as invisible as possible. Lay-in ceilings are

adaptable to several attractive diffusers that can establish a pattern with the light fixtures.

- Locate outlets to provide as uniform an air distribution as possible.
- Size the outlet correctly, using the manufacturer's current catalog, and taking into consideration the outlet cfm, the throw, the discharge pattern, the pressure loss, and the sound level.
- Air outlets for VAV applications must be selected to accommodate the minimum supply cfm while still maintaining the required throw.

Figure 13.1 is a guide for supply outlet selection.

The most popular types of supply outlets are listed in Figure 13.2. The most common applications of these outlets are also listed in this figure. Figure 13.3 is an illustration of the various air outlets.

Supply outlet performance characteristics are described in Figure 13.4. The type, mounting location, discharge direction, and cooling and heating characteristics of the various outlets are given.

Return Air Inlets

Return air inlets must not be located in the direct path of air outlets. Return air inlets should be accepting room air after the 50 fpm throw velocity has been attained. Returns should be located to eliminate a stagnant air condition in the room.

Figure 13.5 gives the recommended return air inlet face velocities. The velocities listed are for the gross inlet area, not the free (grille) area. Figure 13.6 lists the various types of return and exhaust air inlets and their applications. Several of the frequently used accessory devices such as dampers, equalizing grids, blank-offs, etc. that are used with air terminals are listed in Figure 13.7.

Guide for Selection of Supply Outlets	
Type of Outlet	cfm/S.F. of Floor Space
Grilles and registers	0.6 to 1.2
Slot diffusers	0.8 to 2.0
Perforated panel	0.9 to 3.0
Ceiling diffuser	0.9 to 5.0
Perforated ceiling	3.0 to 10.0

(Courtesy of SMACNA)

Figure 13.1

Supply Outlet Types	
Type of Outlet	**Applications**
Adjustable double deflection bar grille	Preferred grille for sidewall installation. Provides both vertical and horizontal air deflection.
Fixed bar grille	Long perimeter grille installations.
Curved vane grilles	Ceiling and high sidewall installation.
Stamped plate grilles	Architectural use only. No adjustment.
Slot diffuser	High sidewall or ceiling installation.
Air light fixture slot diffuser	Ceiling installation. Must be ordered to match light fixture.
Round ceiling diffuser	Install in center of area served.
Square diffuser	Install in center of area served or safe off louvers where airflow is not desired. Size diffuser accordingly.
Adjustable pattern square or round ceiling diffusers	For exposed duct applications where there is no ceiling and it is more difficult to control discharge pattern.
Multi-pattern square and rectangular ceiling diffuser	Install in center of area or adjacent to partitions. Set air pattern according to needs.
Half round diffuser	Install in ceiling adjacent to partition or high sidewall.
Supply and return concentric diffuser	Install in center of area served.
Perforated face diffuser	Install in center of area served or control discharge pattern when installed off center of area served.
Variable area diffuser	Use with VAV to minimize throw variation. Diffuser discharge area can be varied.
Air distributing ceiling	Ceiling system round holes or slots distribute air when ceiling is supply plenum. Use for cooling only. Suitable for large zone areas.
Egg crate grille	Ceiling or sidewall return air. No pattern adjustment.

Figure 13.2

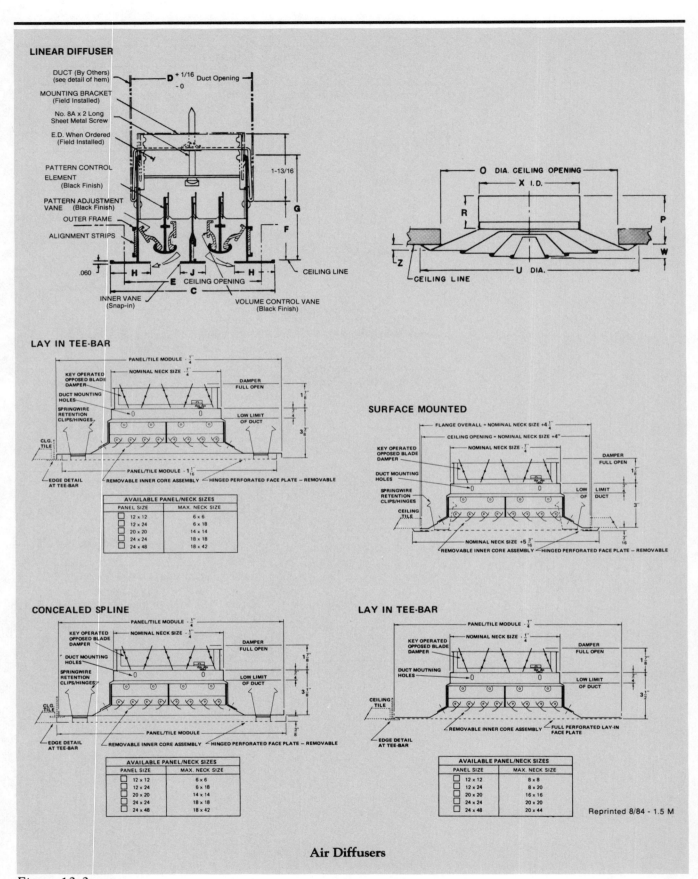

LINEAR DIFFUSER

DUCT (By Others)
(see detail of hem)

MOUNTING BRACKET
(Field Installed)

No. 8A x 2 Long
Sheet Metal Screw

E.D. When Ordered
(Field Installed)

PATTERN CONTROL
ELEMENT
(Black Finish)

PATTERN ADJUSTMENT
VANE (Black Finish)

OUTER FRAME

ALIGNMENT STRIPS

INNER VANE
(Snap-in)

VOLUME CONTROL VANE
(Black Finish)

CEILING LINE

CEILING OPENING

O DIA. CEILING OPENING

X I.D.

CEILING LINE

U DIA.

LAY IN TEE-BAR

PANEL/TILE MODULE

NOMINAL NECK SIZE

KEY OPERATED
OPPOSED BLADE
DAMPER

DUCT MOUNTING
HOLES

SPRINGWIRE
RETENTION
CLIPS/HINGES

CLG. TILE

DAMPER FULL OPEN

LOW LIMIT
OF DUCT

EDGE DETAIL
AT TEE-BAR

PANEL/TILE MODULE

REMOVABLE INNER CORE ASSEMBLY — HINGED PERFORATED FACE PLATE – REMOVABLE

AVAILABLE PANEL/NECK SIZES	
PANEL SIZE	MAX. NECK SIZE
12 x 12	6 x 6
12 x 24	6 x 18
20 x 20	14 x 14
24 x 24	18 x 18
24 x 48	18 x 42

SURFACE MOUNTED

FLANGE OVERALL = NOMINAL NECK SIZE +4 1/4"

CEILING OPENING = NOMINAL NECK SIZE +4"

KEY OPERATED
OPPOSED BLADE
DAMPER

DUCT MOUNTING
HOLES

SPRINGWIRE
RETENTION
CLIPS/HINGES

CEILING
TILE

NOMINAL NECK SIZE

DAMPER FULL OPEN

LOW LIMIT OF DUCT

NOMINAL NECK SIZE +5 3/16"

REMOVABLE INNER CORE ASSEMBLY — HINGED PERFORATED FACE PLATE – REMOVABLE

CONCEALED SPLINE

PANEL/TILE MODULE

NOMINAL NECK SIZE

KEY OPERATED
OPPOSED BLADE
DAMPER

DUCT MOUNTING
HOLES

SPRINGWIRE
RETENTION
CLIPS/HINGES

CLG. TILE

DAMPER FULL OPEN

LOW LIMIT
OF DUCT

EDGE DETAIL
AT TEE-BAR

PANEL/TILE MODULE

REMOVABLE INNER CORE ASSEMBLY — HINGED PERFORATED FACE PLATE – REMOVABLE

AVAILABLE PANEL/NECK SIZES	
PANEL SIZE	MAX. NECK SIZE
12 x 12	6 x 6
12 x 24	6 x 18
20 x 20	14 x 14
24 x 24	18 x 18
24 x 48	18 x 42

LAY IN TEE-BAR

PANEL/TILE MODULE

NOMINAL NECK SIZE

KEY OPERATED
OPPOSED BLADE
DAMPER

DUCT MOUTNING
HOLES

CEILING
TILE

DAMPER FULL OPEN

LOW LIMIT
OF DUCT

EDGE DETAIL
AT TEE-BAR

REMOVABLE INNER CORE ASSEMBLY — FULL PERFORATED LAY-IN FACE PLATE

AVAILABLE PANEL/NECK SIZES	
PANEL SIZE	MAX. NECK SIZE
12 x 12	8 x 8
12 x 24	8 x 20
20 x 20	16 x 16
24 x 24	20 x 20
24 x 48	20 x 44

Reprinted 8/84 - 1.5 M

Air Diffusers

Figure 13.3a

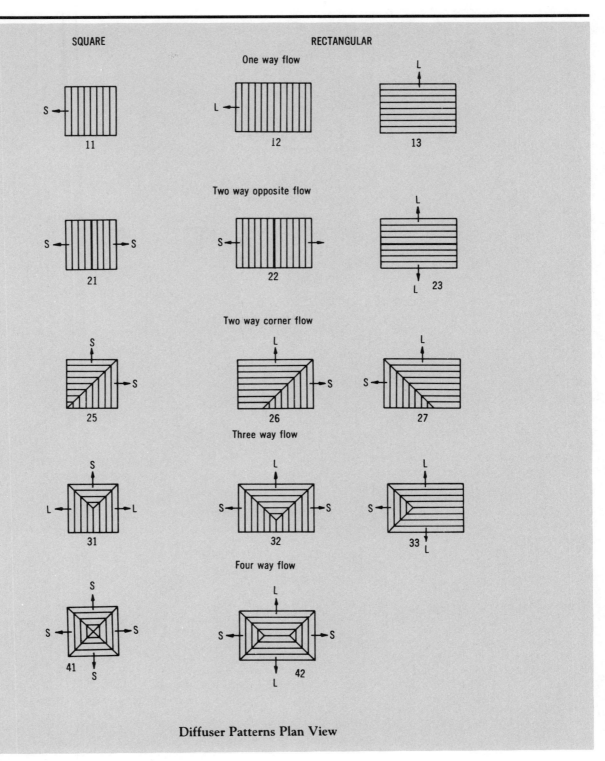

Diffuser Patterns Plan View

Figure 13.3b

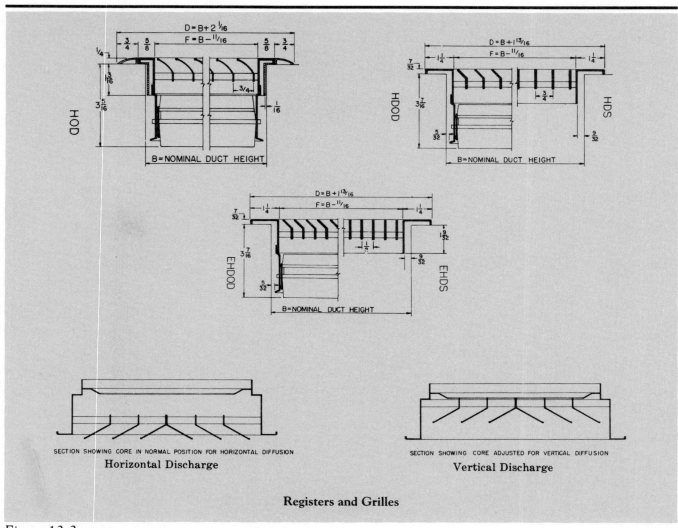

Registers and Grilles

Figure 13.3c

Supply Air Outlet Performance				
			Characteristics	
Type	Mounting	Discharge	Cooling	Heating
High Sidewall Grilles, Sidewall Diffusers, Ceiling Diffusers, Slot Diffusers	Ceiling, High Sidewall	Horizontal	Good mixing with warm room air. Minimum temperature variation within room. Particularly suited to cooling applications	Large stagnant air near floor. In interior zones where loading is not severe, stagnant air area is practically nonexistent
Floor Grilles, Baseboard Units, Fixed Bar Grilles, Linear Grilles	Floor, Low Sidewall, Sill	Vertical Non-Spreading Air Jet	Small stagnant air area generally above occupied zone	Smaller stagnant air area than above-type air outlets
Floor Grilles, Adjustable Bar Grilles, Linear Diffusers	Floor, Low Sidewall	Vertical Spreading Air Jet	Larger stagnant air area than above-type outlets	Smaller stagnant air area than above-type outlets–particularly suited to heating applications
Baseboard Units, Grilles	Floor, Low Sidewall	Horizontal	Large stagnant area above floor in occupied zone. Not recommended for comfort cooling	Uniform temperature throughout area. Recommended for process applications
Ceiling Diffusers, Linear Grilles, Grilles, Slot Diffusers (Vertical Flow), Sidewall Diffusers	Ceiling, High Sidewall	Vertical	Small stagnant air area near ceiling. Select for cooling only applications	Good air distribution. Select for heating only applications
Variable Area Grille, Variable Area Diffuser	Ceiling, High Sidewall	Horizontal, Specially adapted for variable air volume systems	Maintain design air distribution characteristics as air volume changes	Maintain design air distribution characteristics as air volume changes

Figure 13.4

Figure 13.5

Recommended Return Inlet Face Velocities	
Inlet Location	**Gross Inlet Area Velocity fpm**
Above occupied zone	800 and up
Within the occupied zone, but away from the seats	600 – 800
Within occupied zone	400 – 600
Door or wall louvers	200 – 300
Undercut doors	200 – 300

(Courtesy of SMACNA)

Figure 13.6

Return and Exhaust Air Inlet Types	
Type	**Applications**
Fixed bar grille	Return, exhaust and transfer grilles
Adjustable bar grilles	Return, exhaust and transfer grilles. Grilles have a vane pattern to match the air outlets.
V-bar grille	Site-proof and suited for door louvers
Light-proof grille	Used for darkrooms
Stamped grilles	Used for return, exhaust and transfer grilles when matched to supply outlets
Ceiling diffusers	Used for return, exhaust and transfer grilles when matched to supply outlets
Slot diffusers	Used for return, exhaust and transfer grilles when matched to supply outlets
Air light fixture	Used for return, exhaust and transfer grilles when matched to supply outlets
Perforated face inlet	Used for return, exhaust and transfer grilles when matched to supply outlets
Egg crate grille	Used for return, exhaust and transfer grilles when matched to supply outlets

(Courtesy of SMACNA)

Air Terminal Accessory Devices	
Device	**Applications**
Opposed blade volume dampers	Located behind the grille of damper to adjust air volume.
Multi-shutter damper	Used to adjust air volume only when air stream deflection is not objectionable. Parallel blade damper will deflect the airstream when damper is partially open.
Slot diffuser flow equalizing vanes	Adjust discharge pattern of slot diffuser.
Multi-louver round diffuser damper	Series of parallel blades adjust air volume.
Opposed blade round diffuser damper	Pie-shaped blade damper adjusts air volume.
Diffuser splitter damper	Hinged plate at duct branch adjusts air volume. Must be used with equalizing device.
Diffuser equalizing device	Individual adjusted blades provide uniform flow to diffuser.
Diffuser blank-off baffle	Blank-off section of diffuser reduces flow in a given direction.
Diffuser anti-smudge rings	Minimizes ceiling smudging.
Air light fixture slot diffuser plenum with damper flex collar inlet	Attached to light fixture, controls air connection to light fixture slot diffuser.
Linear grille blank-offs	Caps to blank of inactive sections of continuous linear grille.
Linear grille plenum	Attach to linear grille to connect air supply.
Adapter	To adapt square or rectangular diffuser necks to round duct connection.
Plaster frames	Secondaray plaster frame installed prior to plastering. Permits removal of grille without damaging plaster.

(Courtesy of SMACNA)

Figure 13.7

Chapter Fourteen:

VARIABLE AIR VOLUME SYSTEMS

Variable air volume systems became popular with the arrival of the energy crisis of the mid 1970s. Constant volume reheat systems, high and medium pressure dual duct systems, and induction systems were pretty much the state of the art at that time.

Constant volume reheat systems are not energy efficient. Air is cooled to 55°F and supplied to all zones in the system. A steam, hot water, or electric reheat coil installed in the duct serving a zone is activated if the zone is overcooled. Energy is expended to cool the entire system air volume, and additional energy is used to reheat this same air.

Variable air volume, VAV (Figure 14.1) was promoted as a method of eliminating reheat and saving fan energy. VAV accomplishes this by supplying only that volume of cold air to a space that is needed to satisfy the cooling load in the space. Figure 14.2 shows that the size of the 55°F "cloud" of cool air supplied to a room is varied to match the cooling load in the room.

Fan energy is saved when the volume of air handled by the fan is reduced. This is accomplished by installing an automatic damper in the supply duct to each zone. As the room temperature becomes satisfied, the thermostat signals the automatic damper operator to move the supply air zone damper toward the closed position. Figure 14.1 is a schematic layout of a VAV system.

When the zone dampers are throttled, the static pressure in the supply duct is increased. The static pressure sensor located in the supply duct senses the static pressure increase, and adjusts the air volume by:
- Positioning a supply air duct discharge damper.
- Operating supply and return fan variable inlet vanes (VIV).
- Controlling supply and return fan speed.
- Changing vaneaxial fan blade pitch.

Discharge Damper Control

There are more discharge damper air volume systems installed than any other type of VAV system. This popularity is due primarily to their low installation cost. Damper control can be accomplished in two ways.

1. *Let the supply duct static pressure increase as the VAV devices throttle.* No static pressure control or controller will be used. This is the most widely used of all VAV control systems. When the room thermostat throttles the VAV damper, the supply duct static pressure (sp) will

rise. All VAV devices in the system will have to operate against a higher sp differential than when all VAV dampers are open. The greater the static pressure across the VAV, the greater the noise level. Figure 14.3 shows that a 10″ VAV device will generate an NC of 27 delivering 1,200 cfm at 0.50″ sp difference, and will produce an NC of 39 at 1.5″, and an NC of 48 at 3″ sp difference. An NC of 35 is considered acceptable for general offices. This type of VAV control is appropriate for systems with very low supply duct sp, and when an increase in noise level is acceptable.

2. *Have the supply duct static pressure controller operate a damper located at the supply fan outlet.* This is one step up in the VAV ladder. The increase in duct sp will be absorbed by the damper at the supply fan outlet. As Figure 14.4 shows, a damper positioned at 45° in a duct with air traveling at 1,800 fpm, will produce an octave band level of 60.

Figure 14.5 is a set of Noise Criterion Curves. The curves show that the damper generates an NC-60, which can be very objectionable in an air distribution system. Noise travels at a speed of 66,000 feet per minute, while air is traveling in the duct at 1,600 feet per minute. Noise will travel in every direction in an air distribution system.

Damper control VAV systems do not provide the performance and energy savings that other VAV control systems can produce. Chapter 18 explains

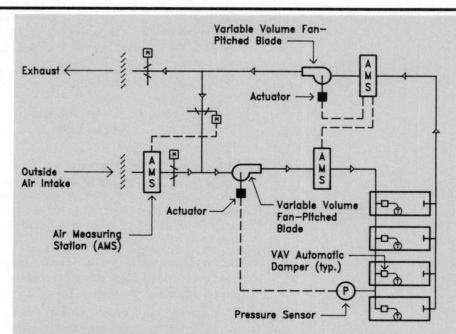

1. The Air Measuring Station (AMS) measures the Supply Fan Air Volume, and adjusts the Return Air Volume to the correct proportion.

2. The AMS located in the Outside Air duct maintains the correct Outside Air volume.

3. Unbalance between Supply and Return Air systems and room over or under pressurization, are eliminated.

Variable Air Volume System

Figure 14.1

that damper-controlled VAV systems can only modulate the air volume about 30-35% with forward-curved fans, and 10-15% with backward-curved fans. Smaller damper-controlled VAV systems with backward-curved fans can perform satisfactorily, as long as fan energy savings and very quiet systems are not important.

Selecting Fans With VAV Damper Control

Figure 14.6 shows fan curves for three different fans. If the system is designed for 40,000 cfm at 3″ sp, the size 98 fan will meet those conditions, operating at 81% efficiency, with 21 bhp. This is an excellent selection for a constant volume application. Operating under damper control, this fan will reach the surge line at 35,000 cfm. The 12-1/2% air modulation is not worth the effort. If one size smaller fan was selected, the size 89 fan would operate at 78% efficiency with 23 bhp. This fan reaches the surge line with damper control at 30,000 cfm, a 25% modulation. The size 80 fan is two sizes smaller than the 98 fan, and would operate at 74% efficiency with 24 bhp at peak load. Surge is not reached with this fan until the air volume is reduced to 27,000 cfm. A 33% modulation is outstanding for a damper-controlled VAV system.

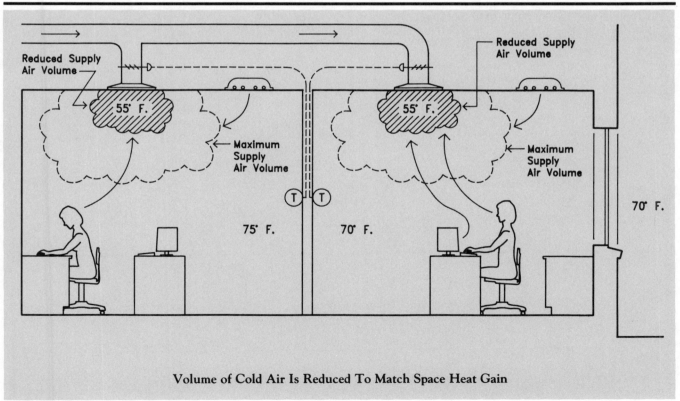

Volume of Cold Air Is Reduced To Match Space Heat Gain

Figure 14.2

The VAV fan is to provide the necessary modulation to make the system perform. The small increase in bhp at peak load is minor compared to the avoidance of reheat for temperature control. Selecting a fan two sizes smaller than the one at maximum efficiency will produce a better system, and reduce construction costs. This is the reason that most existing backward-curved fans used in constant volume systems have performed poorly, when the systems were retrofitted to VAV.

Variable Inlet Vanes (VIV)

Variable inlet vanes are a series of operable vanes located at the inlet to a fan. The vanes give a spinning motion to the air entering the fan. The spinning air rotates in the same direction as the fan impeller, and reduces the fan's performance. The fan efficiency at reduced air volume is maintained, resulting in lower cfm and bhp. Figure 14.7 shows centrifugal fan performance with inlet vanes.

The system effect of wide open inlet vanes installed at the inlet to the fan must be taken into consideration. When fans are selected for inlet van

Variable Air Volume Box Selection Data

	Standard Range		High Range		Discharge NC* at Pressure Difference (p.d.)			Radiated NC* at Pressure Difference (p.d)		
Size	CFM	Min p.d.	CFM	Min p.d.	.50″	1.5″	3.0″	.50″	1.5″	3.0″
10	600	.13	—	—	23	38	47	—	25	30
	800	.23	800	.03	24	38	46	—	28	36
	1000	.36	1000	.05	27	40	47	22	31	40
	—	—	1200	.06	27	39	48	23	34	41
	—	—	1500	.11	31	39	47	24	35	43
12	800	.11	—	—	21	36	46	—	27	34
	1000	.17	—	—	22	36	46	—	29	37
	1200	.25	1200	.03	23	36	45	—	32	39
	1400	.34	1400	.05	24	37	45	20	33	41
	—	—	1800	.07	26	36	45	23	35	44
	—	—	2400	.13	30	37	45	26	38	46
14	1000	.09	—	—	24	37	48	—	30	38
	1400	.18	1400	.02	25	36	46	20	32	43
	1800	.30	1800	.04	26	37	46	22	34	45
	—	—	2400	.07	26	36	47	27	38	47
	—	—	3200	.12	31	38	49	32	41	51

*NC = Noise Criteria

(Adapted from Anemostat)

Figure 14.3

application, the manufacturer's fan curves for fans with inlet vanes should be used. Inlet vanes add a system effect of 0.4 to 0.5 in wg to the system.

The inlet guide vanes change the performance characteristics of the fan. A new pressure curve is established with each percentage of inlet van closure. Note that at 10,000 cfm, only 12 bhp is required. This is a 43% reduction in fan energy when delivering 37% less air.

Inlet vanes obstruct the airstream and slightly reduce fan efficiency. They also generate noise when they throttle to reduce airflow. Adjustments to damper linkage are needed from time to time. The installation costs of inlet vane fans are also considerably higher (ten to twenty times) those of damper control.

Controlling Supply and Return Fan Speed

Speed control is the most efficient method of providing fan air modulation. At each change in fan speed, a new fan pressure curve is established with no change in fan efficiency. AT 10,000 cfm, the fan is turning at 1,000 rpm and drawing only 4.9 bhp. At 10,000 cfm, damper control for the fan in Figure 14.5 will be operating at 17 bhp (12 bhp with

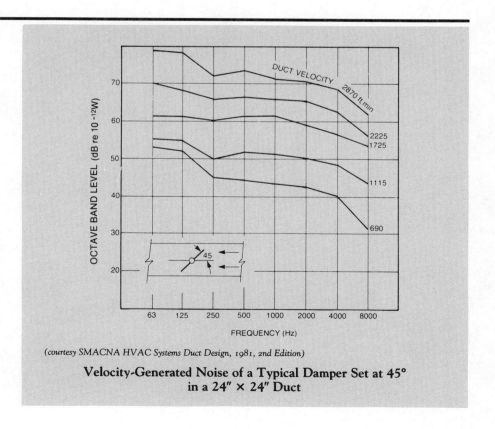

(courtesy SMACNA HVAC Systems Duct Design, 1981, 2nd Edition)

Velocity-Generated Noise of a Typical Damper Set at 45°
in a 24″ × 24″ Duct

Figure 14.4

VIV control and 4.9 bhp with speed control). The installed cost of inlet vanes is approximately one half that of a variable frequency drive speed control.

Changing Vaneaxial Fan Blade Pitch

Vaneaxial fans reduce the air volume by changing the pitch of the fan blades. Figure 14.8 shows the new fan pressure curve that is produced with the change in blade pitch.

For VAV application, the blade pitch must change while the fan is in motion. This is accomplished with an operator similar to those used for VIV. Note that the fan bhp at the air volume reduction to 10,000 cfm is slightly more than the speed control bhp (2%), but much less than the VIV bhp.

The major advantages of vaneaxial fan use for VAV are that they are convenient for direct drive operation, and blade pitch change is easily accomplished. The installed cost of vaneaxial, pitch in motion fans is about 15% less than that of centrifugal fans with speed control, and 25% more than centrifugal fans with VAV.

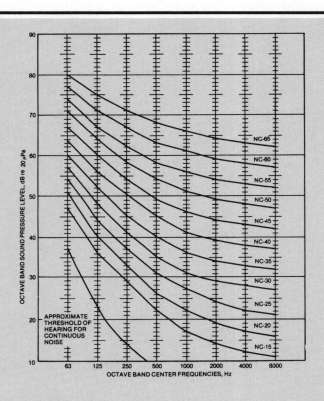

NC (Noise Criteria) Curves for Specifying the Design Level in Terms of the Maximum Permissible Sound Pressure Level for Each Frequency Band

Figure 14.5

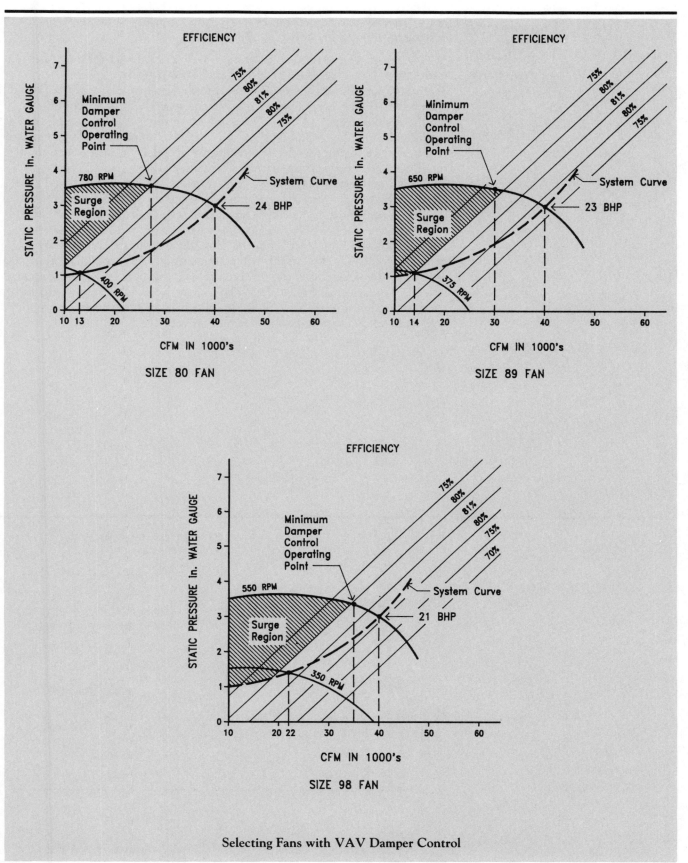

Selecting Fans with VAV Damper Control

Figure 14.6

VAV Design and Installation Requirements

For a VAV system to perform properly, the following items must be designed and installed correctly.

VAV Air Terminal Devices (VAV Boxes) Must Be Selected for the Required Air Modulation.

The main purpose of the variable air volume system is to supply only that volume of air that is necessary to satisfy the space heat gain (cooling requirements). Figure 14.3 is a VAV selection table. Let us assume that a maximum supply air volume of 1,300 cfm, and a minimum of 800 cfm are required. The instinctive decision would be to select from the "standard" range column, and to make a conservative selection. The 12″ box maximum delivery is 1,400 cfm. There may be some concern that 1,300 cfm may be too close to the maximum box capacity, which brings us to the 14″ box, a more comfortable selection, because it has an 1,800 cfm maximum capacity. This theory is fine with boilers, chillers, pumps, etc., but it is not valid for VAV. The 14″ box has a minimum air volume capability of 1,000 cfm, which will provide only a 23% modulation of 300 cfm. A better selection is the 12″ box, which will permit 38% modulation. The 10″ box selection, from the "high" range, is an even better selection. It provides the same 38% modulation, and is less expensive. Construction cost is saved and system performance improved.

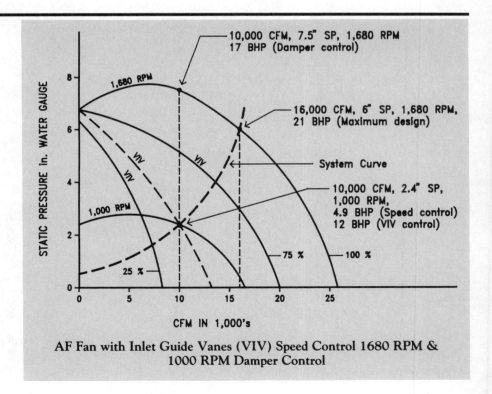

Figure 14.7

AF Fan with Inlet Guide Vanes (VIV) Speed Control 1680 RPM & 1000 RPM Damper Control

The VAV box maximum and minimum air volumes are measured by velocity sensors that are installed at the VAV box inlet. See Figure 18.4 in Chapter 18. The minimum velocity that these controllers can measure is 0.07 in wg. This is equal to 1,050 fpm. Once the velocity at the VAV box inlet falls below 1,050 fpm, the box can no longer control air volume. The only difference between the standard range VAV box and the high

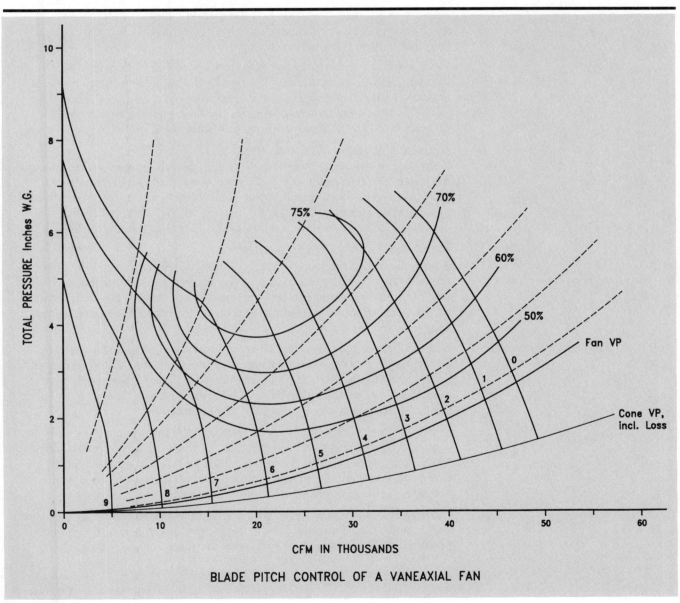

BLADE PITCH CONTROL OF A VANEAXIAL FAN

Figure 14.8

range VAV box is at the box inlet. The high range box has a "donut" installed at the inlet to increase the inlet velocity and allow a lower box air volume.

VAV Box Minimum Air Volume Should Not Create a Stuffy Environment.

The VAV box contributes to a lack of air movement in two ways: First, setting the minimum air volume controller too low, as explained previously, will cause loss of box volume control. Loss of control will fully close the VAV box damper. Occupants claim discomfort when the air circulation in the occupied space falls below four air changes. A maximum supply air volume of 0.75 cfm/SF relates to five air changes. A minimum of four air changes allows only a 20% air volume reduction. If the VAV box minimum cfm setting is less than 20%, complaints from occupants could result.

Second and most common are VAV stuffiness complaints that result from shut off VAV installations. If zero minimum air volume is set in the velocity controller, the VAV box damper will shut completely when the space temperature is satisfied. Zero air circulation will certainly bring complaints from occupants.

VAV systems are used in cooling systems. Cold air is delivered to a space in quantities to satisfy the cooling needs. If a 20-30% minimum cfm is set at the velocity controller, and the space heat gain does not require 20-30% of the maximum air volume, the space will be overcooled. Overcooling can be avoided in the following ways:

Reduce the supply air volume.

If the supply air temperature remains fixed, and the minimum cfm overcools the space, the minimum supply air volume will have to be reduced to maintain proper room temperatures. The VAV box may not accept a lower minimum air volume. This leaves shutting off the VAV (with the associated lack of ventilation and air movement) as the alternative.

Raise the supply air temperature.

This is the better solution to overcooling. The air handling unit supply air temperature can be reset upward, as the space temperatures approach their setpoint. This resetting can be implemented by a return air thermostat, or by the room thermostat sending signals to a microprocessor to reset the supply air temperature.

Resetting the supply air temperature upward during reduced loads can present some problems. Some zones may require more cooling than can be provided by the higher supply air temperature. The higher supply air temperature raises the space relative humidity, because less dehumidification is accomplished by the cooling coil.

A compromise in the supply air temperature dilemma is partial supply air reset and fine tuning room temperature control with some reheat. Good VAV performance occurs when the return air reset schedule is set for a lower maximum supply air temperature, and final temperature control during light loads is refined with the reheat coil.

Fan-powered VAV boxes provide a comfortable environment by reducing the volume of cold air, and raising the supply air temperature. The fan-powered VAV is available in two configurations:

Series fan-powered VAV box. Figure 14.9 shows that the series fan-powered box has a fan, a return air opening, a reheat coil, and a VAV damper controlling the primary cool air supply.

The fan operates continuously during the occupied mode to provide constant air delivery. The VAV problem of lack of air movement is thereby overcome. The room thermostat controls the primary air damper to supply only that volume of cool air that is needed to satisfy the space cooling load. The fan takes a mixture of cool primary air and return air from the room, and delivers the mixed air to the room air terminals. If heating is required, the primary supply air damper is closed, and only recirculated room air is heated. Whenever less than peak load conditions prevail, a higher temperature supply air is delivered to the room.

An additional benefit results during the morning warm-up period. The primary air damper is closed, so that no cool primary air or outside air is heated. Warmer air from the ceiling plenum is heated and delivered to the room.

The series fan-powered VAV box is used in low temperature supply air applications. The cold 45°F supply air is mixed with the 75°F room air to produce a 55°F supply air distribution to the space. Series boxes use more fan energy than parallel fan-powered boxes because the fan operates continuously. Even though the operating fan of a series box is noisier than a parallel box during the cooling mode, the constant fan operation does tend to make the series box more tolerable than the cycling fan noise of the parallel box.

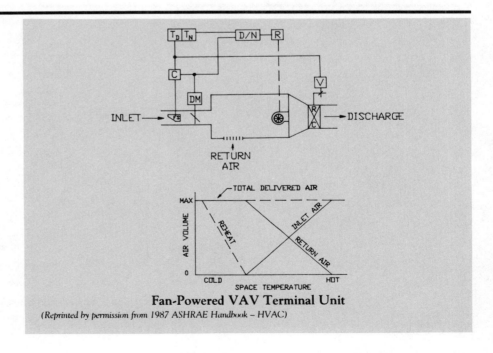

Fan-Powered VAV Terminal Unit

(Reprinted by permission from 1987 ASHRAE Handbook – HVAC)

Figure 14.9

Parallel flow fan-powered boxes, Figure 14.10 shows that the room thermostat controls the VAV damper to regulate the volume of cool air supplied to the room. The fan is not operated during the cooling mode. When the preset minimum primary air volume is reached, the fan is started, and a mixture of primary air and room air is supplied to the room. During the heating mode, the primary damper is closed before the reheat coil is activated. The disadvantage of fan-powered boxes is the added maintenance of fans having to be serviced within the occupied space.

VAV Box Inlet Duct Connection Must Permit Box Control.

Uniform airflow must be attained at the VAV box inlet for proper velocity controller operation. As shown in Chapter 18, a section of straight duct must be installed at the inlet to the box to produce the uniform flow.

Fans Must Be Selected for the Required Modulation.

This subject has been addressed earlier in this chapter and in Chapter 18.

Minimum Outside Air Ventilation Air Volume Must Not Decrease

A decrease in fan air volume propagates reduced outside air for ventilation purposes. Chapter 18 describes the methods for maintaining minimum ventilation air volumes.

Supply Diffusers Must Be Selected To Avoid "Dumping."

Chapter 18 describes how diffusers selected for constant volume can "dump" cool air onto occupants during minimum air volume conditions.

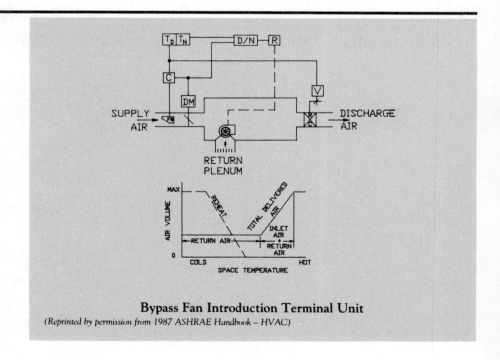

Bypass Fan Introduction Terminal Unit
(Reprinted by permission from 1987 ASHRAE Handbook – HVAC)

Figure 14.10

Minimum VAV Box Volume Must Not Result in Overcooling the Space.

This issue has been covered in an earlier section of this chapter.

All Components of the System Must Be Selected to Minimize Noise.

Noise is generated in VAV systems by:

- *Fans operating in surge.* This situation occurs more frequently in damper-controlled systems, as discussed above and in Chapter 18.
- *High static pressure differentials at VAV and discharge dampers.* This condition has been addressed in the section on damper-controlled VAV systems earlier in this chapter.

VAV Boxes Must Be Accessible for Maintenance.

VAV boxes have damper operators, velocity controllers, and possibly reheat coils, all of which require some attention. VAV devices are located in occupied spaces. Much of the floor spaces are covered with furniture, office cubicle dividers, filing cabinets, etc. Ceilings where VAV boxes are located may have light fixtures blocking access to the box. Care must be taken in locating the VAV box. See Chapter 18 for more information.

Each VAV Box Minimum and Maximum Air Volume Must Be Set.

This issue has been explored in the preceding text.

Static Pressure Controller Should Be Set for the Minimum Operating Point.

The static pressure control at the end of the supply duct determines the operating point of the system. The control should be set for the lowest static pressure that will permit proper VAV operation. A lower static pressure will save fan power, produce a quieter system, and shift the system pressure curve away from the fan surge line. See Chapter 18.

Return Air Fan Air Volume Should Be Proportional to the Supply Fan Air Volume.

There are two methods of controlling VAV fans: open loop control and closed loop control.

Open Loop Control

Probably 98% of VAV fan control is open loop control, because it is the least expensive system to install. Figure 14.11 shows an open loop control system.

In the open loop control, the static pressure control sends the same signal to the supply fan and the return fan. This type of control creates a varying building pressure, either positive or negative. If the return fan is handling more air than it should, the building will be under negative pressure. If the return air is handling less air than it should, the building will be under positive pressure. Open loop control is not the best system because:

- The supply and return air fan curves are not the same.
- The supply and return air system curves are not the same.

Figure 14.12 shows that a static pressure signal calling for two thirds speed will have each fan delivering different percentages of their air volume. In this case, the return air fan will not deliver the desired cfm in order to maintain a stable building pressure.

Closed Loop Control

Figure 14.13 shows the closed loop method of fan control. The static pressure control regulates the supply fan air volume. An air measuring station (AMS) in the main supply duct measures the supply cfm. The supply air AMS signals the return air fan to produce the correct return air volume, through the return air AMS. The building pressure remains stable, because the return air fan is in balance with the supply air fan. Microprocessors provide excellent tools for fan control. They are capable of taking the outside air cfm from the outside air AMS, deducting it from the supply air AMS, and using the calculated return air volume to control the return air fan.

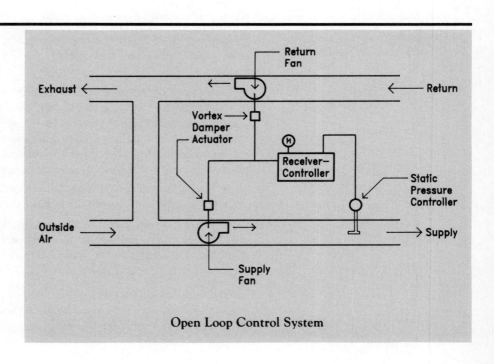

Open Loop Control System

Figure 14.11

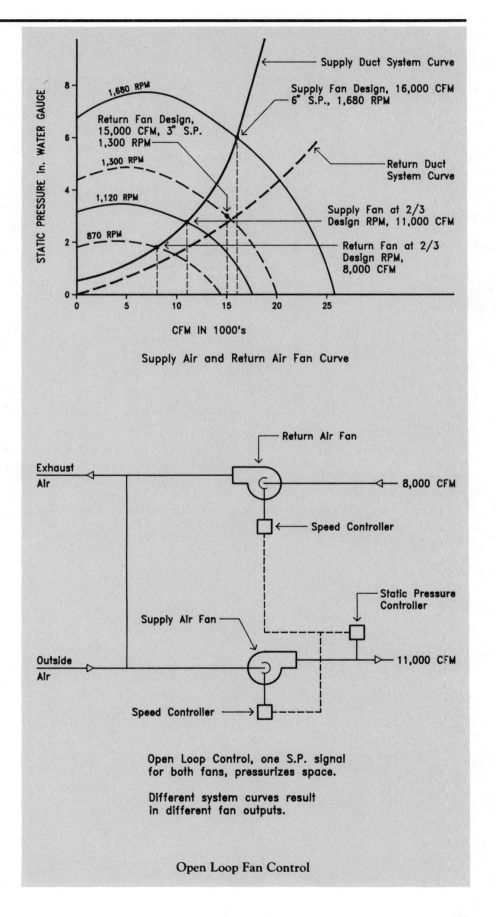

Supply Air and Return Air Fan Curve

Open Loop Fan Control

Figure 14.12

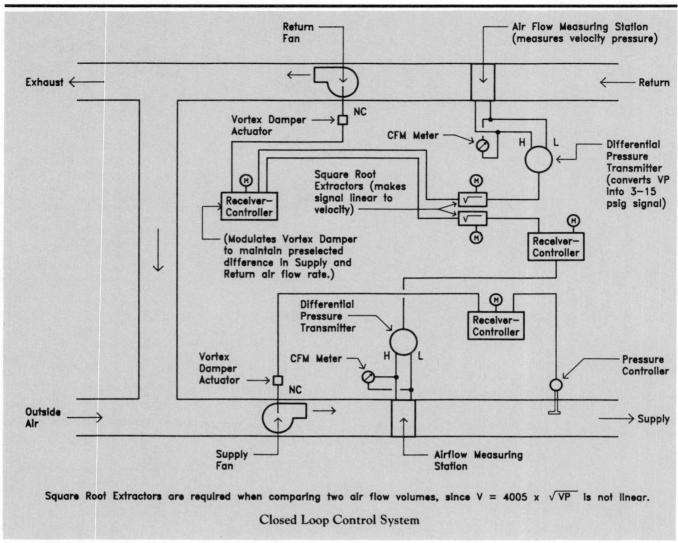

Square Root Extractors are required when comparing two air flow volumes, since $V = 4005 \times \sqrt{VP}$ is not linear.

Closed Loop Control System

Figure 14.13

Chapter Fifteen:

CONSTANT VOLUME REHEAT

Before the energy crisis of the mid 1970s, constant volume reheat, dual duct, and high pressure induction HVAC systems were the norm. Constant volume reheat systems consist of air handling units that deliver a fixed volume of cool air to the conditioned space. Zone thermostats control the reheat coils serving the zones. The supply air temperature is regulated to satisfy the space thermal demands. Figure 15.1 is an illustration.

Constant volume systems produce the best comfort conditions of any multizone system. Air is supplied at 55°F to 58°F to each zone reheat coil. The zone thermostat controls the reheat coil to maintain the desired room temperature.

Most energy codes restrict the use of simultaneous heating and cooling for reheating or recooling the supply air serving a common system. Constant volume reheat systems can be employed in conditions where:

- Recovered energy is used for reheat.
- Humidity must be controlled.
- The cold air supply is automatically reset to the highest temperature level that will satisfy the zone requiring the coolest air.

Recovered Energy Used for Reheat

An excellent source of recovered energy can be found in the refrigeration cycle. Rather than rejecting condenser heat to a cooling tower or air-cooled condenser, as much of that heat as possible may be captured to heat water for reheat purposes. Figure 15.2 shows how refrigeration waste heat is transferred from a heat recovery condenser to the reheat water.

Heat Exchangers

To be effective for reheat loads, the reheat condenser must remove latent heat as well as sensible heat from the refrigerant. De-superheater refrigerant-to-water heat exchangers are designed for heating domestic hot water. By removing only sensible heat, the de-superheater permits hot refrigerant gas to enter the condenser, where the refrigerant's latent heat is removed. The condenser transforms the refrigerant gas to a liquid. Figure 15.3 shows the refrigerant process for heat recovery.

De-superheater heat exchangers are used to avoid problems with hot gas refrigerant piping that conveys liquid refrigerant, as well as possible oil transport velocity problems.

Figure 15.4 shows how full condenser heat recovery can be arranged. A head pressure control valve diverts hot refrigerant gas to an air-cooled condenser whenever heat recovery needs are low. When the reheat fluid does not remove sufficient heat from the hot gas, the increase in saturated condensing pressure positions the head pressure control valve to bypass hot gas to the air-cooled condenser.

Controls   Higher head pressures and condensing temperatures produce higher temperature reheat water. Water temperatures entering the heat recovery

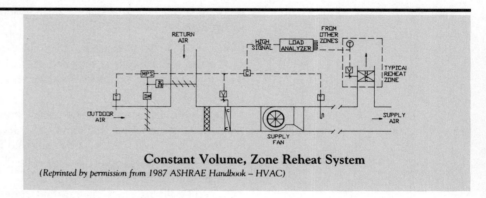

Constant Volume, Zone Reheat System
(Reprinted by permission from 1987 ASHRAE Handbook – HVAC)

Figure 15.1

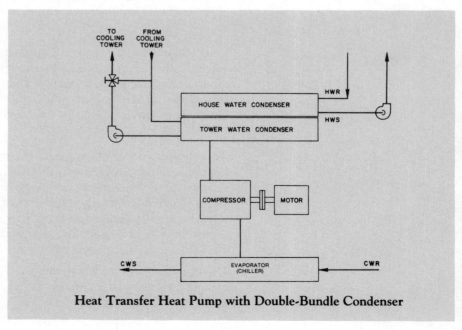

Heat Transfer Heat Pump with Double-Bundle Condenser

Figure 15.2

condenser at over 120°F produce excessive head pressures. Control must be installed to divert water above 120°F around the heat recovery condenser (bypass).

Refrigerant-saturated condensing temperatures below 100°F can also create problems involving the expansion valve and low suction temperature. In this case, water regulating valves should be installed to regulate low condensing temperatures. Reheat water temperatures of 110°F to 115°F are common, and 120°F is possible at a higher head pressure.

It is not desirable to operate refrigeration systems at head pressures over 100°F. The higher the head pressure, the greater the kw/ton (See Chapter 2). Provisions should be made for lowering the condensing temperature when the demand for heat recovery is low.

Air-cooled condensers are controlled to provide a minimum condensing pressure. The air-cooled condenser control must be incorporated into the entire condensing temperature control scheme, when heat recovery condensers are applied. Factory-installed heat recovery condensers eliminate many of these concerns. The field-installed heat recovery condenser is appropriate for retrofit applications.

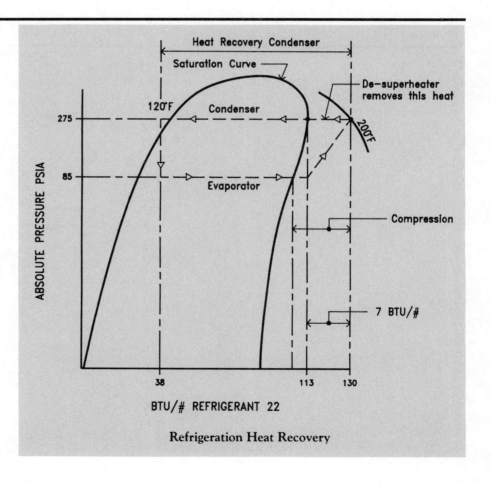

Refrigeration Heat Recovery

Figure 15.3

Appropriate Applications

Constant volume reheat systems are still designed and used for applications that require close temperature and/or humidity control. These include hospitals, museums, research laboratories, and manufacturing plants.

Reheat provides humidity control in this way: when space humidity increases, the cooling coil is activated to condense moisture from the air. This cooled supply air is reheated to maintain the required supply air dry bulb temperature. The zone reheat coils produce the desired room temperature.

Constant volume reheat systems are hard to beat for producing comfort. Waste heat recovery makes them even more desirable.

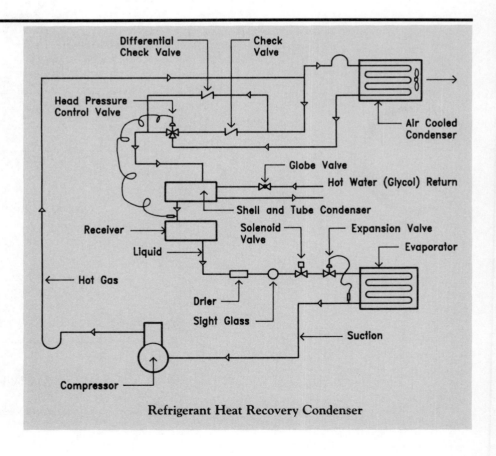

Refrigerant Heat Recovery Condenser

Figure 15.4

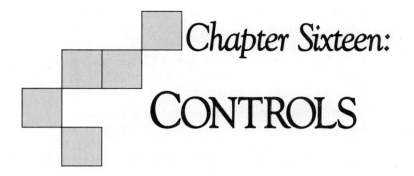

Chapter Sixteen:

CONTROLS

The satisfactory operation of an HVAC system is directly influenced by the selection and functioning of the temperature control system. Temperature control systems have undergone dramatic changes in recent years. The desire to conserve energy and reduce system operating cost has resulted in a demand for more controlled system performance, and better control hardware. The most dramatic change in HVAC control technology is the introduction of direct digital control (DDC) systems. The traditional basic HVAC control systems are pneumatic, electric and electronic.

Pneumatic Control Systems

Pneumatic control systems use compressed air to supply energy for the operation of valves, damper operators, and other control equipment. The pressurized air circuits that operate the pneumatic control devices are small diameter copper or plastic tubes. The tubing varies in size from 1/8″ to 5/16″ diameter for control, and rarely exceeds sizes of 1/2″ diameter for main air. Figure 16.1 shows the basic elements that make up a pneumatic control system.

To be operational, a pneumatic control system requires the following elements:

A source of clean, dry, compressed air. The control air source is usually a dedicated electrically driven air compressor. The control air is cleaned by a high efficiency filter that removes 99.9% of the particulates that are 0.01 microns or smaller. The filter must also remove oil that is introduced to the compressed air stream by the compressor. This is accomplished by a coalescing type air filter. The tiniest droplet of oil or water will render a pneumatic controller, with its .005″ orifice, useless. Water is removed from the control air system by a refrigerated drier. The refrigeration drier condenses moisture from the compressed air supply, and deposits the water in a drain. (See Chapter 17 for more on compressed air filters.)

Pressure reducing valve. The pressure reducing valve reduces the 80 psi (receiver air pressure) to 20 psi for distribution to the system. Large systems with long piping runs may distribute main air at 30-80 psi to make it possible to utilize smaller pipe sizes.

Air lines. As mentioned above, the air lines are usually copper or plastic tubing.

Sensing elements. These measure what is to be controlled, such as temperature, humidity, and pressure. Some of the regulating devices are thermostats, humidistats, pressure regulators, transducers, relays, switches, and sensor controllers. These devices determine how much control air will be passed into the branch line to the operator. The operator is a piston-like device that operates the control valve or damper.

Control devices, such as valve and damper operators. Devices, such as valves or dampers, are positioned by the operator. The operators position the valve or damper in an action similar to that carried out by a piston. Air pressure on the operator diaphragm will move the piston to open, close, or modulate a valve or damper.

Air pressure. Single temperature systems operate at 18 psi. Dual temperature systems operate at 13 psi and 18 psi. A pneumatic system operates by a controller regulating the positioning of a controlled device (valve or damper operator). This is done by taking pressurized air from the main, and delivering it through a branch to the valve operator. The pressure in the branch to the valve operator is varied by the sensing element in the controller. The controller acts as a pressure regulator that varies the pressure, as the controlled condition changes. (See Figure 16.1 for an illustration.)

Direct acting controllers. If the branch line pressure increases in response to an increase in the measured condition (e.g., the temperature rises at the sensing element), the bi-metal thermostat element closes the bleed port to

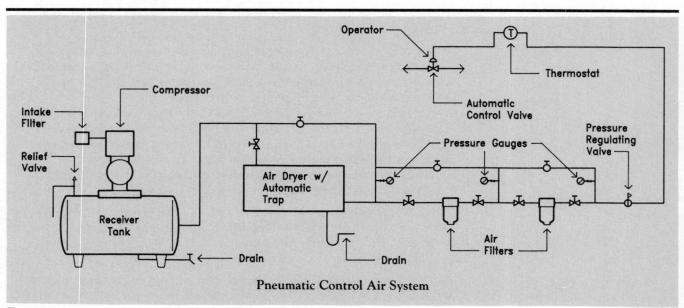

Pneumatic Control Air System

Figure 16.1

increase the air pressure in the branch line to the operator. When the temperature at the sensing equipment decreases, the bi-metal element opens the bleed port, which relieves the compressed air and reduces the air pressure to the controller.

Reverse acting controllers. If the branch line pressure decreases in response to an increase in the measured condition, (e.g., as the temperature at the thermostat's sensing element increases), the bi-metal element opens the thermostat bleed port to release control air, and reduce the pressure in the branch line to the valve operator. When the temperature at the thermostat sensing element decreases, the sensing element closes the bleed port, to increase the pressure in the branch line to the valve operator.

Spring ranges. Springs are installed on the pneumatic operators to provide a full movement of the operator from open to closed, or vice versa, in response to a change in the measured condition through the full 3 to 13 psi throttling range. The spring is usually relaxed at 4 psi and is compressed at 11 psi. Any pressure above 11 psi serves to close off the valve or damper very tightly.

Normally open valves (NO). A normally open valve or damper is in the open position when no air pressure acts on the valve or damper operator diaphragm. The operator spring pressure will keep the valve or damper operator in the open position. Normally open valves are used when a "fail-safe" condition requires full flow through the valve or damper. (Fail-safe refers to what happens when the control system fails.) In cold climates, heating valves are usually normally open. Any loss of control air pressure will permit the spring to open the valve. In warm climates, cooling coil chilled water valves are usually in the normally open position. Return air dampers are usually in the normally open position as well.

Normally closed valves. A valve is normally closed (NC) when the operator spring pressure keeps the valve in the closed position if there is no air pressure on the operator diaphragm. Normally closed valves are used when the "fail-safe" condition of no control air pressure on the operator exists. Normally closed valves are usually used on:

- Steam to water heat exchangers, where lack of water circulation can cause excessive heat and damage the exchanger element.
- Chilled water valves in cooler climates.
- Hot water heating valves in warm climates.
- Condenser water control valves. Valve should be closed when condenser pump is off.
- Chiller automatic valves. Valve is closed when chiller is off.
- Outside air dampers, smoke dampers, back draft dampers, exhaust air dampers, which should be closed when the fan is off.

Advantages

Extremely durable
Pneumatic operators last almost as long as the building. Some operators are still performing after 40 years of service. Thermostats, pressure regulators, operators, and relays can enjoy longevity if the control air supply is clean.

Simple operation
Most mechanical systems operators find simple pneumatic systems within their expertise. Air pressure operated devices tend to be easily understood.

Maintainability
This is the advantage that systems operators promote vehemently when they are consulted about the design of the control system. Most mechanical

maintenance and service people are familiar with pneumatic controls. Controls can readily be checked with an air gauge, and calibrated or replaced as required.

Operators available for larger systems

Pneumatic operators are quite powerful, and can operate very large valves. At the present time, electric operators are not produced in as large sizes as their pneumatic counterparts.

Generally lower installed cost on medium and large projects

In general, pneumatic control systems are still 10 to 20 percent less expensive to install on medium- and large-sized projects, than any other type of control system. An exception is control systems that require many logic conversions such as square root extraction, which could be less expensive to install with DDC than as a pneumatic control system.

Disadvantages

Subject to water and oil contamination

The ports in many pneumatic devices are extremely small. Any contaminants, dirt, water, or oil blocking the port prevent the device from performing its task. It is almost impossible to completely remove the pollutants from the pneumatic system, especially from the control device. The contamination usually originates in the air compressor. Oilless air compressors are not used very often in temperature control work, because they are too expensive, require more maintenance, and do not last very long. Often, control systems that become contaminated must be replaced in their entirety, as the probability of successful cleaning is very low.

More hardware required for the initial control system and for any future modifications

Some pneumatic control schemes require considerably more hardware to implement than is required by a DDC system. Figure 16.2 gives an example.

Modifications to an existing pneumatic control system also involve adding substantial hardware. Control sequences can be changed in DDC control systems merely by modifying the program. For instance, if the system shown in Figure 16.2 were to be changed from open loop to closed loop, pneumatic hardware would have to be added.

Greater deviation than DDC controls

Many pneumatic controls are installed as simple proportional controls. Proportional controls may not pass a high enough air pressure into the branch to the operator to properly control a valve under low or light load conditions. Low load means low air pressure is passed into the branch to the operator to throttle the valve. Proportional plus integral (PI) controllers may correct the condition, but will involve hardware changes. Pneumatic controllers operate with a 5% proportional setting. Controllers with a temperature range between −40°F and 160°F, have a 200°F span. They require a temperature change of 5% of 200°F, or 10°F to vary the branch pressure from 3 to 13 psi. A 10°F temperature change at the sensor is too much for energy efficiency. Sensors with a smaller span are available, but also require a hardware change.

Requires more calibration of regulating devices

The design or construction of pneumatic controllers is such that the springs that control the instruments will demand calibration from time to time. Electronic controls require almost no calibration.

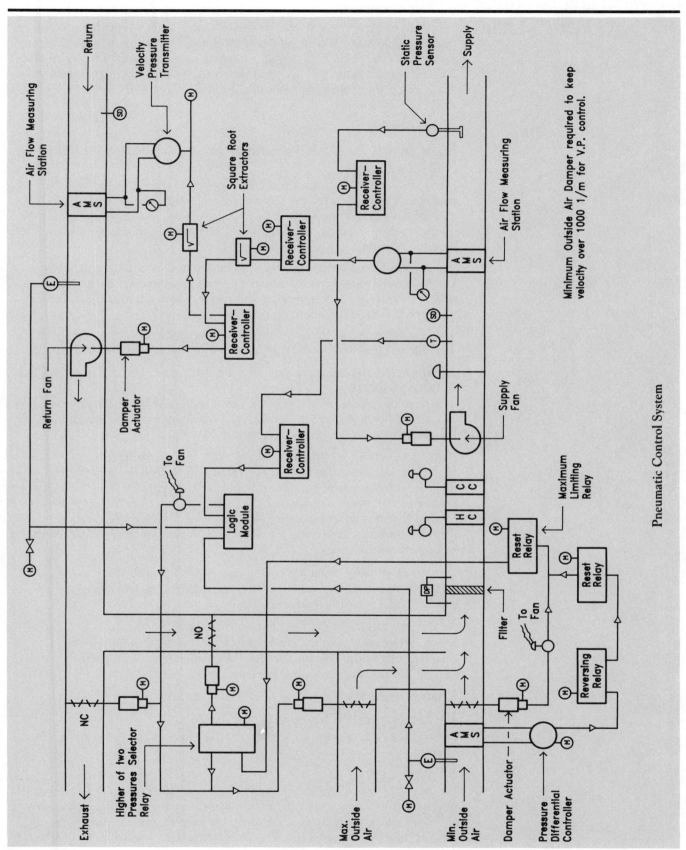

Figure 16.2

Pneumatic Control System

199

Electric and Electronic Control Systems

Electric and electronic control systems use electricity as the power source, and electric wiring to transmit the power from the controller to the operator. Electric controls can be two-position, such as starting a pump or opening a damper, or floating-type controls for reversible motors. Floating controls allow a motor to stop at a position determined by the controller. Electric control operators (motors) are available in 220, 110, and 24 volts A.C. Electronic control systems operate on 24 volt A.C. *input* and 2-22 volt D.C. *output*.

Advantages

Logic panels available for similar projects, or larger standard control systems

Many standard HVAC control sequences can be implemented by using packaged logic panels for the sequences, and wiring the controllers and control devices to the panels. The logic panels are available to perform the following tasks: heating, cooling, and economizer operation for commercial air-conditioning and heat pump equipment.

Electric/electronic controls provide space demand-oriented control of HVAC multizone systems. They receive signals from zone thermostats, select the zone of highest heating and cooling demand, and control three stages of heating and cooling.

The discharge air controller maintains an average discharge air temperature in VAV systems. It utilizes an economizer for free cooling when available.

Inexpensive installation for smaller projects, or larger standard control systems

The installation cost of electronic control systems is less than for pneumatic or DDC control systems for smaller projects. Pneumatic systems usually become cost effective at about 50 tons of HVAC.

Remote control of space sensors easily accommodated

Electronic controls offer versatility in that temperature sensors can be located within a space, and control point adjustments for the sensor can be remote from both the sensor and the operator. Examples: It is possible to control thermostat set points for patient rooms from a nurses' station. Temperature control settings for shops and offices can be regulated by the manager from his office.

Controls readily available

Most electronic controls are available from HVAC supply houses, which can be found in most urban areas.

Many installation and service contractors

Many independent control contractors are capable of installing and servicing electronic control systems. Relatively few are equipped to install and service pneumatic control systems.

Disadvantages

Higher operator hardware cost

Electronic operators cost approximately five times as much as pneumatic operators.

Shorter service

Electronic operator service life is approximately 15-20 years if no electrical problems intervene. Electronic controls burn out due to voltage problems, short circuits, etc. Pneumatic controls, on the other hand, usually last as long as the HVAC system itself.

Higher operator maintenance

Electronic operators must be replaced from time to time, whereas pneumatic operators have a longer life expectancy.

Operators not available for very large diameter (pipe size) valves

Electronic operators are not available for very large control valves. Pneumatic operators, on the other hand, are available for any pipe size automatic valve.

Higher installed cost for larger projects/conduit requirements

Electric/electronic controls have a higher installed cost for larger projects and for those projects where local codes require that wiring must be installed in conduit. Except for the small projects, the installed cost of electronic control systems is usually 15-25% more than for pneumatic control systems.

Direct Digital Control Systems (DDC)

Direct digital control is accomplished by a digital computer. Most DDC systems employ microprocessors with software programs to maintain variable processes. Among the variables are temperature, humidity, and flow rates. The control sequences are entered into the computer by software. The set points are input to the computer by keyboard entries. The computer continuously monitors the controlled variable, which could be temperature, static pressure, airflow, etc., and compares the monitored value to the value input to the computer. The computer transmits the difference in values as a digital pulse that is converted to a modulating signal by a transducer. The transducer sends a pneumatic or electric signal to the controller until the set point is reached. The computer output signal is isolated from the control signal by an interface. Figure 16.3 shows a direct digital control system.

Advantages

Simplicity of changing control sequences and set points

Changes to the control sequences can often be made without adding hardware to the system. For example, suppose a system includes a fan-powered VAV box with a reheat coil. Increased space heat gain has caused occupant discomfort. The following changes can be made from the room thermostat: the maximum box airflow can be increased, the cooling proportional band can be narrowed, the thermostat dead band can be reduced, the morning warm-up or cool-down start can be reset, system shut-down can be enacted later, fan assist started sooner, and the proportional plus integral feature can be installed. Climbing a ladder to remove a ceiling tile, and bodily contortions are no longer necessary to get at the control panel for standard service.

Central control point

One central point can monitor the entire facility and make changes in set points for system controls. In conditions such as strong cold winds, or failure of the chilled water circulating pumps or a chiller, the supply air discharge temperature can be raised, and the morning cool-down and evening shutdown hours can be reset, without setting foot in the mechanical room. Space temperatures, mixed air, outdoor air, return air, discharge temperatures, and static pressures can be monitored in order to diagnose operational problems.

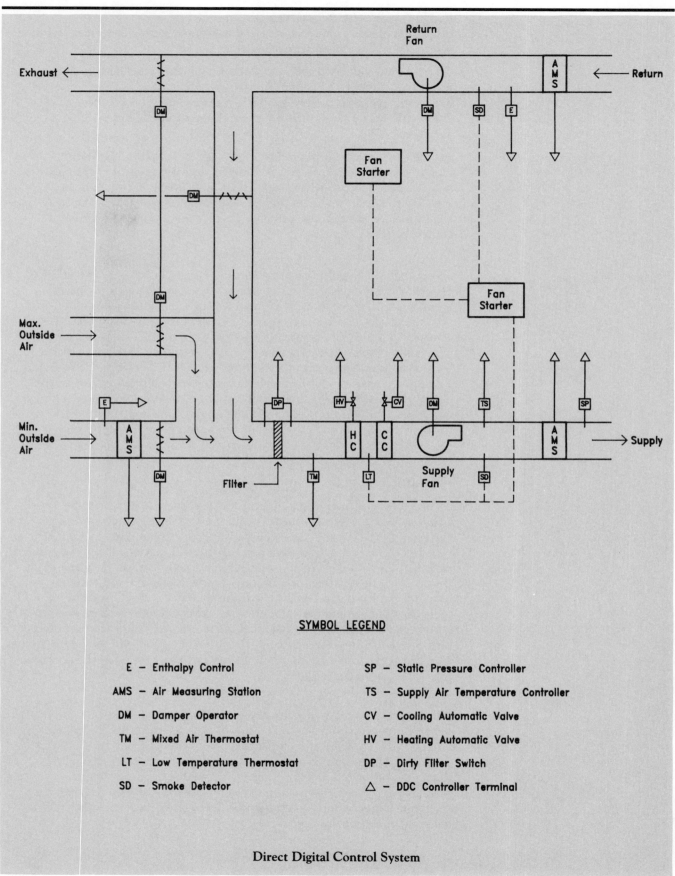

SYMBOL LEGEND

E — Enthalpy Control

AMS — Air Measuring Station

DM — Damper Operator

TM — Mixed Air Thermostat

LT — Low Temperature Thermostat

SD — Smoke Detector

SP — Static Pressure Controller

TS — Supply Air Temperature Controller

CV — Cooling Automatic Valve

HV — Heating Automatic Valve

DP — Dirty Filter Switch

△ — DDC Controller Terminal

Direct Digital Control System

Figure 16.3

Another advantage of the central control point is the improvement in security. Unauthorized people are not likely to tamper with the system, especially if they do not have the access code.

Remote monitoring and operation
Campus-type facilities can use telephone modems to monitor and control several building systems from a central point. The modem also assists operational personnel. A telephone modem can bring the problems to be solved "to the expert" for his or her technical assistance.

Energy efficiency
Dead bands, when no heating or cooling is enabled, can be programmed by keyboard action. The following changes can all be programmed into the computer:

- Precooling the unoccupied building with outside air when recent outdoor temperature history and present outdoor temperatures permit
- Resetting discharge temperature from return air
- Proportional plus integral control
- Closing outdoor dampers
- Closing chilled and hot water valves prior to entering the unoccupied mode
- Closing outdoor air dampers during morning warmup

Occupied and unoccupied schedules are modified easily. Zones can be placed in the unoccupied mode when they are unoccupied for a period of time (e.g., lunch, conferences, seminars). This feature can be obtained with little or no additional hardware.

Precise control
Very precise control is available. The digital computer can be set so that the control point and set point are equal. The change from proportional control to proportional plus integral, is a simple keyboard operation.

Control Components

Automatic Dampers
Automatic dampers are installed to control the flow of air. Control dampers are arranged as opposed blade dampers or parallel blade dampers. Opposed blade damper blades rotate in opposite directions when they open or close. The blades close pointing toward each other. Parallel damper blades rotate in the same direction when they close, so that the blades are always parallel to each other. For descriptive purposes only, the exhaust air damper in Figure 16.4 is shown as an opposed blade damper, and the return air and exhaust air dampers are shown as parallel blade dampers.

Aside from VAV air terminal units, the mixed air damper control system is the most common in HVAC. Figure 16.4 shows a mixed air control system.

The control sequence is to close the return air damper when the outside air and exhaust air damper are opened by the mixed air thermostat. The linear flow changes of all three dampers is important in order to maintain both constant flow and a mixed air condition in the system. Figure 16.5 shows the performance curves of parallel blades and opposed blade dampers.

The numbers from 1 to 200 represent the ratio of the damper pressure drop to the pressure drop of the remaining system. Linear performance is achieved when the percent of the damper stroke is equal to the flow percent. The parallel blade damper provides linear flow when the damper represents 3% of the total pressure drop, and the opposed blade damper provides linear flow when it represents 10-20% of the duct pressure drop. When automatic dampers are installed in mixing boxes, the velocity across the damper is not great enough to produce a pressure drop that is 3-15% of the system pressure drop. The low damper pressure drop through the wide open damper means that the damper will not have an impact on airflow control when it is partially open. The damper will only begin to control when it is almost closed.

Automatic dampers should be smaller than the ducts they are installed in if they are to produce the necessary pressure drop required for good control. When an 8″ valve is handling 5 gpm, the valve must be cracked open a hair before it will control flow. The damper pressure drop must be a significant percent of the system pressure drop in order to control flow.

Automatic Valves

Automatic valves are installed to control the flow of water or steam. Ideal control produces linear *control*, not necessarily linear flow. With the valve actuator at 50% stroke, 50% of the heat transfer is desired.

The capacity of an automatic valve is not influenced by supply pressure, but by pressure *differential*. Valve capacities are compared by the flow

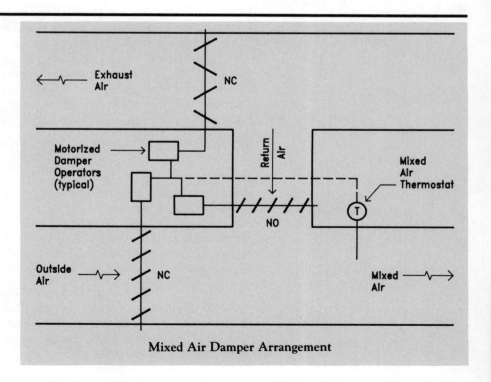

Mixed Air Damper Arrangement

Figure 16.4

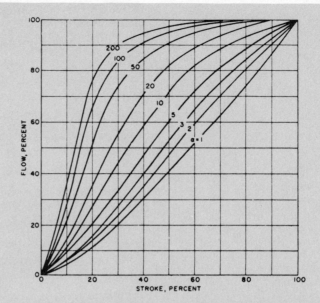

Installed Characteristic Curves of Parallel Blade Dampers

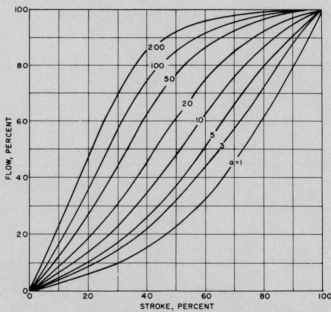

Fig. 15   Installed Characteristic Curves of Opposed Blade Dampers

Installed Characteristic Curves of Opposed Blade Dampers

(Reprinted by permission from 1987 ASHRAE Handbook—HVAC)

Figure 16.5

coefficient Cv, which is the flow in gpm through a valve that will cause one psi pressure drop across the valve.

Control valve performance is determined by the valve's characteristics. These are:

Quick opening: Maximum flow is rapidly approached as the valve begins to open.

Linear: Valve opening and flow are directly proportional.

Equal percentage: Each equal increment of valve opening increases flow by an equal percentage over the previous value.

Figure 16.6 shows the various valve flow characteristics.

Heat emission for hot water coils does not follow a linear pattern. Figure 16.7 shows that 50% of a heating coil's output is obtained with only 10% of the flow, and 90% of the coil's output is produced with a 50% flow.

Linear control of a hot water coil is acquired with a combination of the "quick opening" characteristic and the "equal percentage" valve characteristic, similar to the linear result shown in Figure 16.6.

Automatic control valves for water use should be selected for a significant pressure drop in order to maintain design control. For single coil systems where the coil represents a considerable pressure drop, the valve pressure drop should equal the coil pressure drop. In multiple coil systems, the control valve should be sized for one third the system pressure drop, excluding the valve.

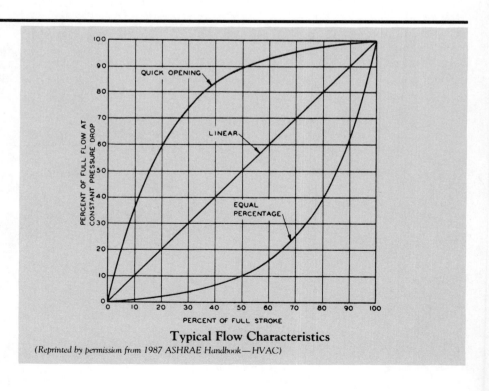

Typical Flow Characteristics
(Reprinted by permission from 1987 ASHRAE Handbook—HVAC)

Figure 16.6

Proportional Control

Figure 16.8 is a diagram of proportional control. The control device (valve) is positioned proportionally in response to slight changes in the controlled variable (temperature). The valve does not run its complete stroke, but immediately assumes a position in proportion to the system requirement. If the temperature at the thermostat drops, indicating heat loss, the thermostat will position the valve in proportion to the drop in temperature, and keep the valve at that position until the thermostat senses another change in the load. This action produces a sine wave-type curve as the room temperature moves from the set point to the top of the throttling range (referred to as *offset*) and then down to the bottom of the throttling range. Proportional control is adequate for most HVAC systems.

Proportional Plus Integral Control (PI)

Control is similar to proportional control, except that when offset (difference between set point and the throttling range) occurs, the automatic reset shifts the control point back to the set point. The control point is the temperature at the thermostat at that moment. The offset is corrected so that the thermostat will control closer to its set point. See Figure 16.9.

DDC accomplishes PI control by adding an integrated error (the continual summation of the space set point minus the actual space temperature during a time period) to the existing proportional error. The proportional error is the set point minus the actual space temperature.

Suppose the actual space temperature is 3°F above the set point, and the error lasts 4 minutes. An integration error is added to the proportional error of 3°F. This causes the controller to operate the cooling for a longer period of time, because it thinks that it is warmer than it is at the thermostat.

The integration control feature is designed to produce a tight system control, not to achieve economy of operation. It is not appropriate for

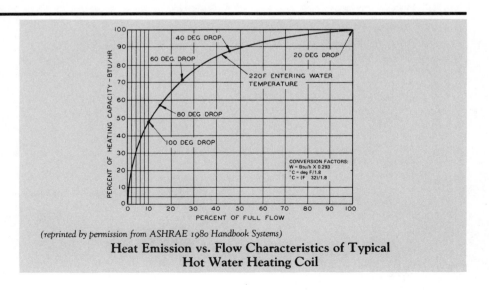

(reprinted by permission from ASHRAE 1980 Handbook Systems)

**Heat Emission vs. Flow Characteristics of Typical
Hot Water Heating Coil**

Figure 16.7

systems with large fluctuations in load such as auditoriums, classrooms, and gymnasium systems which will not benefit from PI control.

PI control can harvest operating savings from chilled water production. If 45°F chilled water supply is needed at peak load, the proportional chilled water controller with a 4°F throttling range will have a set point of 43°F and will cycle the chiller between 41°F and 45°F. During the maximum cooling demand, the chilled water temperature will be 45°F, and during light loads, the chilled water temperature will be 41°F. If there were no cooling load, the chiller would still make 41°F chilled water, two degrees below set point. These are the instructions the chilled water controller is conveying to the chiller.

The chiller will cool water 2°F colder than it should be at any load condition other than at full peak load. PI control eliminates the offset (2°F+ and 2°F-) so that 45°F water will be generated at all times, regardless of the cooling load. Two degrees of chilled water production can represent a significant operational savings.

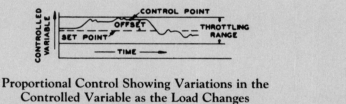

Proportional Control Showing Variations in the Controlled Variable as the Load Changes
(Reprinted by permission from 1987 ASHRAE Handbook—HVAC)

Figure 16.8

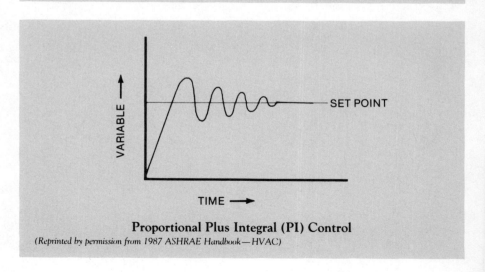

Proportional Plus Integral (PI) Control
(Reprinted by permission from 1987 ASHRAE Handbook—HVAC)

Figure 16.9

Proportional Plus Integral Plus Derivative

This, the highest level of control mode, adds a derivative function to PI control. Derivative action opposes any change and acts proportionally to the rate of change. As quickly as a thermostat senses temperature change, the derivative correction action is taken. PID will bring the control point back to the set point faster than PI control. The derivative function controls the rate of change, which is difficult to establish for each system. Unless the correct rate time setting is set, there is no benefit of PID control.

Fume hood face velocity control is one of the few HVAC control applications that are appropriate for PID control. The rate change setting of the known condition and the desired hood face velocity can be programmed into the microprocessor control.

Part Two:
AVOIDING AND RESOLVING OPERATIONAL PROBLEMS

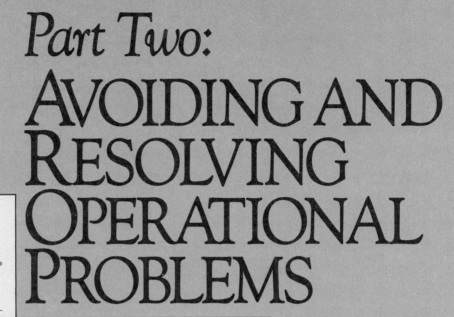

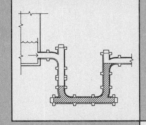

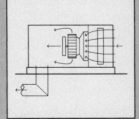

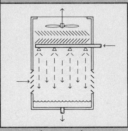

Part Two, "Avoiding and Resolving Operational Problems," focuses on common problems that occur in various HVAC systems and within the system components, and offers strategies to prevent and correct such occurrences. Maintenance is also covered, including general requirements, recommended procedures, and frequency of inspections.

Chapter Seventeen:

CONTROL SYSTEMS

This chapter covers maintenance and operational problems and issues that apply to all control systems, such as moisture and oil removal in pneumatic systems, the advantages of different types of valves, and control system location. The chapter begins with a section on the special operational considerations of Direct Digital Control (DDC) systems, pneumatic control systems, and electric and electronic control systems.

Direct Digital Control Systems (DDC)

The first and most basic question to ask about direct digital control systems is: Should a computer-operated system be installed that the plant operators will not use? The fact is that even experienced HVAC system operators are sometimes reluctant to learn to operate a computer system keyboard. They may be still more hesitant to learn to program the system.

Facility operators may be accustomed to adjusting and repairing pneumatic or electric control systems. On the other hand, most direct digital control systems (energy management systems) are microprocessor-based. Operators must be trained to use the central control console which is a part of these systems. Normally, one or two skilled technicians should be trained to maintain this element in each facility. The various control system components, such as thermostats, control valves, damper operators, discharge sensors, and receiver/controllers, are still pneumatic or electric devices that the console operators will not have time to calibrate and maintain.

It may be too difficult for the (in-house) operating staff to modify or adjust direct digital control systems. Microprocessor-based controls must be programmed with the correct parameters for each application. Each mode must be installed and calibrated. It can be difficult to find and train sufficient technical personnel to efficiently maintain large control systems. If a microprocessor loses its memory because of electrical or radio frequency interference, it may take a technician hours to reprogram the device, if he can do it.

Direct digital controls are being developed to accomplish so much more than can now be performed by pneumatic or electric devices, that they will be essential in order to have a state of the art system. It will be possible for an operator to balance a Variable Air Volume (VAV) system from the computer console. The computer display screen will give the HVAC system operator access to every room in the entire building, and allow him to not only set the maximum and minimum airflows for each VAV box, but also test the performance of each box and the entire system. Rebalancing to satisfy design load changes can be accomplished without entering the rooms and climbing into ceilings.

At least some direct digital control for HVAC systems should be incorporated into every new building, to prepare owners for future advancements. One way to gradually introduce DDC to a building is to install it for one air handling unit in the building. This will enable the operating staff to gain knowledge of the direct digital control system without being overwhelmed by it. As their knowledge increases, so may their desire to expand use of direct digital control.

Avoiding DDC Operational Problems

The power supply to the microprocessor should be in the form of a dedicated circuit, separate from the power to other equipment. It should also be noted that fluorescent lighting located near microprocessors can scramble the controller's memory when the light's ballast is energized. A small voltage regulator wired into the controller's low voltage power supply will open the electric circuit to the controller whenever the incoming voltage is above the controller's limit. See Figure 17.1.

Radio frequency interference from hand-held radio transmitters can also put a computerized control system out of service. Transmitting radios must be kept out of computer control rooms, or the room should be shielded with

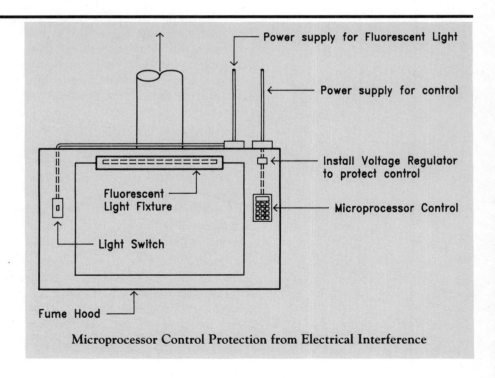

Figure 17.1

Microprocessor Control Protection from Electrical Interference

radio frequency screening. Microprocessor controls should not be located near variable frequency drive (VFD) controllers, as VFD disturbance will cause loss of microprocessor memory. Either locate the microprocessor control in a separate room from the VFD, or provide shielding for the microprocessor.

Sensor wiring must be shielded cable, properly grounded, and not run in the same conduit as power wiring. The interference of power wiring will upset the control performance. Shielded cable should be 22 gauge for runs less than 100', 20 gauge for runs between 100-200', and 18 gauge for runs over 200'.

Pneumatic Control Systems

Pneumatic controls must be calibrated on a regular basis. Otherwise, good control is lost. Uncalibrated thermostats set at 65°F in an 85°F room can open automatic reheat valves to add more heat to the room. Calibration of pneumatic controls every four months is desirable. DDC thermostats, on the other hand, require almost no calibration. Once a year is usually sufficient.

Moisture or Oil Contamination and Removal

The pneumatic control system can become contaminated with water or oil from the control air compressor. The following sections describe how this contamination occurs, and how the oil or water can be removed.

Moisture Removal

Atmospheric air entering the compressor intake contains water vapor. As the air is compressed, water vapor is condensed into liquid droplets. The hot compressed air can hold more water than the incoming air. As the compressed air cools to ambient conditions, water vapor in the compressed air is condensed to liquid droplets inside the compressed air line. If the moisture is not removed from the air, it will contaminate the controls.

To remove the moisture from the compressed air system before it can be condensed by the air surrounding the piping system, the compressed air must be cooled below the dewpoint temperature of the air surrounding the system. A refrigerated air dryer will remove the moisture from the compressed air after the air leaves the receiver tank by reducing the air's moisture content to 33-39°F dewpoint. See Figure 17.2.

The air-leaving side of the receiver tank is the best location for the dryer because of the additional cooling and pulsating dampening that takes place in the receiver. A dryer installed at this location can be sized based on the downstream air demand. If, on the other hand, it is installed *before* the receiver tank, it must be sized for the greater maximum compressor capacity.

Oil Removal

Air compressors are normally splash system lubricated. While force-fed lubricated and non-lubricated cylinder air compressors are also on the market, most temperature control air compressors are splash system lubricated.

Lubricated positive displacement air compressors consume some oil. It is normal for some oil to carry over into the air stream. Because of this oil migration, it is essential that aftercoolers, dryers, and coalescing air filters be utilized on temperature control compressed air systems. See Figure 17.3.

High efficiency air filters that will remove 99.9% of all particles as small as 0.01 microns are available for about $250 as of January 1, 1990. (See *Means Mechanical Cost Data 1990*.) These filters remove 99% of all solid particles and droplets of oil and water by mechanical and centrifugal action. The precleaned air then passes through multi-layered coalescing elements which remove smaller particles and mists.

Another benefit of high efficiency filters is the instrument quality air that they help to produce. Instrument quality air is important for temperature control. If oil were to find its way into a pneumatic control device, such as a thermostat, the device would malfunction. Very little oil is needed to block the extremely tiny control ports. If air cannot pass through the control ports at the correct pressure, the control is disabled. Correcting a contaminated pneumatic control system is almost impossible.

Passing refrigerant through the control tubing may remove accumulated oil, but tubing replacement is usually required. Even if cleaning is successful, the control instruments must be replaced, as instrument cleaning is not feasible. When control instruments are replaced with new ones, filters should be installed at the control air supply to each device. High efficiency coalescing air filters will prevent future contamination. A severely contaminated pneumatic temperature control system that must be replaced may be the best reason to consider installation of a partial or complete direct digital control system.

Control System Locations

Pressure switches, flow switches, and flow meters should be located away from the turbulence created by pipe fittings. Turbulence impedes operation of these devices, since they are set to respond to a particular pressure or

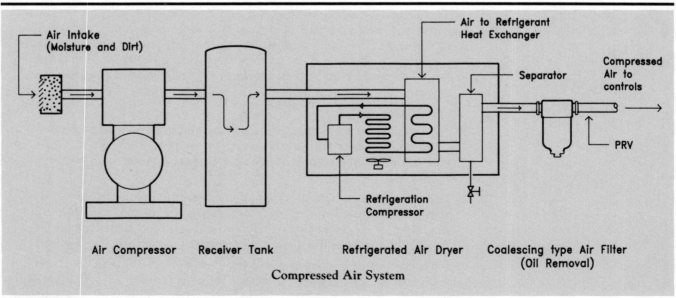

Figure 17.2

216

flow. Approximately three to five diameters of straight pipe are required before and after the device to produce the uniform flow to be measured. See Figure 17.4.

Pressure switches should not be located at the system side of the shut-off valve of the unit it is to protect. In such an arrangement, the pressure switch can be operated even if the unit valve is closed, resulting in an unprotected unit. See Figure 17.5.

Paddle-type flow switches should not be located on the system side of unit shut-off valves. If the paddle were to separate, the entire piping system would have to be drained to replace the paddle. Locating the paddle switch on the unit side of the unit shut-off valve permits paddle replacement with minimal loss of system water.

Automatic Valve Problems in Control Systems

Three-Way to Two-Way Conversions
Converting three-way valves to two-way valves will result in operational problems. Three-way valves are constructed to handle a relatively low seating pressure differential. Water flow is not intended to be shut off, so there is always water flowing out of the common port ("C" in Figure 17.6).

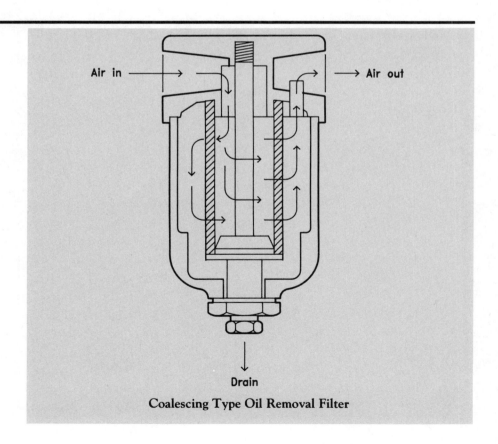

Coalescing Type Oil Removal Filter

Figure 17.3

The 40 psi supply pressure acts on the "A" port. The coil pressure drop reduces the pressure at the inlet "B" to 35 psi. The maximum pressure drop across the valve seat is 5 psi (40 psi minus 35 psi) when the valve is in the coil position ("B" to "C"). Three-way valve actuators are sized for the smaller pressure loads they will overcome.

A two-inch three-way valve maximum seating pressure is rated for 1.8 times the control air pressure. With 15 psi control air pressure, the maximum three-way valve seating pressure is 1.8 x 15 = 27 psi. When the bypass port "A" is capped to make the valve two-way, the pressure entering the valve will increase during minimum pump flow to 48 psi. As Figure 17.7 shows, when the coil water volume (gpm) decreases, the pump pressure will increase.

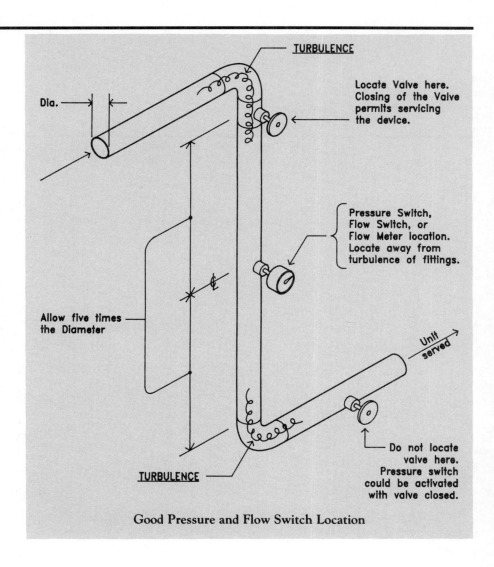

Good Pressure and Flow Switch Location

Figure 17.4

218

Now the modified valve will experience a greater pressure differential (48 psi − 9 psi = 39 psi), which will overpower the actuator and force the valve open. See Figure 17.8.

Normally Open and Normally Closed Automatic Valves

Normally closed automatic valves are maintained in the closed position by spring pressure. If the pressure differential across the valve is greater than the spring pressure, the valve will open. See Figure 17.9.

Normally open automatic valves are kept open with spring pressure. By using 15 psi control air pressure over the proper square inches of diaphragm, enough pressure is developed to overcome the spring and water pressure to close the valve. As the thermostat senses an increase in space temperature, a direct acting thermostat will increase control pressure on the actuator to close a normally open control valve.

Normally open equal percentage control valves and direct acting actuators provide better water flow control for both heating and cooling.

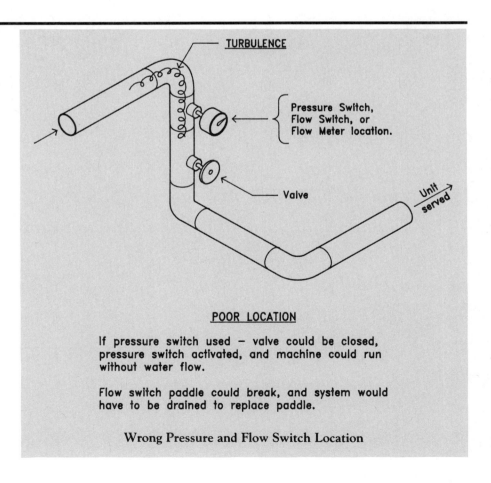

TURBULENCE

Pressure Switch,
Flow Switch, or
Flow Meter location.

Valve

Unit
served

POOR LOCATION

If pressure switch used − valve could be closed, pressure switch activated, and machine could run without water flow.

Flow switch paddle could break, and system would have to be drained to replace paddle.

Wrong Pressure and Flow Switch Location

Figure 17.5

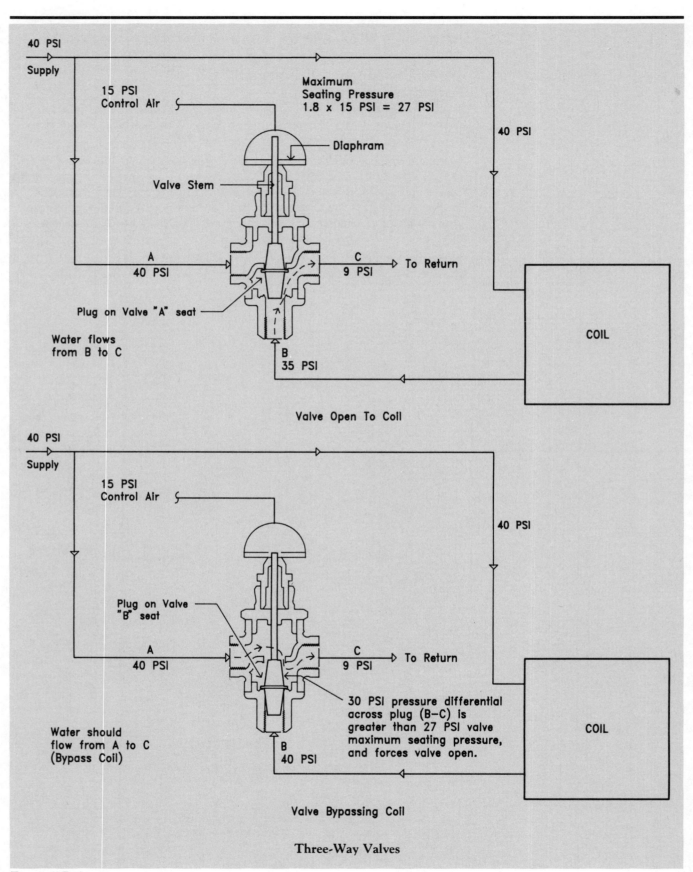

40 PSI
Supply

15 PSI
Control Air

Maximum
Seating Pressure
1.8 x 15 PSI = 27 PSI

40 PSI

Diaphram

Valve Stem

A
40 PSI

C
9 PSI

To Return

Plug on Valve "A" seat

Water flows
from B to C

B
35 PSI

COIL

Valve Open To Coil

40 PSI
Supply

15 PSI
Control Air

40 PSI

Plug on Valve
"B" seat

A
40 PSI

C
9 PSI

To Return

30 PSI pressure differential
across plug (B–C) is
greater than 27 PSI valve
maximum seating pressure,
and forces valve open.

Water should
flow from A to C
(Bypass Coil)

B
40 PSI

COIL

Valve Bypassing Coil

Three-Way Valves

Figure 17.6

220

Sizing Control Valves for Good Control

Automatic control valves must be correctly sized in order to provide good control. Problems often occur when control valves are the same diameter as the supply piping. If a 3″ pipe carries 100 gpm of water to a 3″ control valve serving a coil, the pressure drop through the valve is 1-1/2 psi. If a 4″ pipe delivers 150 gpm to a 4″ control valve, the control valve pressure drop will be less than 1 psi. The control valve will have to close almost completely before it can significantly influence the flow rate. To achieve good control, the valve should provide linear control. The valve movement should be in direct proportion to the flow. (When the valve is half open, it should deliver one-half of the flow, etc.) See Figure 17.10.

In order for a control valve to modulate the flow linearly through the full valve movement, the control must be a substantial portion of the total system pressure drop, even when the valve is wide open. A correctly sized control valve will be several pipe sizes smaller than the pipe. A 2-1/2″ control valve will deliver 150 gpm at an 8 psi pressure drop, and a 2″ control valve will deliver 100 gpm at an 8 psi pressure drop. In both cases, the control valves are two sizes smaller than the supply piping.

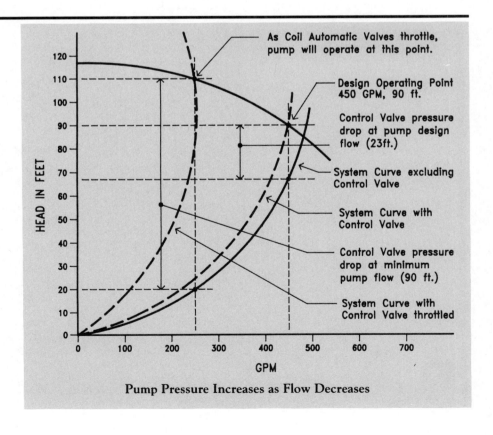

Figure 17.7

Pump Pressure Increases as Flow Decreases

Valve Shut-Off Pressure (Maximum Seating Pressure)

Another common control valve operational problem is water passing through a "closed" control valve, generating a loud chattering noise, and overheating with hot water flow, or overcooling with chilled water flow. Many control valves are rated at 20 psi close-off pressure. In the case of a normally open control valve, the valve operator is sized to overcome the spring pressure in order to keep the valve normally open and to close the valve against a 20 psi pressure difference across the valve. If the pressure difference across the valve exceeds 20 psi, the excessive water pressure will force the valve stem up, allowing water to pass through the valve port. The loud chattering noise is created by the actuator continually pushing the valve stem against the seat as the water pressure raises the stem. See Figure 17.11.

As the load decreases, the pump pressure increases. The increased pump pressure could exceed the valve shut-off rating. Most valve chattering occurs during low load conditions.

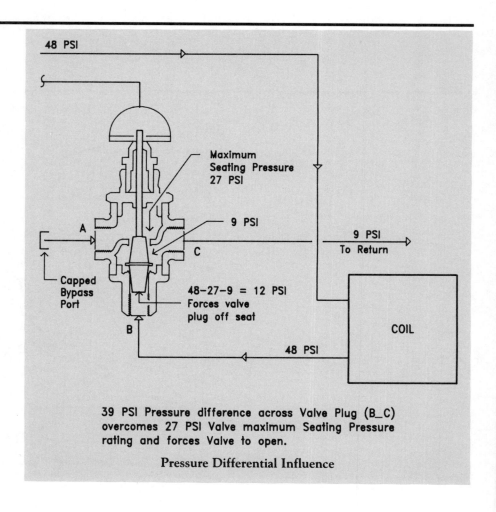

48 PSI

Maximum Seating Pressure 27 PSI

9 PSI

A

C

9 PSI
To Return

Capped Bypass Port

48−27−9 = 12 PSI Forces valve plug off seat

B

48 PSI

COIL

39 PSI Pressure difference across Valve Plug (B_C) overcomes 27 PSI Valve maximum Seating Pressure rating and forces Valve to open.

Pressure Differential Influence

Figure 17.8

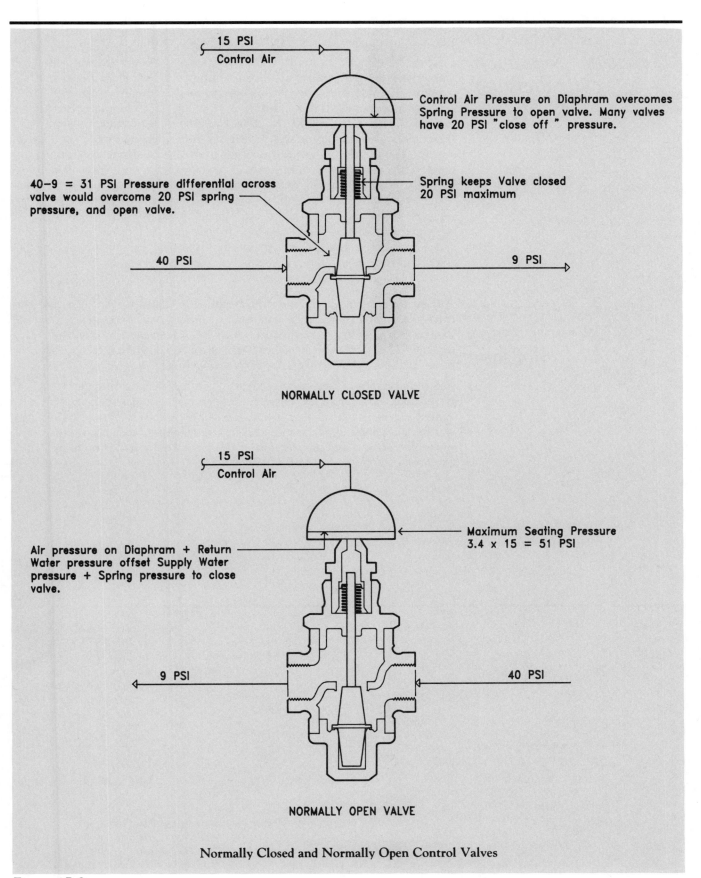

15 PSI
Control Air

Control Air Pressure on Diaphram overcomes Spring Pressure to open valve. Many valves have 20 PSI "close off" pressure.

40−9 = 31 PSI Pressure differential across valve would overcome 20 PSI spring pressure, and open valve.

Spring keeps Valve closed 20 PSI maximum

40 PSI

9 PSI

NORMALLY CLOSED VALVE

15 PSI
Control Air

Maximum Seating Pressure 3.4 x 15 = 51 PSI

Air pressure on Diaphram + Return Water pressure offset Supply Water pressure + Spring pressure to close valve.

9 PSI

40 PSI

NORMALLY OPEN VALVE

Normally Closed and Normally Open Control Valves

Figure 17.9

Automatic control valve shut-off pressure limits should be greater than the maximum pump head. This is usually 25% greater than the pump "design" head. See Figure 17.12.

The system design condition is shown at point "A" in Figure 17.12 at 550 gpm at a 120 foot head. As the load decreases due to valve throttling, the system curve moves up the pump curve to point "C." The pump head is now 160', an increase of 40', or 25%. Since there is not much water flowing through the piping system, there is little pressure loss in the piping system. The control valve must now resist almost the entire 160' pump head. The control valve must have a shut-off rating greater than 160/2.3 = 69.6 psi. A 20 psi shut-off head control valve would sing an unhappy tune. Valves with too low a shut-off pressure will almost always have to be replaced with valves capable of resisting peak pump pressure differential. If the system pressure is a little too high for the installed valves, resetting the system pressure bypass valve to maintain a lower system pressure may reduce the chattering. See Figure 17.13.

Installing Automatic Valves in Existing Buildings

Automatic control valves are often installed to control water or steam flow through existing finned tube radiators, in order to save energy. The energy is saved by preventing overheating of rooms, especially those with a southern exposure. The solar heat gain through southern exposure glass peaks at noon in December, January, and February.

Before automatic valves were used, the uncontrolled heating elements were subjected to very few extreme temperature changes, because the system was either on or off for extended time periods. Heated fluid usually was circulated through the radiators in the morning, and discontinued at night. Expansion and contraction of the finned tube and the piping occurred only once or twice a day.

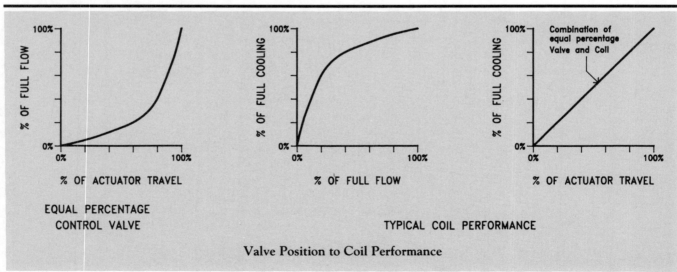

Valve Position to Coil Performance

Figure 17.10

When thermostatically controlled automatic valves are connected to the radiators, the heat through the system is cycled on and off many times a day. Every time the valve opens to heat the radiator, the finned tube and associated piping expands at a rate of 1-1/2" per 100 feet. When the valve closes because the thermostat is satisfied, the finned tube radiator and piping cool and contract. The stress imposed on the piping system by so many expansion and contraction cycles can cause joint failures. See Figure 17.14.

New piping systems must have provisions for relieving thermal stress. It is even more important to avoid stressing old piping systems, because they may have lost structural integrity due to corrosion. Expansion compensators installed at strategic locations and properly anchored and guided will relieve thermal stress. Flexible pipe is not suitable for absorbing expansion. Cycling will fatigue the bellows and produce ruptures. See Figure 17.15.

Microprocessor Control Malfunction

Microprocessors are very sensitive to electrical supply voltage fluctuations. A burst of low supply voltage could cause the microprocessor to lose its memory. Loss of memory results in loss of control, and the controlled device goes to its "fail safe" mode. Automatic dampers and valves will return to their "power off" positions, either normally open or normally closed, and remain there until control is reinstated.

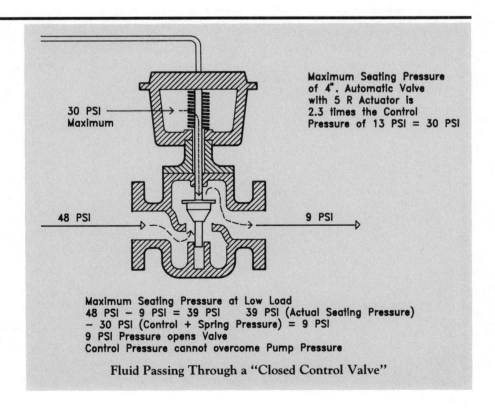

30 PSI Maximum

Maximum Seating Pressure of 4". Automatic Valve with 5 R Actuator is 2.3 times the Control Pressure of 13 PSI = 30 PSI

48 PSI   9 PSI

Maximum Seating Pressure at Low Load
48 PSI − 9 PSI = 39 PSI   39 PSI (Actual Seating Pressure)
− 30 PSI (Control + Spring Pressure) = 9 PSI
9 PSI Pressure opens Valve
Control Pressure cannot overcome Pump Pressure

Fluid Passing Through a "Closed Control Valve"

Figure 17.11

The fluctuation in voltage to the microprocessor could be caused by the activation of a fluorescent light that is connected to the same electric supply as the microprocessor, thereby causing an instantaneous control malfunction due to the loss of memory. The microprocessor should have a dedicated power supply, with no other electrical device connected to its electrical circuit.

If the microprocessor is connected to the same electric circuit as fluorescent lights, control can be stabilized by installing a voltage drop-out relay in the power supply to the microprocessor. The voltage drop-out relay will shut off the power supply to the microprocessor whenever the voltage to the microprocessor drops below the safe point. When voltage returns to the correct level, power is returned to the microprocessor and control is restored.

Radio frequency generated by some electrical equipment, such as variable frequency drives for fans, pumps, and chillers, can affect microprocessor operation. Microprocessors should not be located within the same room as the offending radio frequency emitter.

To determine if the control is influenced by radio frequencies, wrap aluminum foil around the control to shield it from the radio frequencies. If control function returns to normal, a permanent shielding enclosure can be installed for the microprocessor.

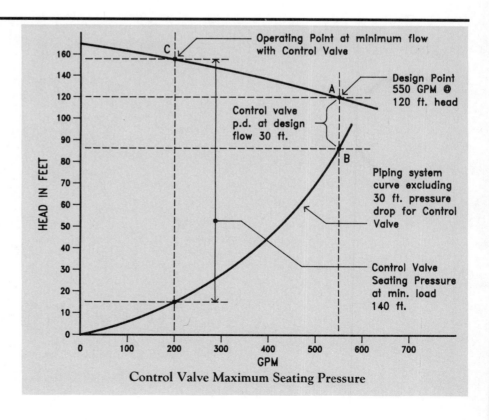

Control Valve Maximum Seating Pressure

Figure 17.12

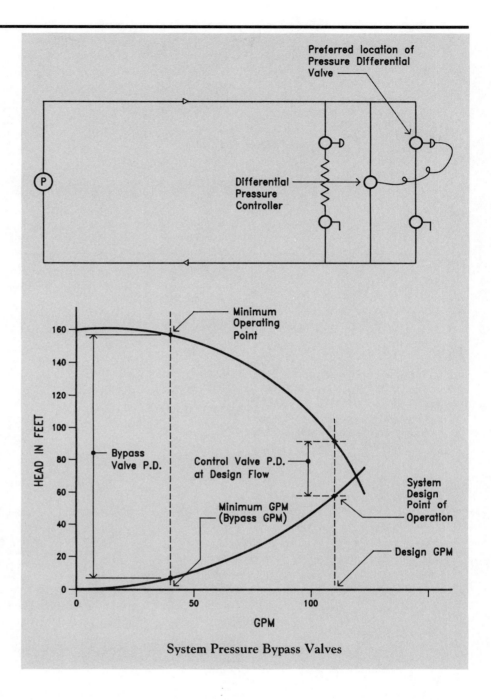

System Pressure Bypass Valves

Figure 17.13

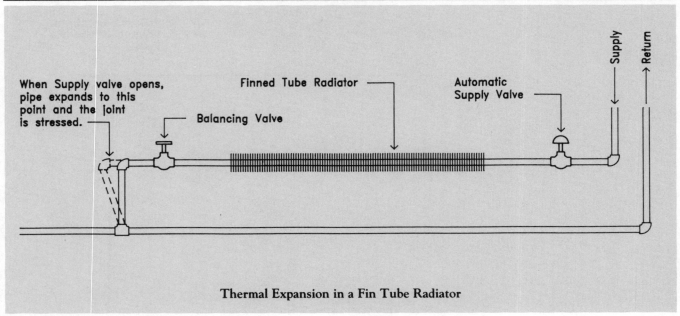

Thermal Expansion in a Fin Tube Radiator

Figure 17.14

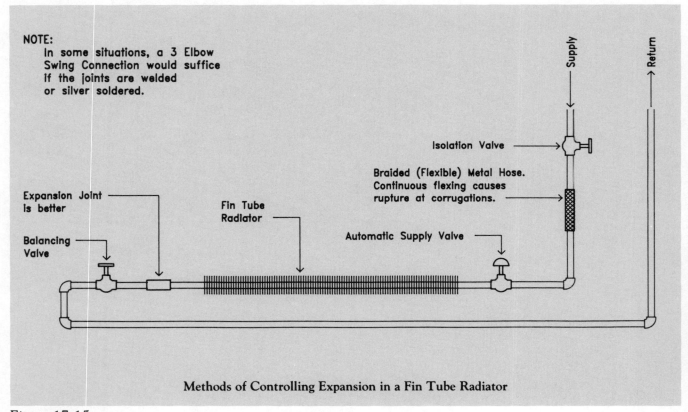

Methods of Controlling Expansion in a Fin Tube Radiator

Figure 17.15

228

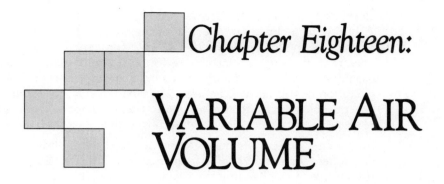

Chapter Eighteen:

VARIABLE AIR VOLUME

Variable Air Volume (VAV) systems are a comparatively new method of air distribution and control. While these systems offer the potential for energy savings, their complexity makes them more susceptible to operational problems than conventional air handling systems. This chapter addresses the most common design and operational problems that occur where VAV systems are used, and suggests methods for avoiding and solving those problems.

Correcting Fan Surge Operation

Fans used in VAV systems should be selected for maximum modulation. If not, they could be operating in their surge region, generating noise, damaging bearings, and possibly causing other physical damage to the fans. Fan surge occurs when the air entering the fan impeller is inadequate to fill the fan blades. Some of the air becomes turbulent and reverses its flow over part of the fan blades.

The basic principle of variable air volume systems is to supply only that quantity of conditioned air that will satisfy the requirements of the spaces served. (See Chapter 14 for VAV selection and installation information.) Fans should be selected to furnish the full range of desired air delivery.

Occupants complain of a "lack of air movement" when the volume of supply air to a space is less than four air changes per hour. In these conditions, there is poor circulation into office cubicles, and dilution of contaminants from carpeting, drapes, furniture, and clothing fabrics does not take place.

Figure 18.1 represents a fan curve for a system that needs 20,000 cfm to satisfy the maximum instantaneous cooling load. In this case, the minimum total supply air volume to satisfy the minimum space heat gain and *the minimum space air changes to maintain comfort* is 8,000 cfm. The supply fan must be selected for a modulation range of 20,000 cfm to 8,000 cfm without creating fan operating problems. This is a 60% reduction of supply air volume. The damper-controlled AF (air foil) fan in Figure 18.1 is operating at 1,200 rpm, 20,000 cfm, and 4.5" static pressure. As the air terminals throttle, the static pressure rises to point "B," 18,000 cfm, and 4.75" static pressure where fan surge operation begins. Only 10% supply air modulation is available before operational problems develop.

Backward-Inclined Fans

A backward-inclined fan selected for maximum efficiency at peak design cfm usually modulates down 10-15% of its design cfm before it begins to operate in the surge or unstable region. Constant volume system fans are usually selected at maximum efficiency. Damper control for VAV operation of this fan will be unsatisfactory. This is because of the noise and possible damage to the fan due to the unbalanced loading of the fan blades caused by operating the fan in the surge area.

Improved VAV performance can be realized by installing a fan speed controller, such as a variable frequency drive (VFD). The VFD reduces the fan speed by changing the frequency and voltage to a standard AC three-phase motor.

Reducing the speed of a backward-inclined fan will permit fan volume reduction of 50% of the design maximum cfm. The backward-inclined fan in Figure 18.2 is operating at 1,200 rpm to deliver the maximum design air volume of 20,000 cfm at 4.5" static pressure (sp).

The static pressure sensor located near the end of the supply duct will maintain a constant 0.50" water gauge (wg) in the supply duct. As the VAV terminal dampers throttle, the supply duct static pressure begins to increase, and the static pressure sensor begins to increase, at which point the static pressure sensor reduces the fan speed to maintain the 0.50" set

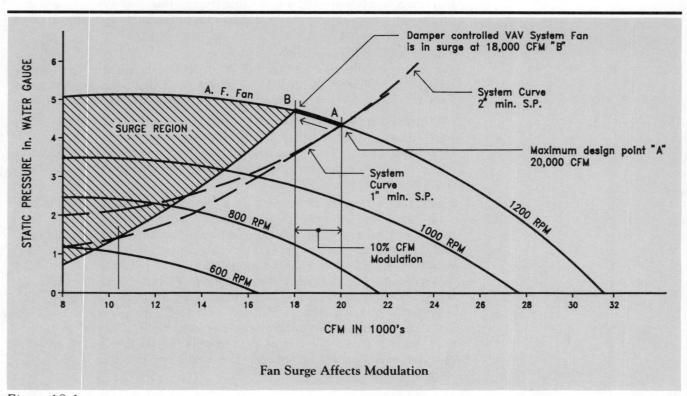

Fan Surge Affects Modulation

Figure 18.1

230

point static pressure. The fan operating point shifts to the intersection of the fan rpm curve and the system curve. The system curve remains fairly constant, since there is little change to the system.

As the fan speed is reduced to 1,000 rpm, the system operates at point "B," 16,500 cfm, and 3.0″ wg. When the fan speed is reduced to 800 rpm at point "C," the system produces 13,800 cfm at 2.0″ wg. At point "D," the fan is turning at 600 rpm, developing 1.2″ sp, and delivering 10,200 cfm; and it begins to operate in minimal surge. Airflow modulation of 50% is available.

Operating an existing damper-controlled backward-inclined fan with a fan speed controller results in a quiet, energy efficient system.

Forward-Curved Fans

VAV damper control is more appropriately used with small forward-curved fans. This is true for air volumes up to 15,000 cfm. Small forward-curved fans under damper control can usually operate with a 35% reduction of volume. Larger forward-curved fans can usually serve damper-controlled VAV systems with an 18-30% reduction of design cfm.

To avoid operating the forward-curved fan in surge, speed control can be added to allow almost any air delivery desired without approaching a surge condition. In Figure 18.3, the forward-curved fan is operating at point

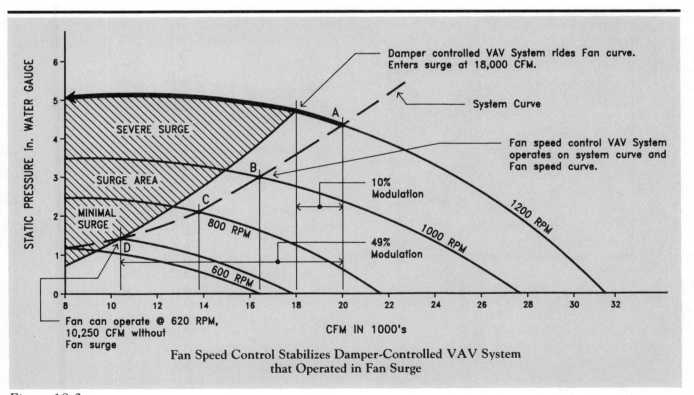

Fan Speed Control Stabilizes Damper-Controlled VAV System
that Operated in Fan Surge

Figure 18.2

"A," 900 rpm, 14,000 cfm, and 3.5″ sp. Throttling VAV dampers raises the sp to point "B," where the fan begins to surge at 11,000 cfm, a 23% reduction of supply air. Fan surge is not reached with the speed reduced as low as 500 rpm and 9,000 cfm.

Air Terminal Retrofit

As stated earlier in this chapter, VAV system air terminal units should be selected for maximum modulation. When an existing VAV air terminal unit fails to modulate the air supply because it was oversized (not selected for the minimum air supply for the minimum cooling load), the result is space overcooling or excessive reheat. In such cases, modifications to the air terminal are necessary to provide correct unit operation. Figure 18.4 shows a metal "donut" installed at the inlet of the air terminal unit to reduce the flow area over the velocity probes.

The donut decreases the area of airflow across the velocity sensor. With a smaller area, the air velocity is increased to a level that will enable the velocity controller to modulate the air volume to the desired minimum level. The donut increases the air terminal unit air pressure drop by approximately 0.18″ wg. The increased sp may require fan speed adjustment.

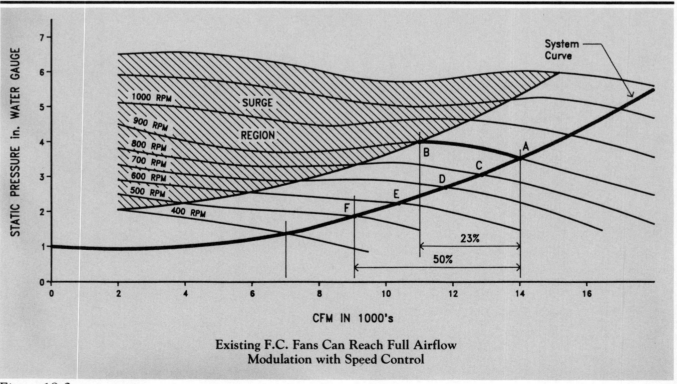

Existing F.C. Fans Can Reach Full Airflow Modulation with Speed Control

Figure 18.3

232

VAV Inlet Connections

A minimum of 2.5 duct diameters of straight duct must be installed at the inlet to the VAV box. Figure 18.5 shows turbulence created by an elbow at the VAV box inlet. A section of straight duct at the inlet to the VAV box will permit the airflow to become uniform when it reaches the velocity sensors.

The air velocity sensors are located at the air inlet to the air terminal units. The velocity sensor measures the air velocity across a fixed area. The air velocity for that area equals the air volume:

area x velocity = air volume

If the airflow is not uniform over the inlet area, the velocity sensor may read a higher or lower velocity at the sensor than the average velocity it was intended to represent. The velocity controller may be conveying a reading of 800 cfm to the air terminal unit damper, when the actual cfm may be 500. Poor cooling performance results from the reduced cool air supply.

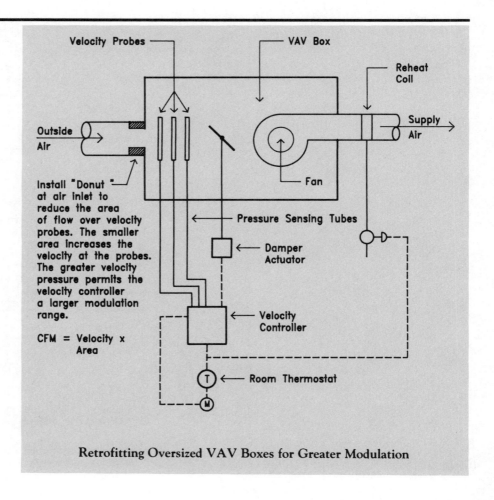

Retrofitting Oversized VAV Boxes for Greater Modulation

Figure 18.4

Access to Air Terminal Units

Velocity controllers and automatic reheat valves must be adjusted during system start-up, and must be serviced during the life of the system. Adjustment knobs, automatic valves, and damper operators must be visually accessible at a distance where the print on the devices can be read. Figure 18.6 shows that these devices must be located so that they can be reached to connect and read gauges and meters required to adjust the velocity controller.

Space must be available at the side of each air terminal unit where the velocity controller, damper operator, and reheat valve are located. The access area must be large enough for a person to get his head, shoulders, and arms close to the devices to be serviced. An 18″ by 18″ access door is the minimum size that is functional. Figure 18.7 demonstrates that mechanical, piping, or electrical conduit must not pass just above the access door, rendering it inoperable.

VAV Operators

Pneumatic operators are usually more reliable than electric or electronic operators. Not only do they usually cost less to purchase and install (see Chapter 14), but they require less maintenance than electric damper motors. Pneumatic operators perform almost indefinitely with little or no attention. Some electric operators, on the other hand, are likely to be replaced during the system life.

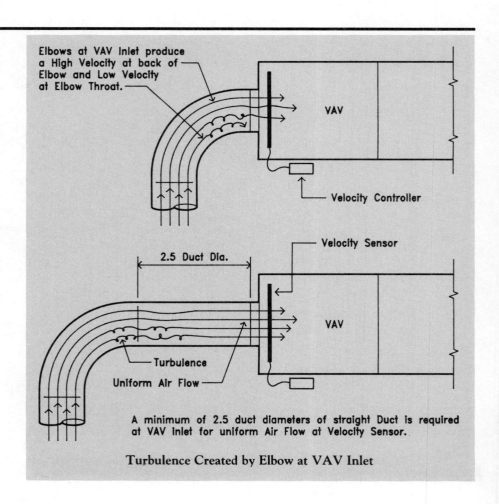

Figure 18.5

Electric Reheat Coils

If electric reheat coils are used, low airflow protection must be provided. Electric heating coils should not be energized unless sufficient air is flowing across the heating surface to remove excessive heat from the heating elements. Lack of minimum airflow over the heating coils will, at best, cause the high temperature cut-out at the heating coil to trip, opening the electrical circuit and shutting off the heat supply. At worst, excessive heat at the coil could cause electrical burnouts and create a fire hazard.

Airflow switches are installed in the air stream at the electric heating coil to measure the velocity pressure at the heating coil. The switches are set for the minimum velocity that relates to the minimum airflow that will permit safe operation of the heating coil.

Commercial airflow switches have a minimum setting of 0.07″ wg. This represents a velocity of 1,050′ per minute. A variable air volume system can reduce the air volume by as much as 40% of maximum flow, and can reduce the velocity to about 400 feet per minute. This low velocity is below the sensitivity of the airflow switch; therefore, the heating coil will be shut down by the airflow switch. To avoid repeated maintenance call-backs, the service mechanic may "jump out" the airflow switch, voiding the protection the switch was to provide.

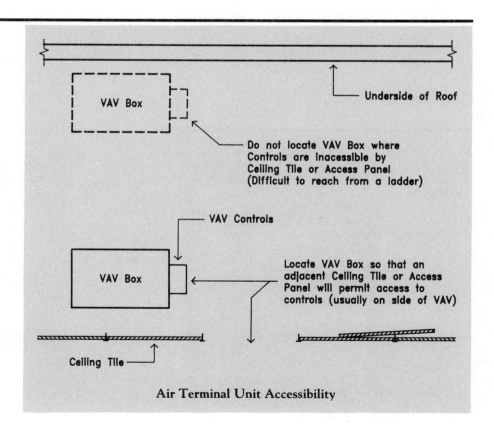

Air Terminal Unit Accessibility

Figure 18.6

An acceptable alternative to the airflow switch is an interlock relay wired in series with the heating coil control circuit. The relay is energized by the supply fan starter, or supply fan pressure differential switch. Should the supply fan starter be disengaged, the electric heater will not be activated. It is possible, however, to have the supply fan starter engaged, but no airflow because of fan belt slippage or motor failure. The airflow pressure differential switch grants a safe electric heater interlock.

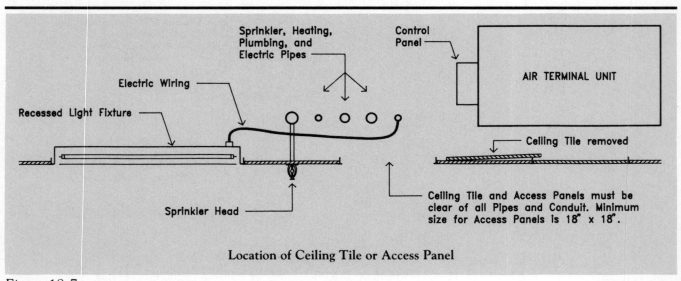

Location of Ceiling Tile or Access Panel

Figure 18.7

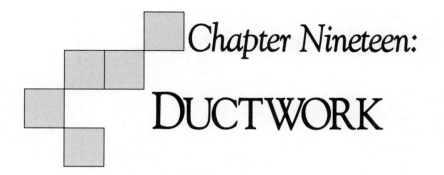

Chapter Nineteen:

DUCTWORK

The single most common problem with most HVAC systems is found in their air distribution systems. For an HVAC system to perform properly, it must deliver the correct volume of conditioned air to a space, and disperse the air evenly without drafts or noise. Unfortunately, this does not always happen. In this chapter, we will address some of the reasons why air distribution systems may be poor achievers.

Leakage: Causes and Prevention

Ductwork must be sealed airtight. SMACNA's (Sheet Metal and Air Conditioning Contractors National Association, Inc.) testing of duct systems using various methods of sealing reveals that unsealed ductwork at 2" of static pressure (sp) could lose 38% of its air supply through duct leakage. In order to effect comfort conditions within the building, the design quantity of conditioned air must reach each area to be conditioned. The correct volume of treated air may enter the supply duct, but not reach its goal. Energy is invested in cooling and moving the air that leaks out of the duct system, yet that "lost" air does nothing to condition the area it was intended to serve. Figure 19.1 shows SMACNA's test results of leakage as a percent of system airflow.

Figure 19.2 shows SMACNA's test results of leakage in cfm of duct surface. Unsealed ductwork at 2" sp will leak 75 cfm for every 100 square feet (sf) of duct surface, 48 cfm/sf at 1" sp, and 30 cfm/sf at 1/2" sp. The greater the static pressure in the duct, the greater will be the air leakage from that duct. Duct systems have been discovered leaking over 50% of the fan output volume. Ductwork should be sealed with silicone or one of the many other commercially available products applied to the joints and seams.

Most VAV air-conditioning duct systems are subjected to operating static pressures from 1.5" to 2.0" of static pressure. SMACNA class C duct sealing is for static pressures up to 2" wg (water gauge). All transverse joints and connections into the walls of the ducts must be sealed under the class "C" category. These include duct collar tap-ins, branch intersections, access doors, and air terminal connections to the ducts. Duct systems with high equivalent lengths due to long duct runs, many fittings, reheat coils, sound traps, acoustic lining, etc. could experience static pressures exceeding 2" wg.

| Leakage Class | Fan CFM Prorated* Per sf | Leakage as % of Flow in System | | | | | |
| | | Static Pressure (in. w.g.) | | | | | |
		1/2	1	2	3	4	6
48	2	15	24	38			
	2-1/2	12	19	30			
	3	10	16	25			
	4	7.7	12	19			
	5	6.1	9.6	15			
24	2	7.7	12	19			
	2-1/2	6.1	9.6	15			
	3	5.1	8.0	13			
	4	3.8	6.0	9.4			
	5	3.1	4.8	7.5			
12	2	3.8	6	9.4	12		
	2-1/2	3.1	4.8	7.5	9.8		
	3	2.6	4.0	6.3	8.2		
	4	1.9	3.0	4.7	6.1		
	5	1.5	2.4	3.8	4.9		
6	2	1.9	3	4.7	6.1	7.4	9.6
	2-1/2	1.5	2.4	3.8	4.9	5.9	7.7
	3	1.3	2.0	3.1	4.1	4.9	6.4
	4	1.0	1.5	2.4	3.1	3.7	4.8
	5	.8	1.2	1.9	2.4	3.0	3.8
3	2	1.0	1.5	2.4	3.1	3.7	4.8
	2-1/2	.5	1.2	1.9	2.4	3.0	3.8
	3	.6	1.0	1.6	2.0	2.5	3.2
	4	.5	.8	1.3	1.6	2.0	2.6
	5	.4	.6	.9	1.2	1.5	1.9

*Typically $\dfrac{\text{Fan cfm}}{\text{Duct Surface Area}}$ will be 2 to 5 cfm/square foot.

% of flow = leakage factor (in cfm/100 at the pressure)

Divided by $\dfrac{\text{Fan cfm}}{\text{s.f. surface}} = \dfrac{\text{cfm}}{100 \text{ s.f.}} \times \dfrac{\text{s.f.}}{\text{cfm}}$

Class 48 is average unsealed rectangular duct. Class 24 and lower are anticipated results for sealed ducts.

(Courtesy SMACNA HVAC Air Duct Leakage Test Manual, 1st Ed.)

Figure 19.1

SMACNA class B sealing is for ductwork exposed to 3″ of static pressure. Class B sealant ducts must have all transverse joints, all connections to the duct walls such as tap-ins, access doors, etc., and all longitudinal seams sealed. See Figure 19.3

Leakage Factor (F) in CFM/100 S.F. Duct						
Pressure W.G.		Leakage Class (C_1)				Unsealed Class 48
$P_{.65}$	p''	Class 3	Class 6	Class 12	Class 24	
.143	.05	.4	.9	1.7	3.4	6.7
.224	.10	.7	1.3	2.7	5.4	10.7
.351	.20	1.1	2.1	4.2	8.4	16.8
.457	.30	1.4	2.7	5.5	11.0	21.9
.551	.40	1.7	3.3	6.6	13.2	26.4
.637	.50	1.9	3.8	7.6	15.3	30.6
.717	.60	2.2	4.3	8.6	17.2	34.4
.793	.70	2.4	4.8	9.5	19.0	38.1
.865	.80	2.6	5.2	10.4	20.8	41.5
.934	.90	2.8	5.6	11.2	22.4	44.8
1	1	3	6	12	24	48
1.30	1.5	3.9	7.8	15.6	31.2	62.4
1.57	2.0	4.7	9.4	18.8	37.7	75.4
1.81	2.5	5.4	10.9	21.7	43.4	86.8
2.04	3.0	6.1	12.2	24.5	49.0	98.0
2.26	3.5	6.7	13.6	27.1	54.2	108.5
2.46	4.0	7.4	14.8	29.5	59.0	118.1
2.66	4.5	8.0	16.0			
2.85	5.0	8.6	17.1			
3.03	5.5	9.1	18.2			
3.20	6.0	9.6	19.2			
3.54	7.0	10.6	21.2			
3.86	8.0	11.6	23.2			
4.17	9.0	12.5	25.0			
4.47	10.0	13.4	26.8			
4.75	11.0	14.3	28.5			
5.03	12.0	15.1	30.2			

(Courtesy SMACNA HVAC Air Duct Leakage Test Manual, 1st Ed.)

Figure 19.2

Longitudinal seams are those where two sections of duct are joined and the seams are parallel to the direction of airflow. Transverse joints are those that connect two sections of duct, with the joints perpendicular to the airflow. See Figure 19.4.

Overcoming Existing Duct Leakage

It must be noted that duct installations suffering from severe leakage are difficult to correct. The duct joints that need to be sealed are seldom accessible for the attention that they need. This is a case where an ounce of prevention is worth a ton of cure. Sealing the ductwork during installation is the right way to *avoid* leakage. However, the following methods may be applied to the problem once it exists.

Sealing Installed Ductwork

All duct joints that can be reached must be sealed. If the ducts are large enough, they can be sealed from inside the duct. As Figure 19.5 demonstrates, sealant should be placed to fill the space *between* the overlapping pieces of metal. Ideally, sealant should be pumped or brushed into the slip connections. While this can only be done during duct

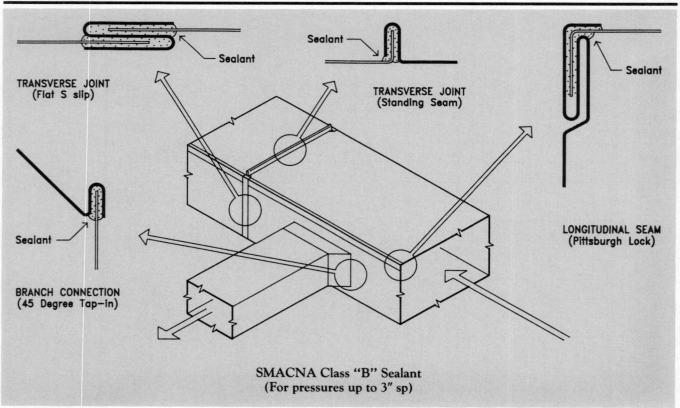

Figure 19.3

assembly, it is still worthwhile to liberally apply the liquid sealant over the existing joints to accomplish as much sealing as possible.

Lowering Duct Static Pressure

If leakage is still excessive after sealing the ductwork, attempts must be made to reduce the static pressure in the duct. It may be possible to relocate equipment with high pressure drops closer to the fan. The HEPA (high efficiency particulate air) filter in Figure 19.6 increases the negative duct pressure by 1″. By relocating the HEPA filter at the fan as shown in Figure 19.7, the 1″ decrease in duct pressure reduces the leakage from 100 cfm/sf to 80 cfm/sf.

Changing Bad Fittings

The preferred method of reducing duct pressure is by replacing the fittings that create the high pressure loss with low resistance fittings. The low pressure loss fittings not only reduce the leakage, but lower the fan power required to move the air. Figure 19.8 shows that by replacing the converging "T" with a converging "Y" fitting, the duct pressure is reduced by 0.32″. Changing the offset around the beam from four 90° vaned elbows to four 45° smooth radius ells will reduce the duct pressure another 0.67″ wg. Relocating the HEPA filter and making two fitting changes will reduce the duct leakage from 100 cfm/sf to 55 cfm/sf.

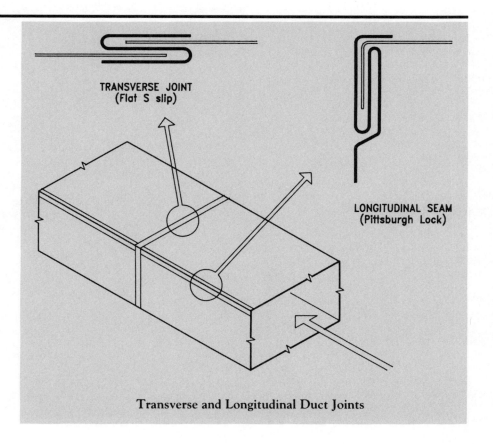

Transverse and Longitudinal Duct Joints

Figure 19.4

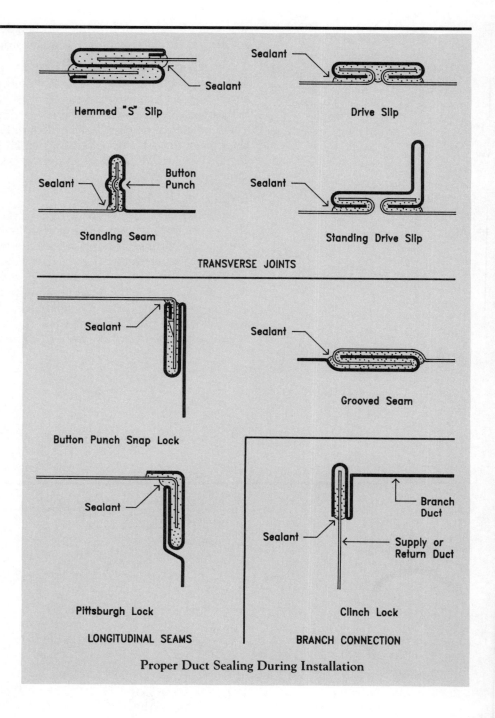

Hemmed "S" Slip

Sealant

Drive Slip

Sealant

Sealant

Button Punch

Standing Seam

Sealant

Standing Drive Slip

TRANSVERSE JOINTS

Sealant

Sealant

Button Punch Snap Lock

Grooved Seam

Sealant

Sealant

Branch Duct

Supply or Return Duct

Pittsburgh Lock

Clinch Lock

LONGITUDINAL SEAMS

BRANCH CONNECTION

Proper Duct Sealing During Installation

Figure 19.5

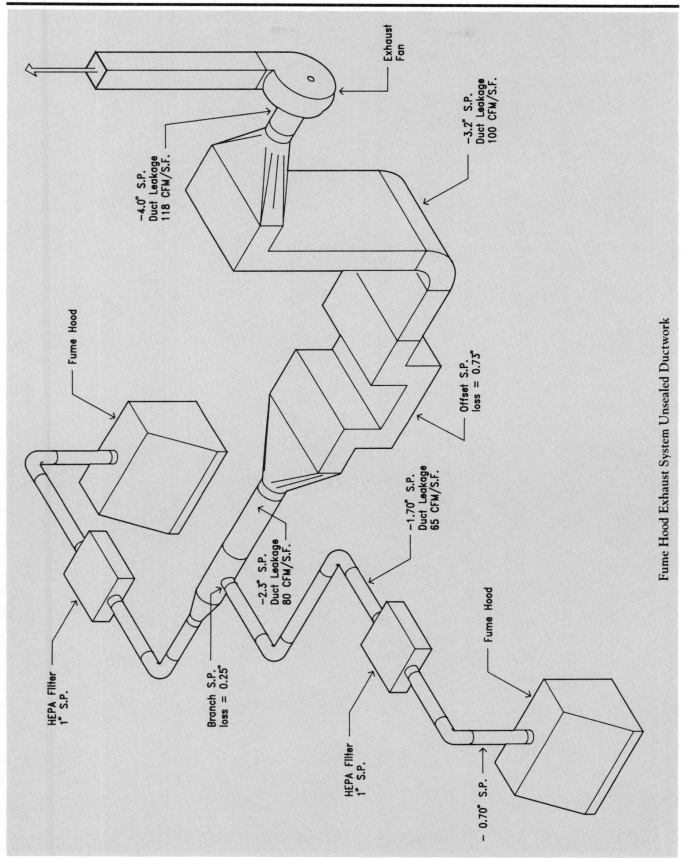

Fume Hood Exhaust System Unsealed Ductwork

Figure 19.6

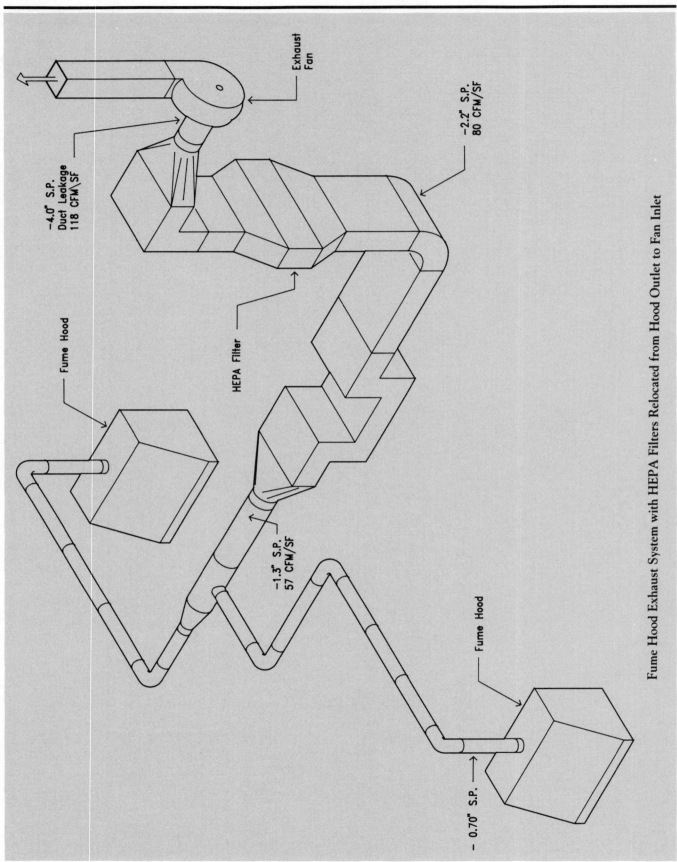

Exhaust Fan

−4.0" S.P.
Duct Leakage
118 CFM\SF

−2.2" S.P.
80 CFM/SF

Fume Hood

HEPA Filter

−1.3" S.P.
57 CFM/SF

Fume Hood

− 0.70" S.P.

Fume Hood Exhaust System with HEPA Filters Relocated from Hood Outlet to Fan Inlet

Figure 19.7

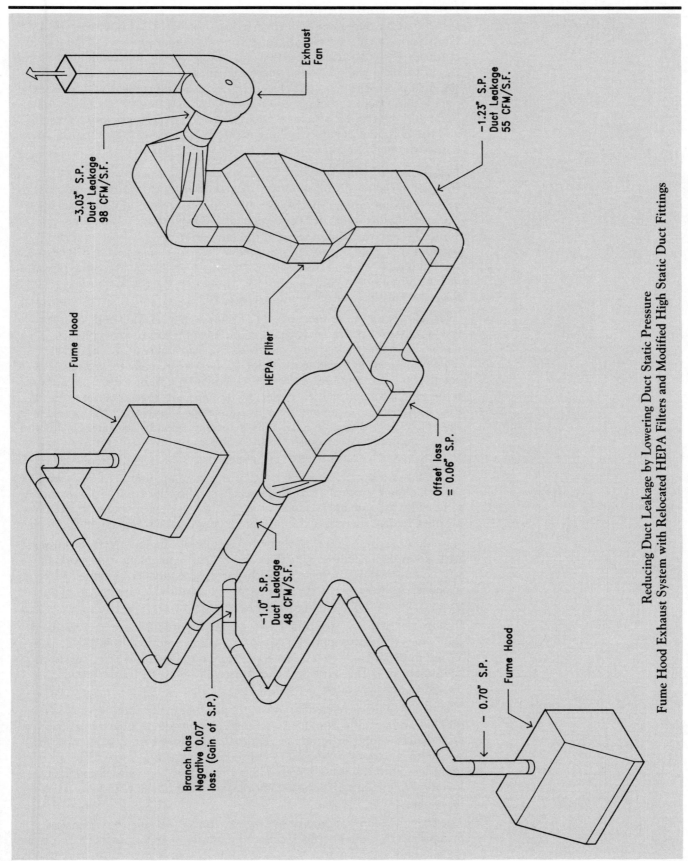

Exhaust Fan

−3.03" S.P.
Duct Leakage
98 CFM/S.F.

−1.23" S.P.
Duct Leakage
55 CFM/S.F.

Fume Hood

HEPA Filter

Offset loss
= 0.06" S.P.

−1.0" S.P.
Duct Leakage
48 CFM/S.F.

Branch has
Negative 0.07"
loss. (Gain of S.P.)

− 0.70" S.P.

Fume Hood

Reducing Duct Leakage by Lowering Duct Static Pressure
Fume Hood Exhaust System with Relocated HEPA Filters and Modified High Static Duct Fittings

Figure 19.8

245

Installing Booster Fans

If the duct pressure must be further reduced, it may be necessary to install a fan in series with the existing fan. If the existing fan shown in Figure 19.6 produces 4″ sp to overcome the system resistance, the maximum duct sp can be cut to 1.0″ by allowing the additional fan to handle the hood loss plus one-half of the common duct loss. This system arrangement will reduce the duct leakage from 118 cfm/sf to 48 cfm/sf. In the case of a laboratory fume hood duct system, care must be taken to avoid creating a positive pressure and the potential of a duct leaking toxic air from the duct system into any part of the building.

Balancing Dampers

In order for an air distribution system to deliver the specified air quantity to each supply air outlet, and to take in the necessary cfm into each air inlet, the pressure loss from the fan to each air terminal must be the same. As Figure 19.9 shows, the pressure at the fan outlet "A" is 2″ wg, and the pressure at the last supply air terminal "F" is zero. The pressure loss between the fan outlet "A" and the last air terminal "F" is 2″ wg. Since the pressure at the fan outlet "A" is 2″ wg and the pressure at the second outlet "E" is zero, the pressure loss of circuit A-E is 2″ wg.

Every supply air outlet must have a pressure loss from the fan discharge to the air outlet of the same 2″ wg. If the total equivalent length of each circuit is not the same, the circuit with the least equivalent length will deliver more air to increase the velocity in order to make its pressure loss equal to the pressure loss of the other circuits. We want the design airflow to be supplied from each air outlet, so we must add resistance, or pressure loss, to the circuits that are lower in equivalent length than the circuit with the greatest equivalent length. This is accomplished by partially closing the branch damper to increase the pressure loss in the circuit. The damper is throttled to deliver the desired air quantity.

The return air duct system is treated in the same manner. The pressure loss of each circuit, from the air inlet in each room to the fan inlet, must be the same. The pressure at the fan inlet is − 2″ wg, and the pressure at the room air inlet is zero; therefore, the pressure loss in each circuit should be 2″ wg.

If we do not install and throttle branch return air dampers in those circuits with lower equivalent lengths than the greatest equivalent run, more air will be returned from the rooms with the lower equivalent lengths. More air is needed in order to increase the pressure loss in those circuits, to make them equal to the larger equivalent length circuit.

Return air balancing dampers are very often omitted because it is assumed that whatever air is supplied into a room will be returned from the room. When more air is taken out of a room than is delivered to the room from the supply air outlet, air from other areas will enter the room through doorways, wall openings, windows, etc.

The negative pressure created in the room could make opening doors very difficult and possibly introduce warm or cold outside air into the space. Unbalanced supply or return air terminals can also be excessively noisy due to the high velocity generated by the excess air passing through the air terminal. Throttling the branch damper in order to equalize the pressure loss in each circuit also generates noise as a result of the high velocity at the damper.

The branch damper should be as far away from the room air terminal as possible, so that the ductwork can attenuate the damper noise. Duct elbows are good noise attenuators. Acoustic lining or sound attenuators may also be necessary to reduce noise. If more than 0.2″ wg must be added to a circuit, it may be necessary to have more than one damper in a branch. If pressure drops of 0.5 or more are required, duct pressure reducing valves (acoustically lined dampers and boxes) may be needed.

Air terminals should not be directly connected to the supply or return air main duct. The short collar connecting the terminal to the main duct does not permit installation of a damper far enough away from the air terminal to attenuate the noise. When the damper at the grille is adjusted to provide design flow at 0.25″ wg pressure drop, the room air terminal sound pressure level is increased by 20 dB. This large increase in sound power level will exceed the space noise criterion. It is not unusual to damper 0.50″ of wg to produce design flow. The more than 50 dB sound pressure increase generated by the action creates an uncomfortable acoustic environment. A branch duct with several elbows is needed to control the noise generated by the balancing damper. See Figure 19.10

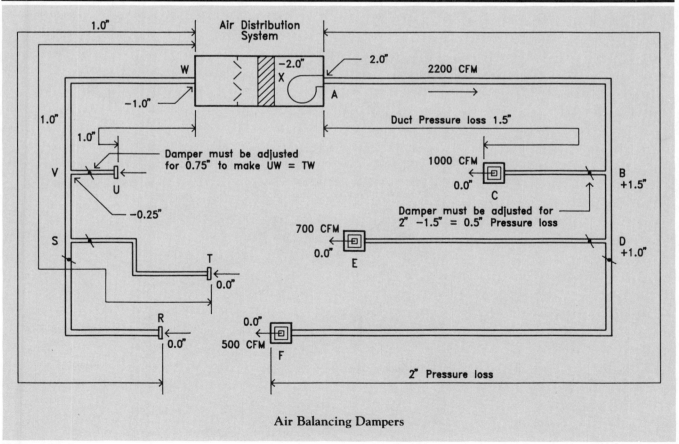

Air Balancing Dampers

Figure 19.9

Splitter Dampers

Splitter dampers create more problems than they solve. They provide little or no air volume control, and their only value is to divert air into the branch duct. See Figure 19.11.

When splitter dampers are used in 45° takeoff fittings, they cause turbulence in the main duct that adds pressure loss to the system. Splitter dampers are more expensive than butterfly dampers and can generate noise if they are not rigid and tight.

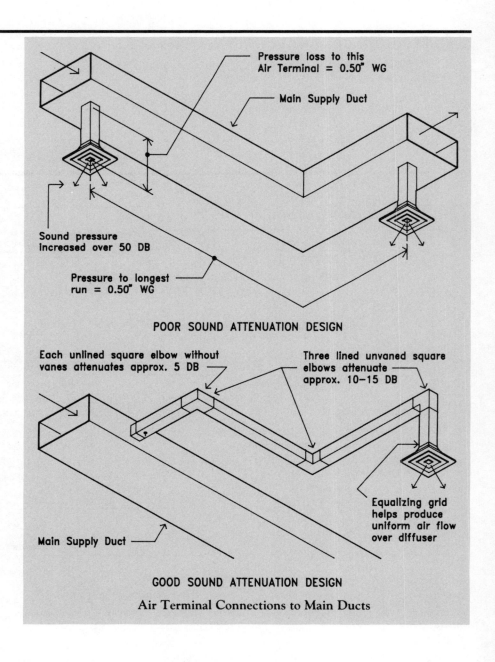

Pressure loss to this
Air Terminal = 0.50" WG

Main Supply Duct

Sound pressure
increased over 50 DB

Pressure to longest
run = 0.50" WG

POOR SOUND ATTENUATION DESIGN

Each unlined square elbow without
vanes attenuates approx. 5 DB

Three lined unvaned square
elbows attenuate
approx. 10–15 DB

Equalizing grid
helps produce
uniform air flow
over diffuser

Main Supply Duct

GOOD SOUND ATTENUATION DESIGN

Air Terminal Connections to Main Ducts

Figure 19.10

248

Volume Dampers

Volume dampers installed in the branch duct near the connection to the main duct produce better balancing results than splitter dampers. They do not add pressure loss to the main duct system, do not normally generate "rattling" noise, and cost less to install than splitter dampers. See Figure 19.12.

Locate the balancing dampers as far away from air terminals as possible. Duct elbows and ductwork attenuate some of the noise generated at the throttled damper. The ideal balancing damper location is in the branch duct connection to the main duct.

The damper must be installed in the branch duct, where it can be reached through a ceiling access door or removable ceiling tile. See Figure 19.13.

Acoustic Lining

The lining of the duct inside surface reduces the aerodynamic noise generated by the airflow through the system, and provides thermal insulation. Acoustic lining should be double-walled near fans, where the greatest velocity and turbulence occur, and should be metal-nosed elsewhere. Care must be taken not to damage the lining during installation. Lining edges at the cut sections may otherwise protrude into the air stream. Portions of the lining "peel off" and collect at fittings, restricting airflow. Fiberglass fibers that pass into the air stream are then discharged into the conditioned space and can be inhaled by the occupants. See Figure 19.14.

Acoustic lining doubles the friction loss of the metal duct that it covers. The acoustical benefit of more than 15 feet of lined duct is minimal. For

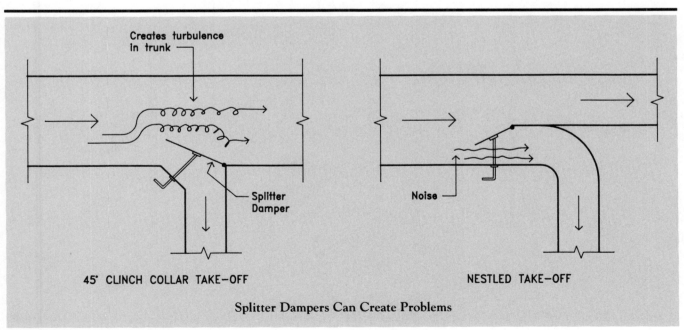

Splitter Dampers Can Create Problems

Figure 19.11

these reasons, lining ductwork should be limited to 10 – 15' from the noise source. Double-wall construction consists of a "sandwich" type duct, made up of an outside standard duct, the acoustic lining, and an inside perforated sheet metal surface. This construction will maintain the acoustic lining stability, especially at the highest duct velocity, near the fan. See Figure 19.15.

Metal Nosing

Metal nosings are sheet metal caps installed over the transverse lining sections. The sheet metal caps prevent the raw lining edges from separating from the lining sheets. See Figure 19.16.

Metal caps over lining edges are difficult to fabricate and install. They are an expensive lining protection method. A metal strip secured to the lined duct over the lining edges offers an acceptable alternative. See Figure 19.17.

Flexible Duct

The use of flexible duct should be minimized. It is not uncommon to find flexible duct installations with pressure losses five times greater than those

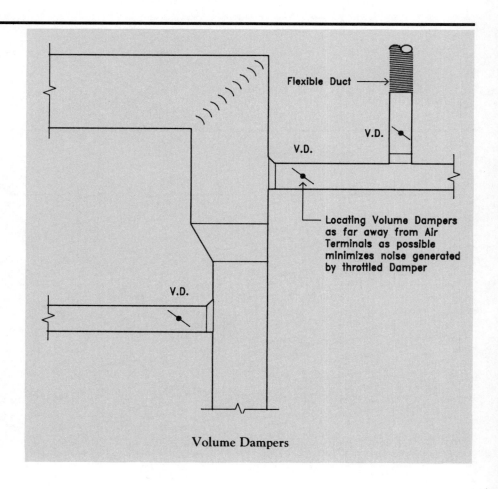

Volume Dampers

Figure 19.12

found in metal ducts. Long flexible ducts are squeezed almost closed at hangers, and flexible ducts used as elbows can crimp, restricting airflow. See Figure 19.18.

When the space above the diffusers is limited, it is better to install a round sheet metal elbow, and connect the flexible duct between the metal elbow and the metal branch duct. See Figure 19.19.

Flexible ducts connecting supply and return air terminals significantly reduce the ductwork installation cost. Flexible duct is easily adjusted to connect to a diffuser located in a ceiling grid. Rigid duct can rarely be installed straight from the branch duct to the diffuser.

Offsets made of sheet metal are expensive to fabricate and install. The offset must be kept to a minimum in order not to generate noise. As Figure 19.20 shows, a 2" offset in a 16" long, 8" round vertical connection

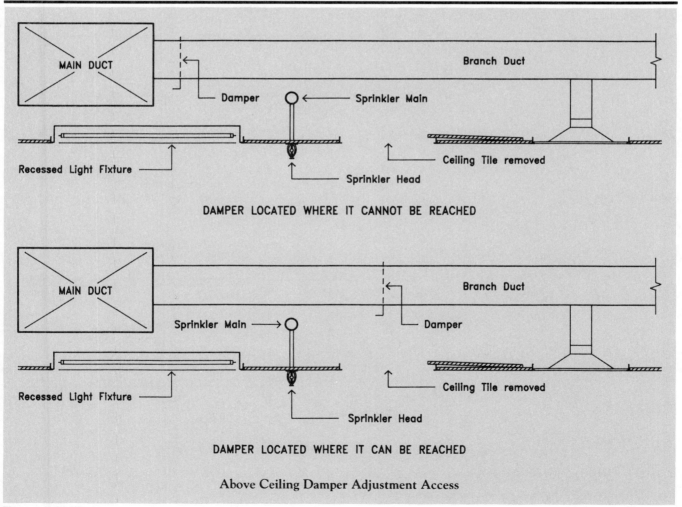

Above Ceiling Damper Adjustment Access

Figure 19.13

to the outlet will not affect the sound power level. A 4″ offset will appreciably increase the sound power level, and an 8″ offset will increase the sound power level by 15 dB.

Flexible duct should be limited to a maximum length of 5′. Flexible duct elbows should be avoided or supported to permit smooth airflow.

Rooftop Ductwork

Ductwork should be installed at least 18″ above the roof in order to permit servicing of the roof. Steps are also desirable to avoid stepping on and damaging the ductwork and its weather protection. The outdoor ductwork

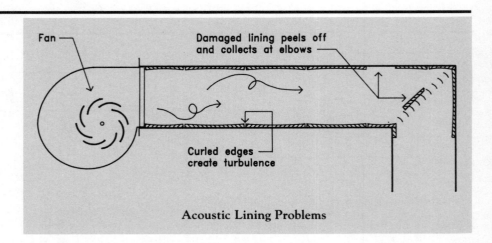

Acoustic Lining Problems

Figure 19.14

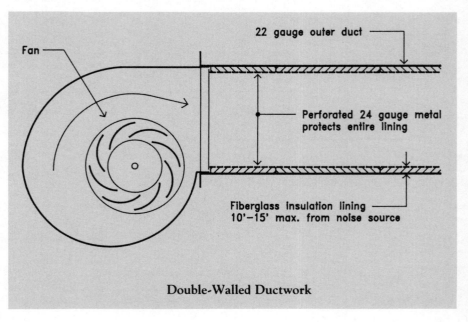

Double-Walled Ductwork

Figure 19.15

must be insulated and made watertight. Exterior duct insulation and an aluminum jacket with watertight joints makes for a good trouble-free installation. See Figure 19.21.

System Effect Factors

Fans are almost never connected to operate under the conditions that the fan was tested in the laboratory. The standard fan testing configuration

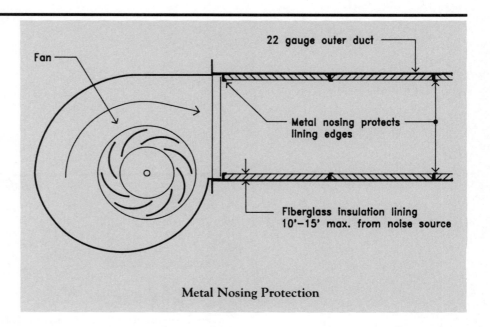

Figure 19.16

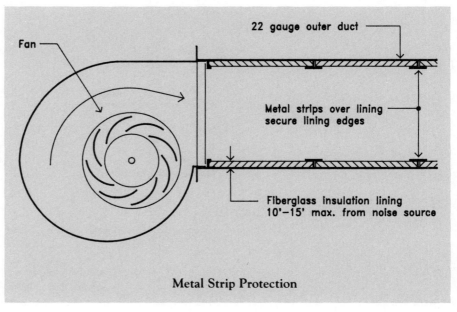

Figure 19.17

shown in Figure 19.22, of 15° transitions, 3.5 fan diameters between the fan and air straighteners, and 10 fan diameters to the test station, are almost impossible to reproduce in the construction world.

Fan manufacturers' fan curves and tables are established from data derived from the test assembly in accordance with AMCA (Air Movement and Control Association) standards. The differences between the fan performance characteristics obtained in the test rig and those occurring in the actual installation are known as *system effects*.

The major differences between the test rig and the actual installed conditions occur at the fan inlet and the fan outlet. When a fan is connected so that the airflow into the fan inlet is disturbed and the velocity at the impeller is not evenly distributed, extra losses result. Figure 19.23 shows examples of fan inlet connections that produce substantial system effect losses.

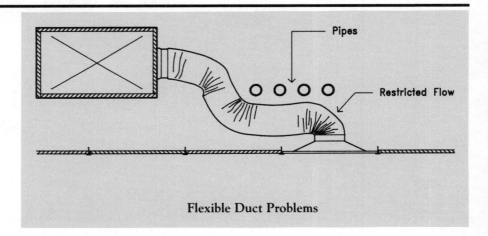

Flexible Duct Problems

Figure 19.18

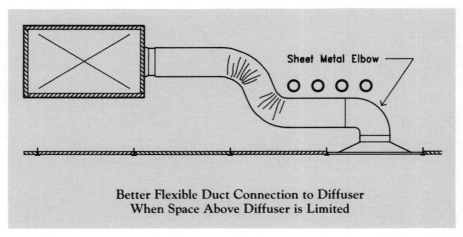

**Better Flexible Duct Connection to Diffuser
When Space Above Diffuser is Limited**

Figure 19.19

The fan connections depicted in Figure 19.24 are very common in actual fan installations. Non-uniform flow into the fan inlet is the most common cause of poor fan performance.

The inlet spin in the same direction as impeller rotation reduces the fan pressure/volume curve in the same manner as inlet vanes installed at the fan inlet. A counter-rotating swirl at the fan inlet will produce an increase in the fan pressure volume curve, along with a substantial increase in fan horsepower. See Figure 19.25.

Inlet spin generated by rectangular elbows at the fan inlet are impossible to tabulate, but they result in capacity losses of 45-50%. Fan performance losses of this magnitude should be avoided by installing round elbows or square elbows with turning vanes, as per Figure 19.26, that have tabulated losses which can be addressed.

Figure 19.27a & 19.27b show typical fan inlet configurations and their system effect factors as tabulated by AMCA. Figure 19.28 is the AMCA chart that converts the system effect factor and fan inlet velocity into the static pressure to be added to the system curve.

Correcting Inlet Spin

The ideal fan inlet condition provides for uniform airflow across the fan inlet. Fans connected as in Figure 19.24 can lose 45% of their capacity. It is not practical to modify the fan inlet connection by reconnecting the fan with ten diameters of straight duct at the fan inlet. However, it may be possible to modify the duct inlet connections as shown in Figure 19.26.

AMCA Figure C in Figure 19.27a represents a square 90° elbow with short guide vanes and a transition. The R/H is (12"/25") 0.5, the duct length is 0, resulting in a system effect factor of "S." Using the system effect factor of "S" and an inlet velocity of 2,000 fpm in Figure 19.28, the increase in

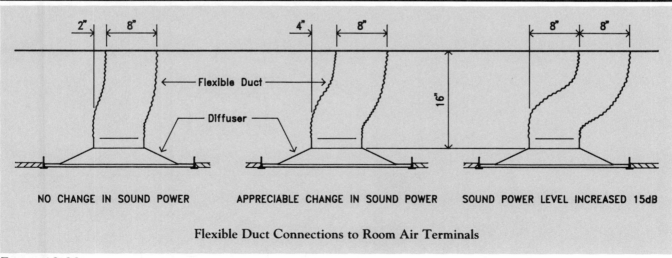

Flexible Duct Connections to Room Air Terminals

Figure 19.20

the system curve is only 0.20" wg, *a remarkable improvement over the 45% loss of fan volume the original fan inlet connection produced.*

If the fan inlet connection was modified with a round to rectangular transition and even a short radius 24" x 12" elbow with short turning vanes, the fan's performance will be dramatically improved.

Fan Suction Boxes

The very bad fan inlet condition displayed in Figure 19.26 can be easily corrected by installing a fan suction box at the fan inlet. The Type 2 fan suction box shown in Figure 19.29 requires minimal duct modification. In most cases, only the section connecting the fan inlet collar needs replacement. The Type 1 fan suction box involves a duct offset to accommodate the 5" long transition between the fan inlet collar and the

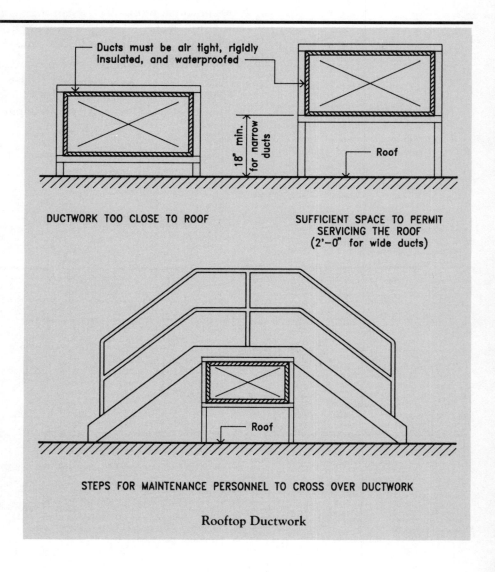

Figure 19.21

256

vertical duct drop portion of the fan suction box. Figure 19.29 demonstrates the fan suction box solution to the common, ineffective fan connection of Figure 19.24.

The Type 2 fan suction box in Figure 19.29 has a system effect factor of "S". The intersection of the "S" system effect factor with a 2,000 fpm velocity in Figure 19.28 results in a static pressure loss of 0.20" wg.

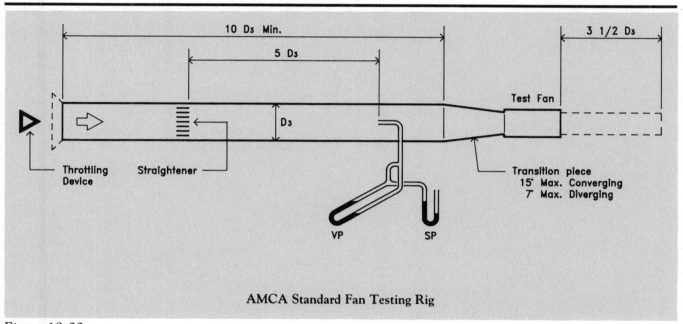

AMCA Standard Fan Testing Rig

Figure 19.22

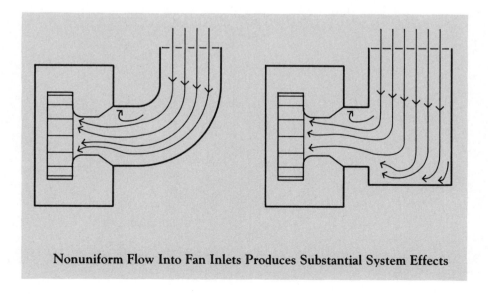

Nonuniform Flow Into Fan Inlets Produces Substantial System Effects

Figure 19.23

The Type 1 fan suction box has a system effect factor of "T." Inserting "T" SEF with 2,000 fpm in Figure 19.28 results in a static pressure loss of an almost insignificant 0.13" wg. The SEF created by fan suction boxes can be offset by a slight increase in fan speed, while ensuring that the fan will deliver the design air volume.

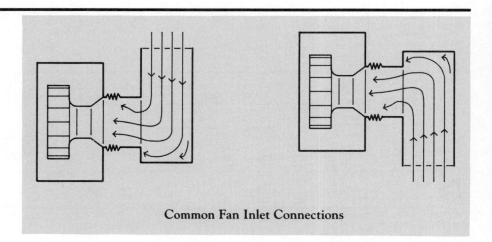

Common Fan Inlet Connections

Figure 19.24

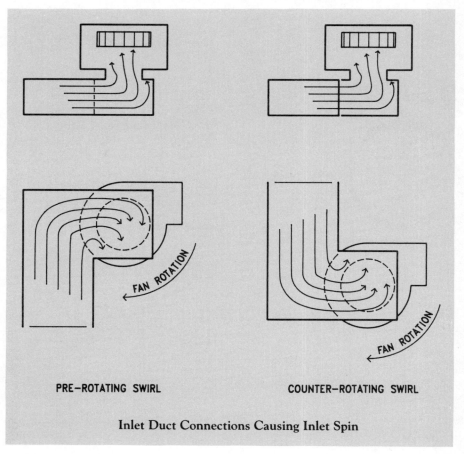

PRE-ROTATING SWIRL   COUNTER-ROTATING SWIRL

Inlet Duct Connections Causing Inlet Spin

Figure 19.25

Fan Outlet Duct System Effect

The performance of a fan is based on the tabulated results of testing the fan in accordance with the AMCA criteria shown in Figure 19.22. Fans can almost never be connected with 3-1/2 diameters of discharge duct. AMCA's detail in Figure 19.30 shows the change in velocity profile with the distance from the fan outlet. No system effect occurs at 2.5 – 3.0 duct diameters of discharge duct length, which is 100% effective duct length. Figures 19.31a – d contain SMACNA data on outlet ducts. Several typical fan outlet duct connections that produce system effect factors are illustrated in Figure 19.31a.

Figure 19.31b provides the system effect factor for various fan outlet connections. In Figure 19.31b, the ratio of the fan blast area to the nominal fan outlet area is given by the fan manufacturer. For the example shown, it is 0.6.

Since the elbow is directly connected to the flexible connection at the fan, the effective duct length is 0%. With the elbow orientation at "A", the SEF is tabulated as "Q." The intersection of "Q" SEF and 2,000 fpm velocity is 0.40″ wg pressure loss. The static pressure loss of ductwork, coils, filter, louvers, grilles, etc. must be increased by 0.40″ for the fan outlet system effect to determine the fan's performance. If the discharge elbow was located in position "C" of Figure 19.31a, the SEF would be "N-0." The intersection of SEF "N-O" 2,000 fpm velocity is a very significant pressure loss of 0.70. The 90° elbow bend configuration "D" in Figure 19.31a creates an "0" SEF. Plotting the "0" SEF on the chart in Figure 19.28 reveals a substantial 0.63 pressure loss.

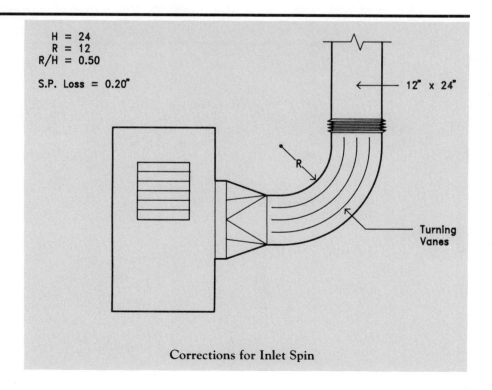

H = 24
R = 12
R/H = 0.50

S.P. Loss = 0.20″

12″ x 24″

R

Turning Vanes

Corrections for Inlet Spin

Figure 19.26

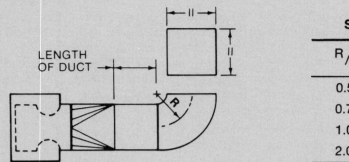

SYSTEM EFFECT CURVES			
R/H	NO DUCT	2D DUCT	5D DUCT
0.5	O	Q	S
0.75	P	R	S-T
1.0	R	S-T	U-V
2.0	S	T-U	V

a SQUARE ELBOW WITH INLET TRANSITION — NO TURNING VANES

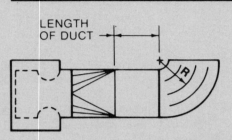

SYSTEM EFFECT CURVES			
R/H	NO DUCT	2D DUCT	5D DUCT
0.5	S	T-U	V
1.0	T	U-V	W
2.0	V	V-W	W-X

b SQUARE ELBOW WITH INLET TRANSITION — 3 LONG TURNING VANES

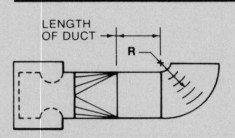

SYSTEM EFFECT CURVES			
R/H	NO DUCT	2D DUCT	5D DUCT
0.5	S	T-U	V
1.0	T	U-V	W
2.0	V	V-W	W-X

c SQUARE ELBOW WITH INLET TRANSITION — SHORT TURNING VANES

$$D = \frac{2H}{\sqrt{\pi}}$$

THE INSIDE AREA OF THE SQUARE DUCT (H × H) IS EQUAL TO THE INSIDE AREA CIRCUMSCRIBED BY THE FAN INLET COLLAR. THE MAXIMUM PERMISSIBLE ANGLE OF ANY CONVERGING ELEMENT OF THE TRANSITION IS 15°, AND FOR A DIVERGING ELEMENT 7½°.

System Effects for Various Duct Elbows

(Reprinted by Permission from AMCA Publication 201, Fans and Systems)

Figure 19.27a

Correcting Poor Fan Outlet Duct Installations

If a 24″ long straight duct were added between the fan outlet and the elbow in Figure 19.32, the 50% effective duct length will reduce the SEF to "S" and the pressure loss would be reduced from 0.70 to 0.20″ wg. This is an impressive static pressure reduction of 0.70 − 0.20 = 0.50″ wg.

Note that if the fan arrangement in Figure 19.32 was reversed to that of position A in Figure 19.31a, the pressure loss would be reduced from 0.70 to 0.40, a static pressure reduction of 0.30″ wg without adding straight duct to the fan outlet.

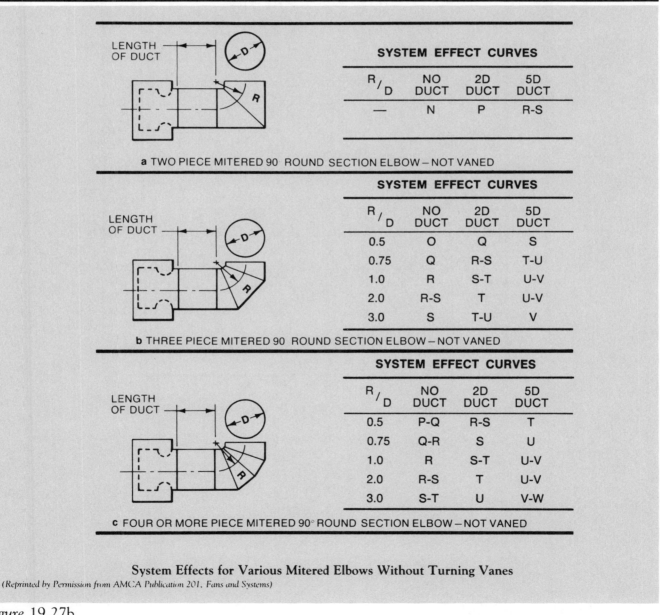

a TWO PIECE MITERED 90 ROUND SECTION ELBOW – NOT VANED

SYSTEM EFFECT CURVES

R/D	NO DUCT	2D DUCT	5D DUCT
—	N	P	R-S

b THREE PIECE MITERED 90 ROUND SECTION ELBOW – NOT VANED

SYSTEM EFFECT CURVES

R/D	NO DUCT	2D DUCT	5D DUCT
0.5	O	Q	S
0.75	Q	R-S	T-U
1.0	R	S-T	U-V
2.0	R-S	T	U-V
3.0	S	T-U	V

c FOUR OR MORE PIECE MITERED 90° ROUND SECTION ELBOW – NOT VANED

SYSTEM EFFECT CURVES

R/D	NO DUCT	2D DUCT	5D DUCT
0.5	P-Q	R-S	T
0.75	Q-R	S	U
1.0	R	S-T	U-V
2.0	R-S	T	U-V
3.0	S-T	U	V-W

System Effects for Various Mitered Elbows Without Turning Vanes

(Reprinted by Permission from AMCA Publication 201, Fans and Systems)

Figure 19.27b

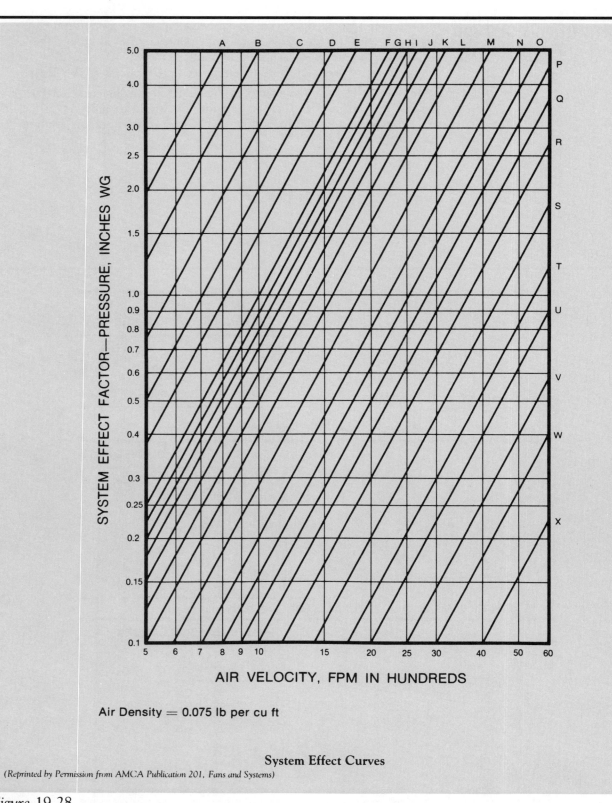

SYSTEM EFFECT FACTOR—PRESSURE, INCHES WG

AIR VELOCITY, FPM IN HUNDREDS

Air Density = 0.075 lb per cu ft

System Effect Curves

(Reprinted by Permission from AMCA Publication 201, Fans and Systems)

Figure 19.28

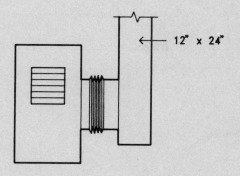

Installed Fan Inlet
45% CFM Loss

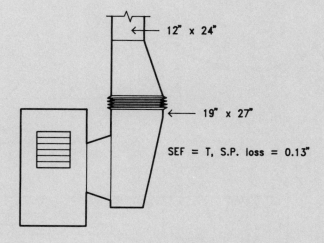

Type I Suction Box
Fan performance is improved from 55% to almost 100%

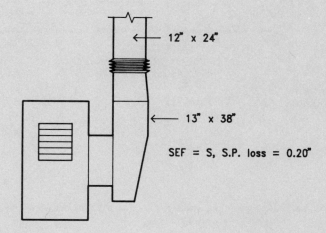

Type II Suction Box

Correcting Poor Fan Inlet Connection with Fan Suction Box

Figure 19.29

263

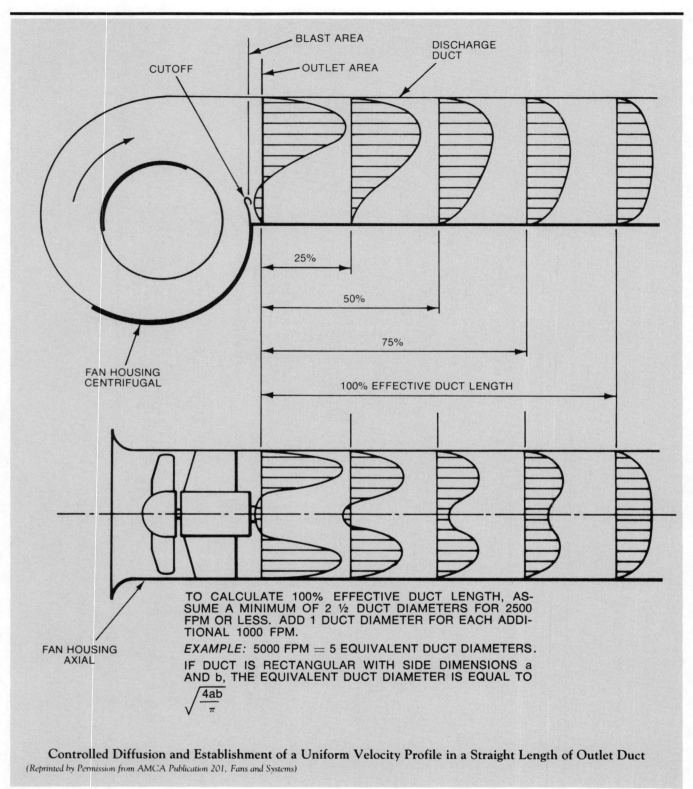

BLAST AREA

CUTOFF

OUTLET AREA

DISCHARGE DUCT

FAN HOUSING CENTRIFUGAL

25%

50%

75%

100% EFFECTIVE DUCT LENGTH

FAN HOUSING AXIAL

TO CALCULATE 100% EFFECTIVE DUCT LENGTH, ASSUME A MINIMUM OF 2 ½ DUCT DIAMETERS FOR 2500 FPM OR LESS. ADD 1 DUCT DIAMETER FOR EACH ADDITIONAL 1000 FPM.

EXAMPLE: 5000 FPM = 5 EQUIVALENT DUCT DIAMETERS.

IF DUCT IS RECTANGULAR WITH SIDE DIMENSIONS a AND b, THE EQUIVALENT DUCT DIAMETER IS EQUAL TO

$$\sqrt{\frac{4ab}{\pi}}$$

Controlled Diffusion and Establishment of a Uniform Velocity Profile in a Straight Length of Outlet Duct

(*Reprinted by Permission from AMCA Publication 201, Fans and Systems*)

Figure 19.30

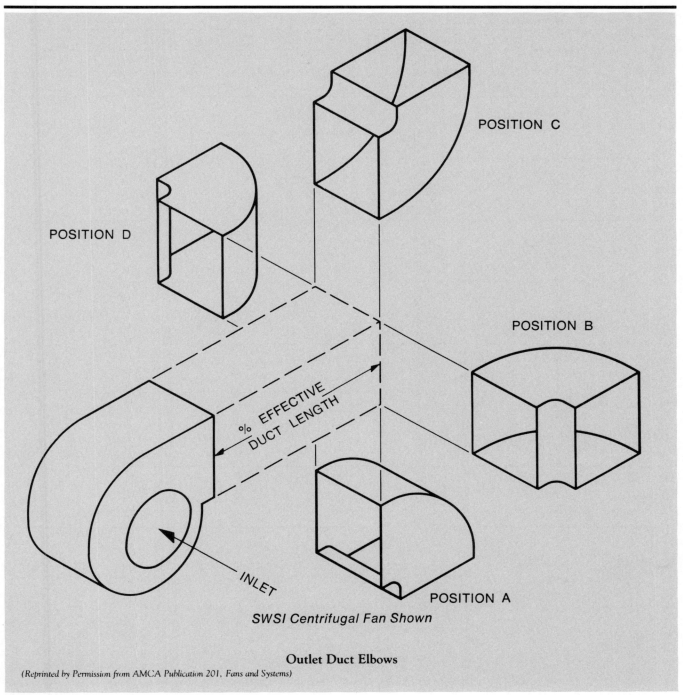

POSITION C

POSITION D

POSITION B

% EFFECTIVE DUCT LENGTH

INLET

POSITION A

SWSI Centrifugal Fan Shown

Outlet Duct Elbows

(Reprinted by Permission from AMCA Publication 201, Fans and Systems)

Figure 19.31a

System Effect Factors for Outlet Elbows						
Blast Area / Outlet Area	Outlet Elbow Position	No Outlet Duct	12% Effective Duct	25% Effective Duct	50% Effective Duct	100% Effective Duct
0.4	A	N	O	P-Q	S	No System Effect Factor
	B	M	M-N	O	R	
	C	L-M	M	N	Q	
	D	L-M	M	N	Q	
0.5	A	P	Q	R	T	
	B	N-O	O-P	P-Q	S	
	C	M-N	N-O	O-P	R-S	
	D	M-N	N-O	O-P	R-S	
0.6	A	Q	Q-R	R-S	U	
	B	P	Q	R	T	
	C	N-O	O-P	P-Q	S	
	D	O	P	Q-R	S-T	
0.7	A	S-T	T	U	W	
	B	R-S	S	T	V	
	C	Q-R	R	S	U-V	
	D	R	R-S	S-T	U-V	
0.8	A	S	S-T	T-U	V-W	
	B	R	R-S	S-T	U-V	
	C	Q	Q-R	R-S	U	
	D	Q-R	R	S	U-V	
0.9	A	S-T	T	U	W	
	B	R-S	S	T	V	
	C	R	R-S	S-T	U-V	
	D	R-S	S	T	V	
1.0	A	R-S	S	T	V	
	B	S-T	T	U	W	
	C	R-S	S	T	V	
	D	R-S	S	T	V	

System Effect Factor Curves for SWSI Fans

For DWDI fans determine system effect factor curve using the above tabulation for SWSI fans. Next, determine system effect factor (ΔP) by using Figure 18. Then apply appropriate multiplier from tabulation below:

Multipliers for DWDI Fans

Elbow Position B = ΔP x 1.25
Elbow Position D = ΔP x 0.85
Elbow Positions A and C = ΔP x 1.00

Refer to Figure 19.31a for elbow position designations.

(Reprinted by Permission from AMCA Publication 201, Fans and Systems)

Figure 19.31b

Example of Fan Performance When System Effect Is Ignored

Assume 2,000 cfm, 1.5″ static pressure, excluding fan inlet and outlet conditions. The inlet velocity is 2,000 feet per minute. The blast area/outlet area is 0.60. The fan is connected as per Figure 19.32.

The 24 x 12 rectangular elbow is connected to the inlet transition as shown in Figure 19.27a. The SEF is "0", and the static pressure loss is 0.40″. The discharge elbow is connected in orientation "C" as shown in Figure 19.31a. The SEF is "N-O" and the static pressure loss is 0.70″ wg. Figure 19.33 shows the calculated system curve with no SEF allowance. The intersection of the size 13.5, 1,080 rpm fan pressure-volume curve at 2,000 cfm and 1.5″ wg at point 1.

The fan inlet SEF pressure loss is 0.40″ and the fan outlet SEF pressure loss is 0.70″ wg. The 1.10″ total SEF pressure loss is added to static

Figure 19.31c

Pressure Loss Multipliers for Volume Control Dampers	
Blast Area / Outlet Area	ΔSP Multiplier
0.4	7.5
0.5	4.8
0.6	3.3
0.7	2.4
0.8	1.9
0.9	1.5
1.0	1.2

System Effect Curves for Outlet Ducts					
	No Duct	12% Effective Duct	25% Effective Duct	50% Effective Duct	100% Effective Duct
Pressure Recovery	0%	50%	80%	90%	100%
Blast Area / Outlet Area	System Effect Curve				
0.4	P	R-S	U	W	—
0.5	P	R-S	U	W	—
0.6	R-S	S-T	U-V	W-X	—
0.7	S	U	W-X	—	—
		V-W	X	—	—
0.9	V-W	W-X	—	—	—
1.0	—	—	—	—	—

Figure 19.31d

pressure at the design volume for a total static pressure of $1.10 + 1.5 = 2.60''$ wg at point 2. Point 2 establishes the actual duct system curve designated by the broken line. The actual duct system curve intersects the fan curve at point 4, $1.70''$ wg. The SEF at actual flow is the difference between the intersection of the fan curve and the actual system curve point 4 ($1.7''$ wg) and point 3 ($0.90''$ wg). The loss of fan performance is the difference between the design calculated system point 1 (2000 cfm) on the fan curve and the actual system point 4 (1600 cfm) on the fan curve. In this case, the fan will deliver 400 cfm or 20% less than required.

Centrifugal Exhaust Fans Should Have Discharge Ducts

It is not unusual to find centrifugal exhaust fans installed as presented in Figure 19.34. The duct velocity is 2,500 fpm,

$$\frac{\text{blast area}}{\text{outlet area}} = 0.7$$

The damper pressure loss with uniform velocity is $0.35''$ wg.

In Figure 19.31c, the pressure loss for the louvre damper is

$$2.4 \times 0.35 = 0.875'' \text{ wg}$$

Figure 19.31d shows the SEF for outlet ducts. The SEF for 12% effective duct length and 0.7 blast area ratio is "U".

Plotting the "U" SEF with 2,500 fpm velocity in Figure 19.28 results in a $0.16''$ wg pressure loss. The total fan outlet loss is $0.16 + 0.875 = 1.035$

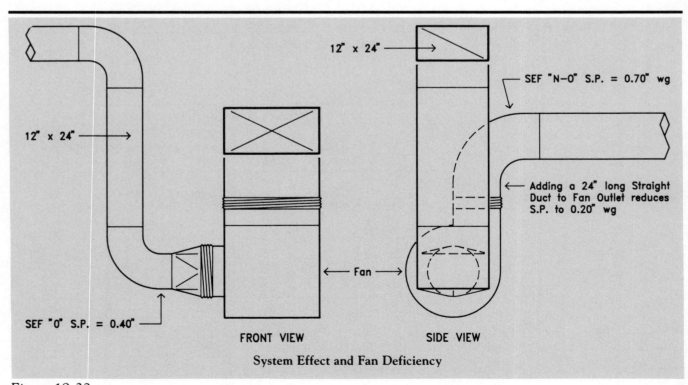

12" x 24"

12" x 24"

SEF "N–O" S.P. = 0.70" wg

Adding a 24" long Straight Duct to Fan Outlet reduces S.P. to 0.20" wg

Fan

SEF "O" S.P. = 0.40"

FRONT VIEW

SIDE VIEW

System Effect and Fan Deficiency

Figure 19.32

The entire outlet loss of over 1″ of static pressure can be avoided by merely adding a duct 2.5 diameters long to the fan outlet. In the case of the 24″ x 12″ duct in the previous examples, the equivalent diameter of the 24 x 12 duct is 19″. Multiplying the 19″ diameter by 2.5 results in a 48″ long discharge duct.

Noise Transfer

Noise transfers through return air grilles to ceiling plenums. Ceiling return air plenums provide an excellent avenue for sound to travel from one office to another. As Figure 19.35 demonstrates, sound passes through the ceiling return air grille, into the ceiling, and out of an adjacent office ceiling return air grille, into the adjoining office. Rooms that require privacy should be constructed with insulated partitions to the underside of the floor slab above. See Figure 19.36.

Acoustically lined elbows installed on top of the ceiling return air grille, and discharging into the ceiling plenum away from the adjacent return air acoustic elbow, will eliminate most of the sound transmission between air grilles. See Figure 19.38b.

A return air duct system will yield a quieter environment than a ceiling return plenum. Ductwork attenuates noise. Lined ductwork is much more

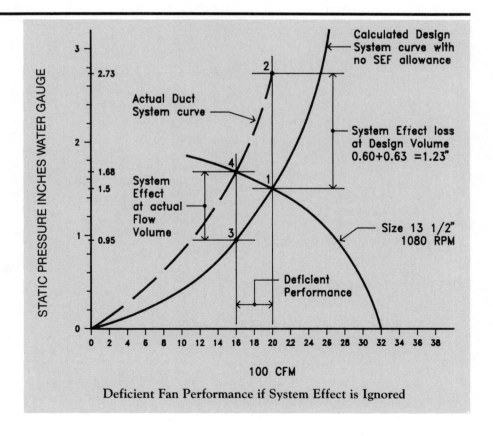

Figure 19.33

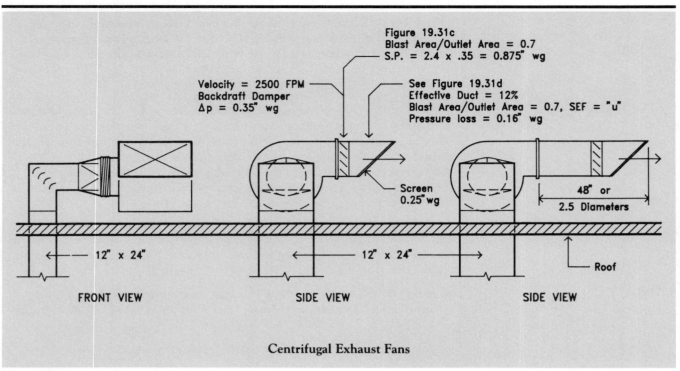

Figure 19.31c
Blast Area/Outlet Area = 0.7
S.P. = 2.4 x .35 = 0.875" wg

Velocity = 2500 FPM
Backdraft Damper
Δp = 0.35" wg

See Figure 19.31d
Effective Duct = 12%
Blast Area/Outlet Area = 0.7, SEF = "u"
Pressure loss = 0.16" wg

Screen
0.25" wg

48" or
2.5 Diameters

Roof

12" x 24"

12" x 24"

FRONT VIEW

SIDE VIEW

SIDE VIEW

Centrifugal Exhaust Fans

Figure 19.34

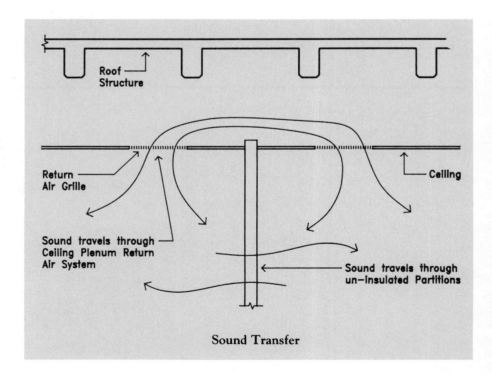

Roof
Structure

Return
Air Grille

Ceiling

Sound travels through
Ceiling Plenum Return
Air System

Sound travels through
un-insulated Partitions

Sound Transfer

Figure 19.35

effective than unlined ductwork in attenuating noise. Research by SMACNA indicates that acoustic lining reaches its maximum attenuation value at 15' of lined duct. Lined duct over 15' long will not attenuate more noise.

Aerodynamic Noise

SMACNA tests reveal that aerodynamic noise is created when the airflow in a duct system becomes turbulent. Turbulence occurs at sharp bends, and at devices that produce a substantial pressure drop. They found that aerodynamic noise was not important when the velocity was below 2,000 feet per minute in the main ducts, below 1,500 feet per minute in branch ducts, and below 800 feet per minute in ducts serving room air terminal devices. When the duct velocities exceed the above or when good airflow duct design principles are not followed, aerodynamic noise is a problem. Figure 19.37 shows examples of duct configurations that SMACNA has found to generate noise.

The use of turning vanes and 45° entry fittings should be used to minimize turbulence. Thin plate butterfly volume dampers should not be used near diffusers. The turbulence generated by the butterfly damper will produce a whistle. Butterfly dampers should not be used in duct systems with velocities over 2,000 feet per minute. Opposed blade volume dampers are more appropriate since they yield less turbulence than butterfly dampers.

Equalizing Grids at Diffusers

Manufacturers publish sound data for their diffusers based on a uniform velocity across the diffuser neck. If diffusers are installed without an equalizing grid to produce a uniform velocity gradient within the diffuser

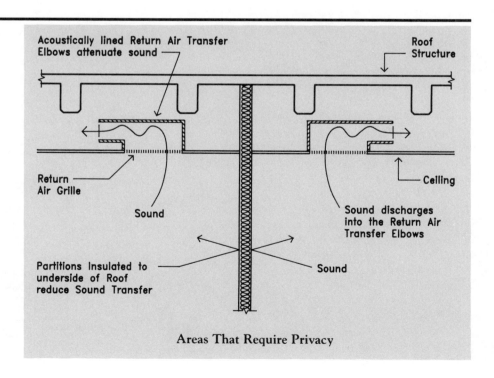

Figure 19.36

271

neck, the sound power level will increase as much as 12 dB. Deleting the diffuser grid (as shown in Figure 19.38) causes the air to flow into only one half of the diffuser neck. This doubles the velocity and increases the sound power level.

Installing equalizing grids in existing noisy diffusers will reduce the sound power level generated by the diffuser.

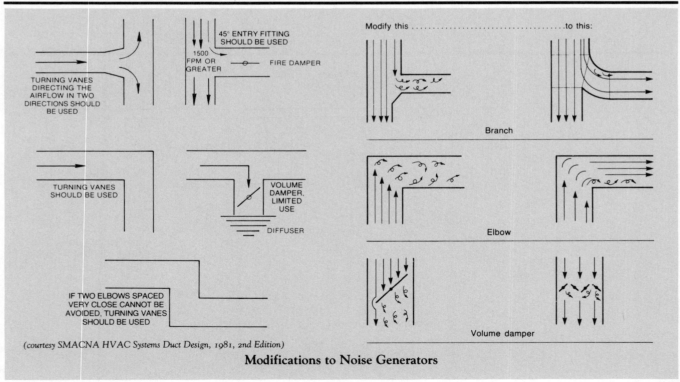

(courtesy SMACNA HVAC Systems Duct Design, 1981, 2nd Edition)

Modifications to Noise Generators

Figure 19.37

Linear Diffusers

Linear diffuser internal deflectors must be set for the direction of the airflow. Each linear diffuser slot is equipped with a moveable deflector that must be positioned to discharge air either left, right, or straight. The deflector must be moved toward the direction of the airflow. Linear diffusers that are installed as they come from the factory could discharge air toward the adjacent wall instead of into the open space. See Figure 19.39.

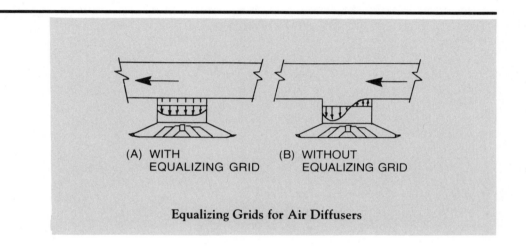

(A) WITH EQUALIZING GRID

(B) WITHOUT EQUALIZING GRID

Equalizing Grids for Air Diffusers

Figure 19.38

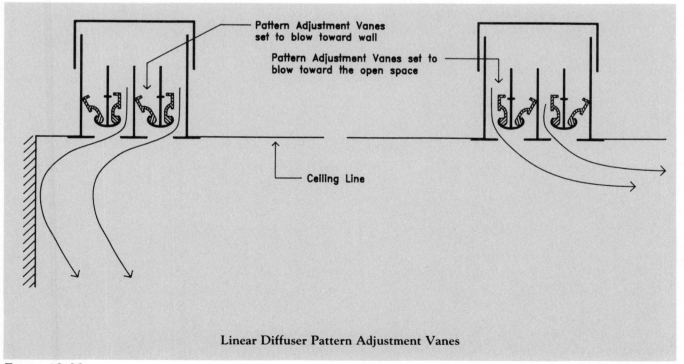

Pattern Adjustment Vanes set to blow toward wall

Pattern Adjustment Vanes set to blow toward the open space

Ceiling Line

Linear Diffuser Pattern Adjustment Vanes

Figure 19.39

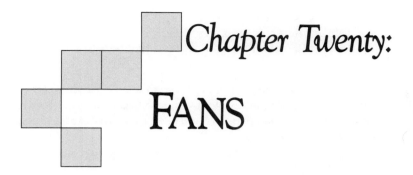

Chapter Twenty:

FANS

Fan Redundancy

HVAC fan failure ceases all system operation. The ideal HVAC design makes provisions for sustaining fan operation by providing redundancy, or back-up. This chapter presents the redundancy alternatives that have been used successfully to extend and ensure air handling performance, and points out common problems and solutions that occur in each of these systems.

Two Fans in Parallel

An alternative to installing one large fan is to install two fans in parallel. Figure 20.1 shows a typical arrangement of a parallel two-fan installation. In such an arrangement, each fan is sized for 50% of the total cfm at the design static pressure. If one fan fails, the other fan, operating alone, will deliver approximately two thirds of the design cfm. As Figure 20.2 shows, the fan motor must be capable of producing the fan brake horsepower of the single fan to deliver two thirds of the design cfm. This is a greater brake horsepower (bhp) than each fan draws when the two fans operate together.

Each of the two 33″ fans in Figure 20.2 are operating at 1,100 rpm and will deliver 24,000 cfm for a total air supply of 48,000 cfm at 4″ static pressure, and draw 23 bhp. When one fan operates alone, the 1,100 rpm fan curve intersects the system curve at 34,000 cfm. One fan operating alone will produce 34,000 cfm, which is 70% of the 48,000 cfm that the two fans deliver. The single 33″ fan demands 25 brake horsepower at 34,000 cfm. A 30 hp motor should be supplied for each fan, in order to handle the increased hp of a single fan operation.

It should be noted that the fan cabinet that houses the two smaller fans will be larger than that for a single larger fan. The additional cost of installing two smaller fans in parallel, including fans, cabinet, wiring and starters is $4,500 for a 16,000 cfm system, or $0.28/cfm.

Many manufacturers of factory-fabricated air handling units will not modify their standard units for a two-fan operation. If the design requires two fans in parallel, it may have to be a custom built or built-up air handling system.

Fans in Series

Two fans operating in series can achieve the same fan redundancy that parallel fan installations produce. However, the installation of two main supply fans in series involves considerable space and expense. See Figure

20.3. Both fans must be sized for the total air volume, but at one half of the total static pressure. Larger diameter fans are required for series fan operation (40″) than for parallel fan operation (33″).

Return Air Fans

An inexpensive method of creating air moving redundancy is adding return air fans. Figure 20.4 shows a 25,000 cfm system with a 40″ air foil plug supply air fan operating at 1,100 rpm, 4.6″ static pressure, and 27.5 bhp (point "A").

The 40″ airfoil plug return air fan operates at 950 rpm, 2.2″ static pressure, and 17.5 bhp (point "B").

The supply system curve is identified as "S." The return system curve is denoted "R," and the combined supply and return system curve is indicated by "C."

If the return air fan fails, the supply fan must overcome the supply and return air distribution systems, represented by curve "C." The 1,100 rpm fan curve intersects the combined system curve at "D," delivering 22,000 cfm. Notice that the bhp at point "D" is still 27.5. Furnishing 88% of the design cfm should suffice for most cooling applications until the return fan can be repaired.

Should the supply fan become inoperative, the return air fan, running along at 900 rpm, intersects the combined system curve at point "E." The return air fan will deliver 18,500 cfm at the same 17.5 bhp. The return air fan operating alone will circulate 74% of the maximum design cfm. This should maintain suitable space conditions until the supply fan is returned to service.

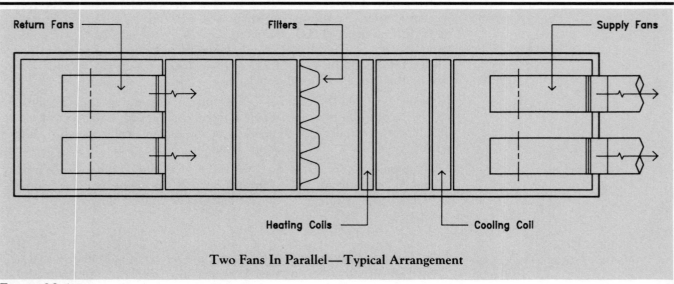

Two Fans In Parallel—Typical Arrangement

Figure 20.1

When the return fan acts as the sole air mover in the system, provisions should be made for the introduction of outside ventilation air to the system. The main system outside air damper must be closed to prevent return air from flowing through the return air damper into the outside air damper in reverse, and discharging through the outside air louver.

A separate, manually-operated outside air damper should be located and designated "emergency outside air damper." This will permit the return air fan to draw outside ventilation air into its intake and mix it with the return air before discharging the mixed air to the filter. By taking "fan back-up" into consideration when selecting the return air fan, fan redundancy can be obtained at almost no additional cost!

Plug Fans (Plenum Fans)

Plug fans are single width, single inlet centrifugal fans without the discharge scroll. (See Figure 20.5 for an illustration.) Plug fans are ideally suited for plenum (a closed chamber used to collect conditioned air) applications. The plug fan discharges directly from the fan blades, in a 360°F pattern, into the supply plenum.

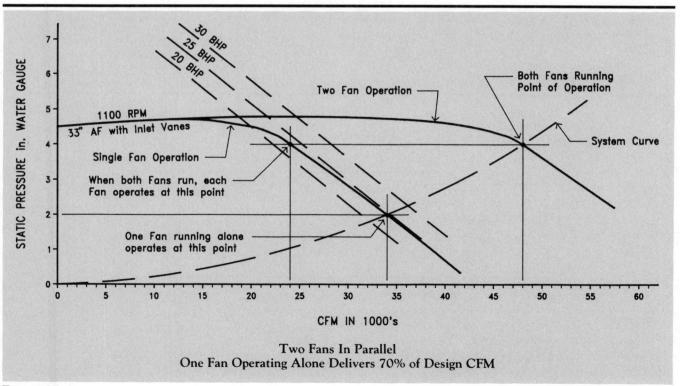

Two Fans In Parallel
One Fan Operating Alone Delivers 70% of Design CFM

Figure 20.2

Advantages of the Plug Fan

More uniform airflow over cooling coil

Since the diameter of the plug fan is greater than a double-width, double-inlet centrifugal fan, a more uniform airflow is created across the cooling coil, into the plug fan inlet. (See Figure 20.5.)

A backward-inclined air foil double inlet 25,000 cfm centrifugal fan has a 36.5" diameter. The comparable plug fan diameter would be 44.5". For a 15,000 cfm system, the double-inlet fan diameter would be 27", and the plug fan diameter 36.5".

As Figure 20.5 shows, air tends to concentrate at the ends of the cooling coil toward the inlets at each end of the double-width double-inlet centrifugal fan, avoiding the center of the coil directly opposite the back of the fan scroll.

Simplified ductwork

Supply ductwork can be simplified by making connections from any side of the plenum. As Figures 20.6 and 20.7 show, plenum fan connections are especially suitable for air distribution from rooftop units.

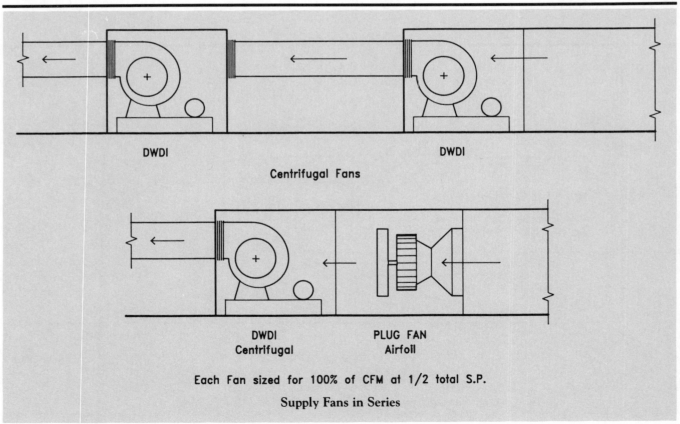

Figure 20.3

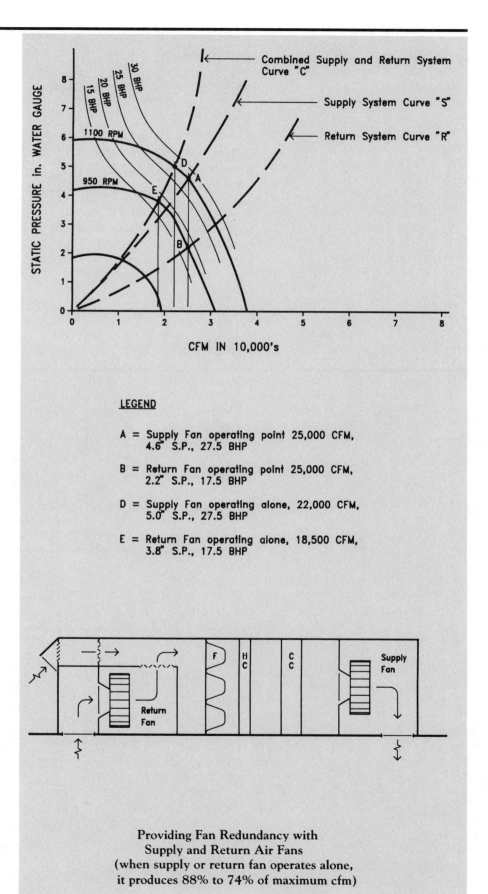

LEGEND

A = Supply Fan operating point 25,000 CFM, 4.6" S.P., 27.5 BHP

B = Return Fan operating point 25,000 CFM, 2.2" S.P., 17.5 BHP

D = Supply Fan operating alone, 22,000 CFM, 5.0" S.P., 27.5 BHP

E = Return Fan operating alone, 18,500 CFM, 3.8" S.P., 17.5 BHP

Providing Fan Redundancy with
Supply and Return Air Fans
(when supply or return fan operates alone,
it produces 88% to 74% of maximum cfm)

Figure 20.4

279

As Figure 20.8 demonstrates, a supply plenum would have to be added to the double-inlet fan unit to permit multiple duct outlets or a final filter. Equalizing plates would also be required for uniform flow through a final filter. These add to the initial cost and contribute a substantial pressure drop to the system. Figure 20.9 shows how the plug fan's uniform air discharge into a plenum eliminates the need for equalizing plates.

Plug fans discharging into an internally-lined plenum result in a quieter operating system than a supply duct connected directly to the fan outlet. The lined plenum helps to attenuate the fan noise.

Disadvantages of the Plug Fan

Longer fan section

A 25,000 cfm plug fan section is 18" longer than a double-inlet fan section.

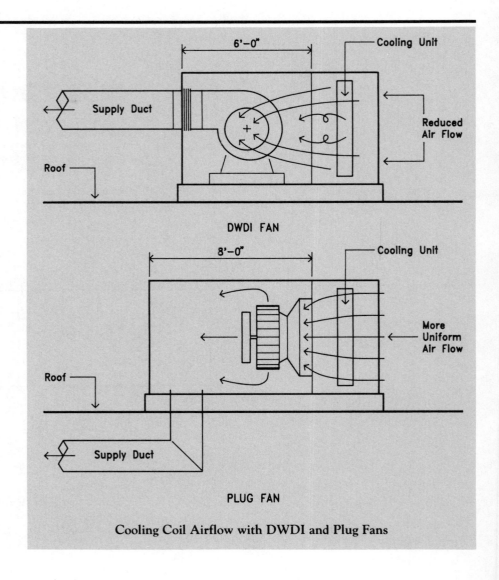

Cooling Coil Airflow with DWDI and Plug Fans

Figure 20.5

If a single supply duct can be arranged from a double-inlet fan without abusive system effect penalty, the double-inlet fan unit could be less expensive because there is an additional cost for the larger supply fan casing.

Greater Starting Torque

The plug fan does not have a scroll to direct airflow to a concentrated outlet area. The large mass of air dispersed to the plenum directly from the plug fan's blades produces a greater starting torque than is the case with a conventional centrifugal fan. This is not a problem when across-the-line motor starters are used. However, if variable frequency drives are being used, the variable drive must be set for a longer time period to reach full design fan speed. Otherwise, it will trip out on overload protection.

Fan Surge

Fan surge occurs when there is insufficient air entering the fan wheel to completely fill the fan blade space. During periods of significant stall, the turbulence generated between the fan blades can cause air to flow in the reverse direction over part of the fan blade. That turbulent flow is fan surge. Fan surge is a problem when the noise and vibration become objectionable. Fans operating in severe surge are not only noisy, but can literally tear themselves apart. Fans operating in surge can destroy bearings, damage fan shafts, break fan blades and propel them through the unit casing.

Fan surge can be avoided by proper fan selection when the system is installed. Many fan manufacturers identify the surge region on their fan curves. The unstable area is to the left of the maximum static pressure and efficiency. This is the point where the fan curve goes down toward the left. In the surge region, two air volumes are possible at the same static pressure. (See Figure 20.10.)

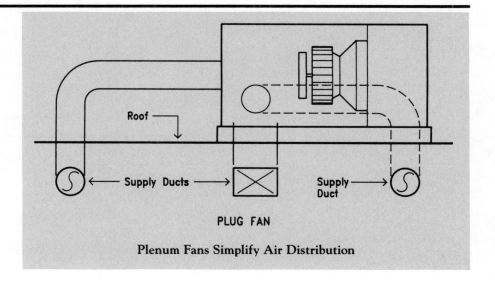

Figure 20.6

PLUG FAN

Plenum Fans Simplify Air Distribution

Note that at 900 rpm and 2.3″ static pressure, the fan can deliver 7,000 cfm or 9,500 cfm. A slight change in system static pressure can shift the operating point, resulting in unstable operation.

In selecting the fan, important criteria include not only maximum operation, but also minimum airflow, to be sure that the fan does not operate in the surge region.

Figure 20.11 shows a typical air foil fan with two system curves plotted. Note that curve B in this figure operates in the surge region, below 10,000 cfm. Curve A does not intersect the surge area. Fan surge is not noticeable at 1″ static pressure. Fan VAV operational problems can be avoided by selecting a fan for the desired modulation range. (See Chapter 14 for information on VAV fan selection.)

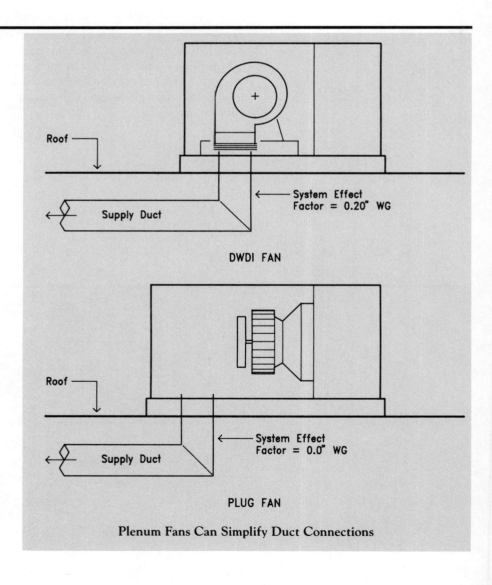

Plenum Fans Can Simplify Duct Connections

Figure 20.7

VAV surge operational problems may be eliminated by taking the following actions:

- Lowering the duct static pressure set point to the lowest possible point that will satisfy minimum load. In Figure 20.11, lowering the static pressure set point from 2" to 1" prevented the fan from operating in surge.
- Raising the minimum air volume delivery by resetting the supply air temperature (upward) as the load decreases. As Figure 20.12 shows, increasing the fan cfm will move it out of the surge region.
- As Figures 20.13 and 20.14 indicate, if the VAV system has a fixed

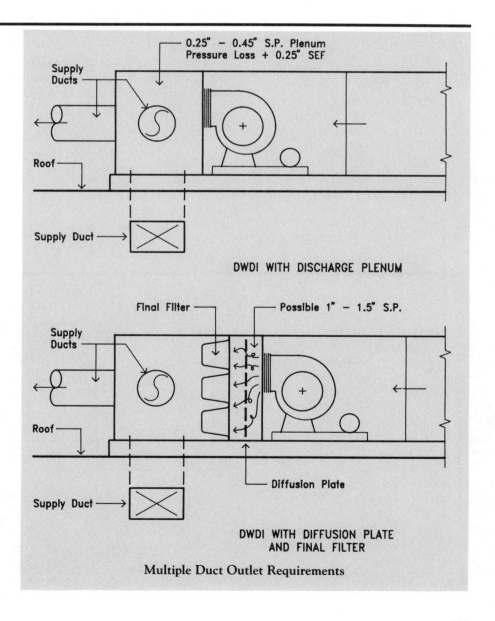

DWDI WITH DISCHARGE PLENUM

DWDI WITH DIFFUSION PLATE
AND FINAL FILTER

Multiple Duct Outlet Requirements

Figure 20.8

fan rpm and is damper-controlled, adding inlet vanes or speed control will move the fan operating point out of the surge area during reduced supply air volume.

- If the VAV fan is inlet-vane-controlled, the installation of a speed control device may avoid surge operation.

Direct Drive Fans

The single greatest fan maintenance task is servicing fan belts. Fans commonly operate at a fraction of their design speed simply because belts are loose. Fan motors may also be found humming away to a sleeping fan, the drive belts having stretched and slipped off. Often, broken fan belts are discovered on the floor of the unit.

To avoid these problems, belts must be properly aligned, tension must be adjusted, and worn and broken belts must be replaced.

The facilities personnel should be prepared and have the right size belts on hand for replacement. They should know that two- and three-belt drives must be matched, or they will not equally share the load. Maintenance personnel should also be acquainted with the belt's horsepower rating and should be able to recognize when belts have too much slack, or are too tight. Noise and premature belt failure will result if belts are not at the correct tension.

One method of avoiding belt problems is by the use of *direct drive fans*. Direct drive fans are constructed with the fan wheel either directly mounted on the motor shaft or connected to the motor shaft by a flexible

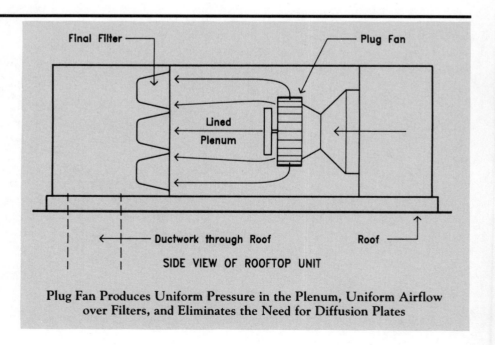

Plug Fan Produces Uniform Pressure in the Plenum, Uniform Airflow over Filters, and Eliminates the Need for Diffusion Plates

Figure 20.9

coupling. Direct drive fan speeds are fixed by the motor speed. The standard motor speed is 1,750 rpm, with 850, 1,150, and 3,550 rpm motors also available.

The use of variable speed drives and two-speed motors offers greater direct drive fan flexibility of operation. Since the rpm level is fixed by the motor speed, if variable speed drives are not used, it may be difficult to change the installed fan's performance. Fan selections are often made on the basis of a standard impeller, but using an impeller width no greater than that needed to satisfy the design criteria. If a 20" diameter DWDI (double-width, double-inlet) fan with a 28" wide impeller is selected, 3" wide metal strips could be placed over the two outer edges of the fan wheel blades to reduce the 20" DWDI fan's capacity to the desired performance level.

If it is necessary to increase the fan's air delivery or static pressure, this can be accomplished by reducing the width of, or removing, the metal strips that have been placed on the fan's impeller. The fan motor hp should be selected for a non-overloading condition.

If the installed fan's capacity is to be reduced, wider metal strips should be placed on the fan's impeller to reduce the fan's capacity to the desired performance level. The fan must be rebalanced whenever modifications are made to the impeller.

Figure 20.10

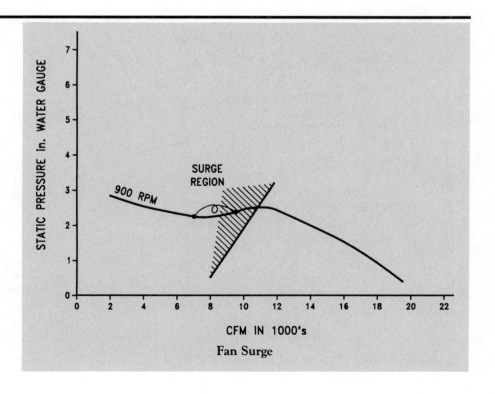

Fan Modulation

Fan modulation is produced by the following:

- Changing fan speed
- Inlet vanes
- Vaneaxial variable pitch in motion

These issues have been addressed in Chapter 18, which covers VAV operational problems and their solutions.

Proper fan performance can be ensured by:

- Correct fan inlet and outlet duct connections to minimize system effect. (See "System Effect Factors" in Chapter 19 for more information.)
- Avoid belt-driven fans if possible. Direct drive fans eliminate belt maintenance.

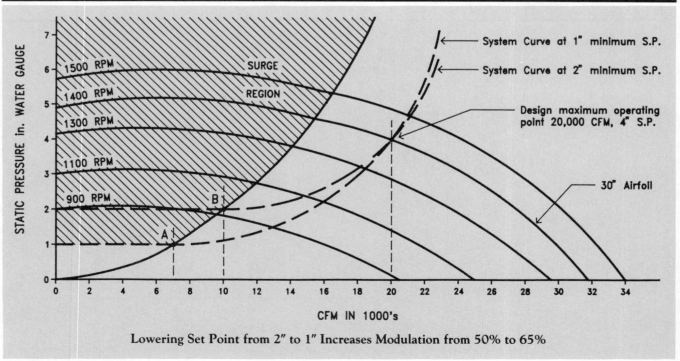

Lowering Set Point from 2″ to 1″ Increases Modulation from 50% to 65%

Figure 20.11

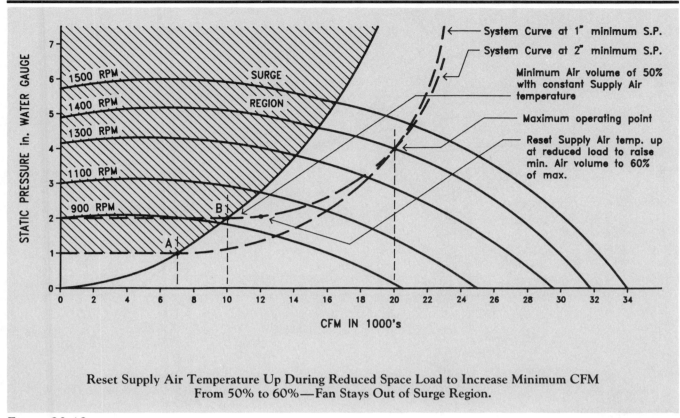

Reset Supply Air Temperature Up During Reduced Space Load to Increase Minimum CFM
From 50% to 60%—Fan Stays Out of Surge Region.

Figure 20.12

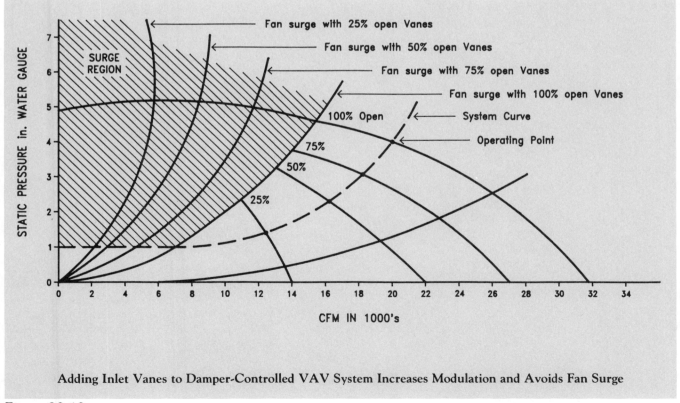

Adding Inlet Vanes to Damper-Controlled VAV System Increases Modulation and Avoids Fan Surge

Figure 20.13

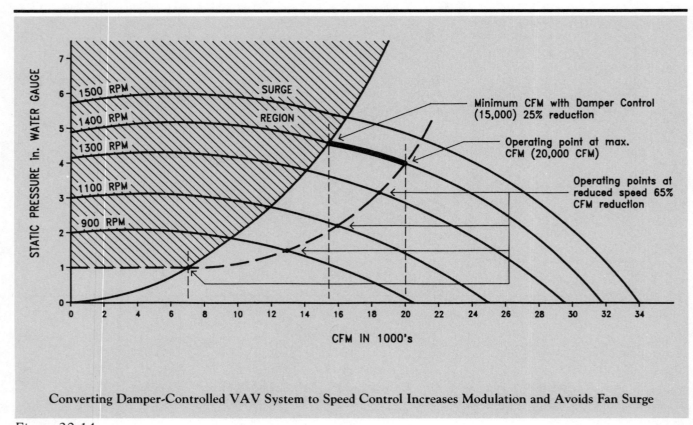

Converting Damper-Controlled VAV System to Speed Control Increases Modulation and Avoids Fan Surge

Figure 20.14

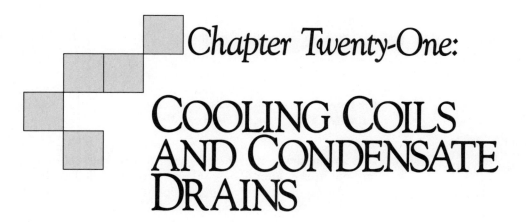

Chapter Twenty-One:

COOLING COILS AND CONDENSATE DRAINS

Cooling coils and condensate drain pans can represent a major expenditure of both dollars and down time in air handling unit maintenance. The cost of replacing a cooling coil and condensate drain pan often equals the original cost of the entire air handling unit. Furthermore, the unit may be out of service for two to three days while the coil and pan are being replaced.

Material and Deterioration

Deterioration of cooling coil casings and fins, as well as condensate drain pans, can accelerate when the air in contact with the cooling coil is contaminated with sulfur dioxide, charcoal, grease, dirt, etc. Cooling coils are further prone to corrosive attack because they are wet with condensate. The condensed water is aerated and incorporates contaminants from the air.

Since there is no practical way to chemically treat the water condensed from the air in order to inhibit its corrosiveness, the material used in the manufacture of cooling coils and casings should be decay-resistant. Stainless steel coil casings and condensate drain pans have proven to be extremely durable. Lead-coated copper fins will not deteriorate. Therefore, if the budget allows it, these are the ideal material choices for these applications. The additional cost for these preferred materials for a 7,000-25,000 cfm range of air handling units is $250 per 1,000 cfm.

Protective Coatings

Protective coating for coil casing, finned tubing, and drain pans effectively protects the metal, if the coating is properly applied. The following coatings are recommended for the specific conditions described.

- *Alkyd resin coatings* are suitable for normal atmospheric conditions. However, they are not effective in resisting acid or alkali (salt) contamination.
- A *vinyl resin coating* is recommended when acid conditions are expected. Vinyl is resistant to moisture and can be air-dried. Proper surface preparation is mandatory for vinyl resin application. Shop-applied coatings are preferable, because shop conditions are better for achieving controlled adhesion.
- *Catalyzed epoxy* is the best coating to use for severe alkali conditions. It is also the more abrasion-resistant coating. Since catalyzed epoxy is a two-component system, it too should be shop-applied.

Applying any of the above coatings to galvanized steel coil casings with a high zinc primer will extend the life of the coating and protect the metal.

Where budget constraints prevail, protective coatings for the cooling coil casing, tubes and fins, and stainless steel drain pans present a superior product to the standard cooling coil and drain pan assembly. As of 1990, the premium cost of coated cooling coils and stainless steel drain pans is an estimated $120 per 1,000 cfm.

Condensate Drain Trap Installation

Condensate drain traps must be sized for the fan suction or discharge pressure in the drain pan. In almost every installed medium-to-large air handling (cooling) unit, the condensate drain system suffers from inadequate drain trap performance. As Figure 21.1 shows, an undersized trap water seal allows air to enter the condensate drainpipe and impede condensate flow out of a draw-through unit.

Draw-Through Unit Condensate Drains

The minimum height of the "U" trap must be greater than the fan inlet static pressure for draw-through units, and the fan discharge static pressure for blow-through units. The height of the trap must be sufficient to permit the negative fan pressure to raise the water up one leg of the trap, and still provide a water seal at the bottom of the trap.

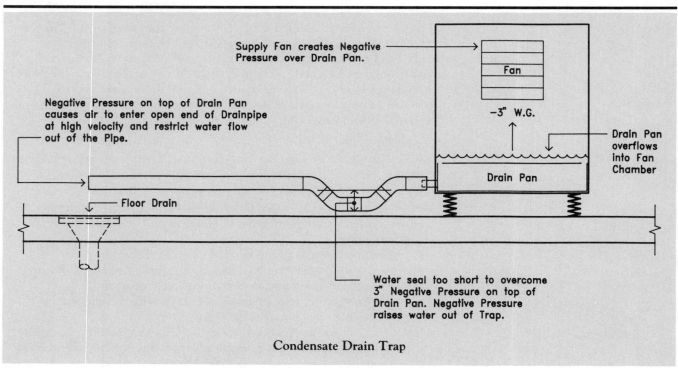

Condensate Drain Trap

Figure 21.1

The water seal is essential to prevent air movement through the drainpipe. If the trap water seal is less than the fan negative static pressure at the drain pan of draw-through units, the negative pressure generated by the fan will induce the water up the drain riser into the drain pan. Once the water seal is lost, air enters the open end of the drainpipe. Since the pressure at the open end of the pipe is higher (atmospheric) than the pressure at the drain pan (negative), water will not properly flow from the drain pan to the floor drain. Water then accumulates in the fan section, rusts the fan, and spills over the edge of the drain pan. When the fan is shut off, water pours out of the condensate drainpipe and the unit, overflowing the floor drain.

Figure 21.2 shows a draw-through air handling unit with 3″ of negative pressure at the fan inlet. The trap exiting vertical leg (A) must be at least 1″ greater than the 3″ fan suction pressure. The fan negative pressure will raise the water up the trap inlet leg (B) 3″. This will leave a 1″ water seal at the bottom of the trap (C) to prevent air passage in the drainpipe. The bottom of the trap horizontal exiting pipe (D) must be at least 1″ below the drain pan to furnish the head necessary to overcome the drainpipe friction pressure loss. The union that enables cleaning of the trap should be installed in the vertical position in order to prevent dirt from accumulating in the union recesses. Brass plugs or caps should be installed to facilitate drain system cleaning. Most codes require a 2″ air gap between the unit drain and the sanitary system. At least 10″ of height must be available between the bottom of the drain pan outlet and the floor in order to properly install a condensate drain trap.

Blow-Through Units

Blow-through air handling unit fans create positive air pressure at the drain pan outlet. Figure 21.3 shows the correct design of a blow-through unit condensate trap.

A positive air pressure of 3″ wg will raise the water in the trap 3″. The trap height must be sufficient to allow water in the entering leg to drop 3″, while maintaining a 1″ water seal and a 1″ vertical head to overcome drainpipe friction. The trap exiting the vertical leg must be able to contain the 1″ of water seal and additional 3″ of water blown in from the trap-entering leg. The trap installation should be similar to the draw-through design.

Condensate Drain Installation

The coil drain pan must be elevated high enough above the floor drain for the condensate trap installation. A 4″ concrete pad and vibration isolators usually will not permit the 10″ height needed for the drain trap installation. A 6″ or 8″ concrete pad, or a 4″ concrete pad and steel blocking may be required to elevate the unit drain pan outlet 10″ above the floor drain inlet.

Drainpipe Sizing

The condensate drainpipe diameter should never be less than the drain pan connection. In most cases, the diameter will be 1″ for up to 2,000 cfm, 1-1/4″ up to 12,000 cfm, and 1-1/2″ up to 24,000 cfm. Units over 24,000 cfm should have two 1-1/2″ drains, or one 2″ drain. Local codes must be checked for minimum condensate drainpipe size requirements. Clean-out connections should be installed at least every 40 feet. Drain piping is usually soldered copper or plastic tubing with drainage fittings. The plugged clean-outs must be threaded brass or plastic for cleaning purposes.

Venting Condensate drain piping with more than a 6′ unvented vertical drop should be provided with a vacuum breaker at the top of the vertical drop to prevent a vacuum from developing in the riser. As the water in a full drainpipe flows down, the negative pressure on the pan and the water seal in the trap help seal the inlet end of the pipe. When the water in the riser is not open to atmospheric pressure, a vacuum is created. Atmospheric pressure at the outlet of the pipe is greater than the negative pressure in the riser, thereby preventing water from exiting the pipe. This is the same process that occurs when you insert a straw into a glass of water and seal

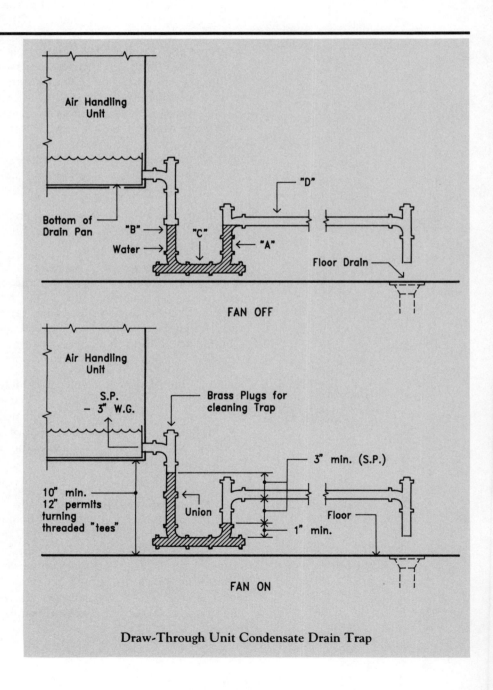

Draw-Through Unit Condensate Drain Trap

Figure 21.2

the straw's open end by placing your finger over the top of the straw. When you withdraw the straw from the glass, keeping your finger on the top end, the water in the straw will not flow out of it, but remains inside, just like the riser. When your finger is removed from the top of the straw, atmospheric pressure is allowed to enter it and water exits through the bottom.

As Figure 21.4 shows, a 1/2″ pipe at the top of the drain riser with its open end 6″ above the drain pan serves as a suitable vacuum breaker.

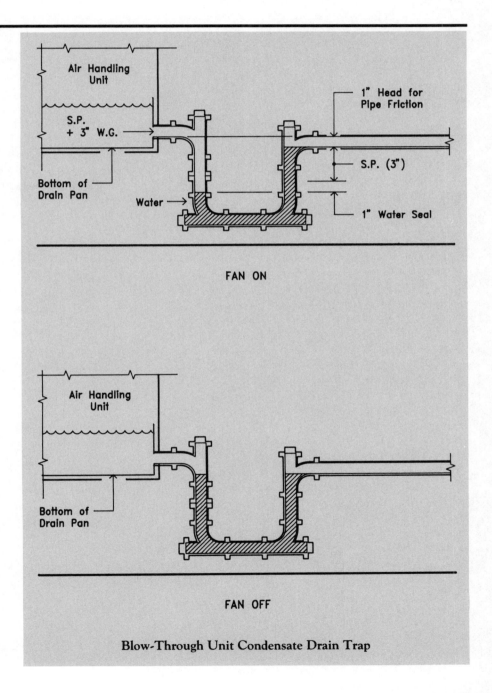

Blow-Through Unit Condensate Drain Trap

Figure 21.3

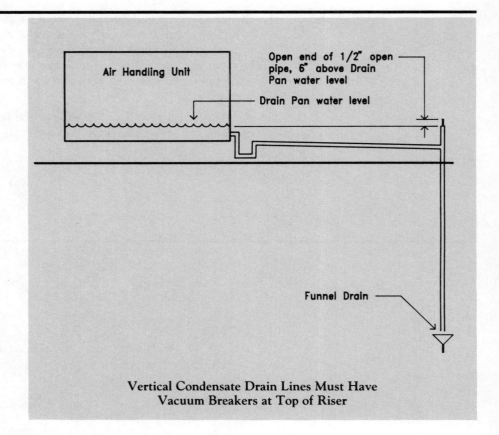

**Vertical Condensate Drain Lines Must Have
Vacuum Breakers at Top of Riser**

Figure 21.4

294

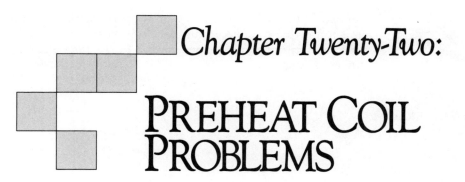

Chapter Twenty-Two:

PREHEAT COIL PROBLEMS

Air handling units that operate in subfreezing weather introduce performance difficulties along with the cold outside air. This frigid air can freeze the fluid in the coils, causing them to rupture or burst. The ensuing building damage, down time, and system replacement costs far outweigh the cost of precautionary design procedures and safety measures to prevent a freeze-up. Some of the more common preheat coil problems and solutions are discussed in this chapter.

Stratification

Stratification occurs when cold outside air does not thoroughly mix with the indoor return air and passes through the preheat coil at a temperature close to the outdoor temperature. (See Figure 22.1.) When the temperature is low enough to activate the freeze-stat which stops the supply fan, operating personnel should be alerted that a problem is building. When subfreezing outdoor air finds a direct path to freeze and rupture a preheat, reheat, or chilled water coil, the operating staff should be immediately involved and the building occupants quickly affected.

Freeze-stats should be installed across the face of coils to avoid freezing and rupturing of the coil, and to prevent the discharge of freezing air into the occupied areas. These problems are avoided by shutting down the fan when the temperature of the fluid in the coil reads a predetermined limit, and by closing the outside air dampers.

Stratification can be minimized by the following methods:

Installing Baffles
Figure 22.2 shows how mixing outside air and return air can be improved by installing a baffle to divert the outside air directly into the return air stream.

Using Parallel Blade Dampers
Figure 22.3 shows that the mixing of outside air and return air can be enhanced by the use of parallel blade dampers set to direct or blend the two air streams with one another.

Connecting the Outside Air Duct Into the Return Air Duct
Figure 22.4 shows the most effective mixing of outside air and return air. Connecting an outside air duct into the return air duct via a "bull head" fitting enhances air stream mixing. The outside air duct should be

connected to the return duct as far away as practical from the air handling unit to provide the opportunity for the two air streams to mix before they enter the unit. The turbulence created by the duct elbows stimulates air stream mixing.

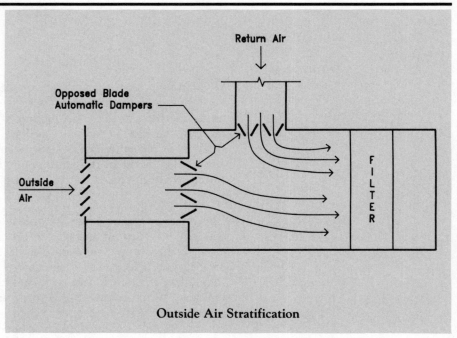

Outside Air Stratification

Figure 22.1

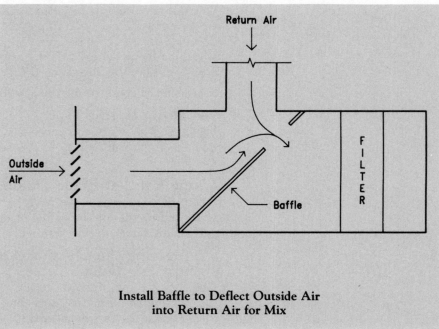

Install Baffle to Deflect Outside Air into Return Air for Mix

Figure 22.2

Preheat Coil Area

Preheat coils should be sized for the same face velocity as the cooling coil in order to avoid stratification. Many preheat coils are sized for 700 feet per minute velocity to facilitate trap installation. Figure 22.5 shows the smaller preheat coil area used to permit the selection of a shorter height coil. The shorter coil height means that the coil return pipe connection can be a greater distance from the floor, producing height for the steam trap to drain. Most of the 700 feet per minute (fpm) velocity leaving the preheat coil enters the cooling coil, carrying moisture on the cooling coil into the air stream. If more trap height is needed, the best solution is to elevate the unit on a stand, or to locate the trap below the floor on which the unit is located. Figure 22.6 demonstrates that when the preheat coil and the cooling coil are the same size, airflow across the cooling coil is uniform at 500 fpm, with no moisture carryover.

Preheat Coil Freeze-ups

The major causes of hot water and steam preheat coil freezing are:

Outside Air Stratification

Avoiding outside air stratification is addressed in the beginning of this chapter. Freeze-stats are installed to prevent freezing.

Steam Coils Under a Vacuum

When steam condenses in a coil, the condensate is removed through a trap, the capacity of which is dependent on the pressure differential between the trap inlet and outlet, and the anticipated load or volume of condensate. When the control valve throttles, the steam pressure in the coil decreases, and the trap capacity is reduced. Figure 22.7 shows that the steam trap capacity reduces as the pressure differential across the trap is reduced.

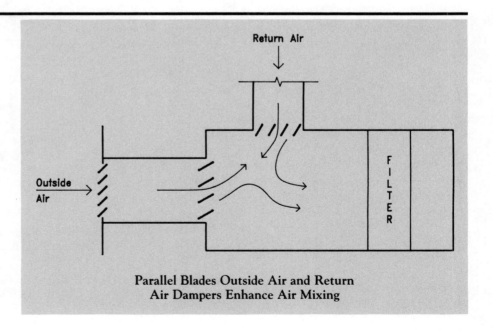

**Parallel Blades Outside Air and Return
Air Dampers Enhance Air Mixing**

Figure 22.3

Figure 22.8 demonstrates that when the steam control valve action and the condensed steam reduces the coil pressure below atmospheric pressure, the trap has no capacity, since there is no inlet pressure available to force the condensate through the trap into the return main. The condensate is held in the coil, and can freeze if the air entering the coil is below freezing. A vacuum breaker should be installed at the steam coil inlet. If the coil pressure falls below atmospheric, the vacuum breaker will introduce air at this point and free the entrained condensate. Point "A" in Figure 22.8 is the initial steam temperature; Point "B" is the entering air temperature; and Point "C" is the leaving air temperature. The coil pressure will become atmospheric at 18°F, and coil freezing is likely to occur.

Incorrect Steam Trap Sizing and Installation

A 30″ hydraulic head at the trap inlet will ensure one psi of trap pressure differential for proper trap performance. Figure 22.9 is a steam preheat coil piping diagram with a vacuum breaker and steam trap.

If 30″ vertical height is not available between the coil outlet and the trap inlet, the trap can be located in the floor below the coil. If the unit location is "slab on grade," the unit must be elevated to produce the required static head above the steam trap. Figure 22.10 is an illustration of this application.

If 30″ vertical space is not available, the trap should not be sized for a 1 psi pressure differential. The trap should be sized for the vertical height available. If there are only 15″ of height between the coil outlet and the trap inlet, size the trap for 1/2 psi pressure differential. If only 7-1/2″ are

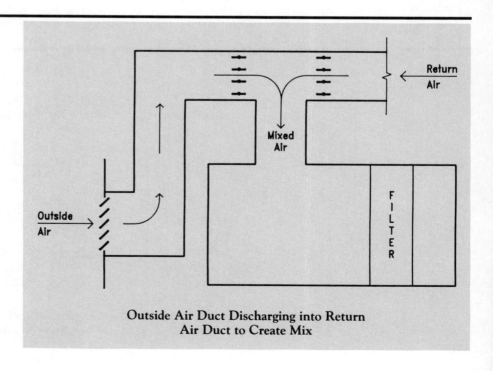

**Outside Air Duct Discharging into Return
Air Duct to Create Mix**

Figure 22.4

available, size the trap for 1/4 psi pressure differential. Figure 22.11 shows the correct steam trap installation in situations where limited height is available.

Freeze Protection

Air handling unit coils are "protected" from freezing by the following devices:

Freeze-stats

Freeze-stats are usually remote bulb thermostats, with a long capillary tube that is spread evenly over the entering face area of the cooling coil. See Figure 22.12 for an example of a freeze-stat installation.

Should air at a temperature below the set point temperature (usually 35°F), come into contact with any section of the capillary tube, the freeze-stat electrical contacts are opened and the supply fan is stopped. As Figure 22.13 shows, sufficient space is needed between the preheat and the cooling coil for the installation of the freeze-stat.

An access space of at least 18″ wide is essential for a technician to erect the mounting supports and to secure the freeze-stat capillary tube. If access space is not furnished, a hole must be drilled into the panel between the preheat and the cooling coil. The capillary tube is inserted into the hole until the entire tube length of capillary is inside the unit. However, most of the capillary tube tends to curl up in one area of the coil space and is therefore unable to sense the air temperature passing through most of the coil face area that it is supposed to protect. Stratification could allow low temperature outside air to pass undetected through the coil and cause a

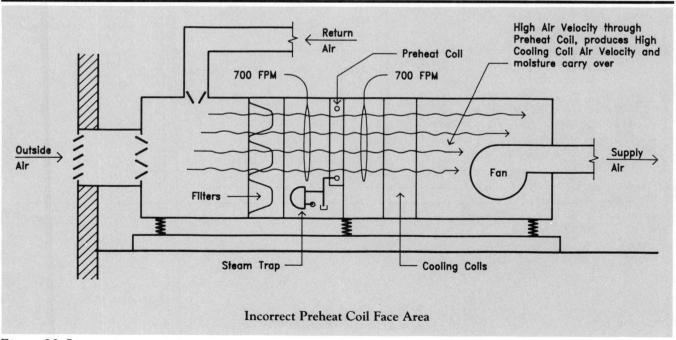

Incorrect Preheat Coil Face Area

Figure 22.5

freezeup, while warmer return air in contact with the freeze-stat element erroneously satisfies the control setting.

Freeze Protection Pump

The basic tenet for freeze protection is to *keep the water moving*. Fast moving mountain streams do not readily freeze in winter. Likewise, circulating water in a pipe or coil will not freeze if it is not exposed to freezing air for a long enough period of time.

Freeze protection pumps connected to preheat coil piping (as shown in Figure 22.14) will afford protection, as long as the pump operates and circulates water. An outside air thermostat starts the freeze protection pump at 32°F ambient temperature.

The disadvantages of freeze protection pumps are:

- The installed cost of multiple freeze protection pumps could be greater than the installed cost of a glycol preheat system.
- The freeze protection pumps require maintenance.
- Pumps are unreliable in certain circumstances. Loss of electrical power, pump motor failure, broken shaft coupling, air bound coil piping, and automatic valve malfunction lead to loss of circulation and freeze protection.

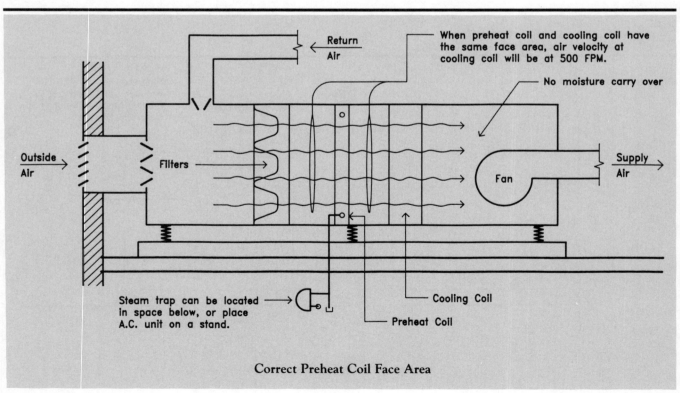

Correct Preheat Coil Face Area

Figure 22.6

Low Pressure Capacities of Straightway Float & Thermostatic Traps

Capacities lbs/hr

Model	Size	Orifice Inches/mm	1/4	1/2	1	2	5
					Differential Pressure, psig		
FT-553	1/2", 3/4"	.157/4	87	123	173	227	389
FT-552	1/2", 3/4"	.110/2.8	41	58	82	107	183
FT-551	1/2", 3/4"	.079/2	21	30	42	55	95
FT10-4.5	1/2", 3/4"	.157/4	102	145	203	287	419
FT10-10	1/2", 3/4"	.106/2.7	49	69	97	137	209
FT10-14	1/2",3/4"	.079/2	24	34	48	68	106
FT10-4.5	1"	.276/7	374	529	748	1058	1566
FT10-10	1"	.205/5.2	163	230	325	459	695
FT10-14	1"	.157/4	109	154	218	309	441
FT10-4.5	1 1/2"	.689/17.5*	975	1378	1949	2756	4388
FT10-10	1 1/2"	.591/15*	608	860	1216	1720	2700
FT10-14	1 1/2	.531/13.5*	394	557	788	1114	1764
FT10-4.5	2"	1.122/28.5*	3196	4520	6392	9040	14332
FT10-10	2"	.807/20.5*	1637	2135	3274	4630	7166
FT10-14	2"	.657/16.7*	780	1102	1559	2205	3418
FT32-4.5	3/4"	.157/4	10	155	219	310	440
FT32-10	3/4"	.126/3.2	65	91	129	183	275
FT32-14	3/4"	.106/2.7	47	66	93	132	203
FT32-21	3/4"	.079/2.0	25	35	50	70	110
FT32-32	3/4"	.063/1.6	16	22	31	44	66
FT32-4.5	1"	.276/7	389	550	778	1100	1655
FT32-10	1"	.205/5.2	202	285	403	570	870
FT32-14	1"	.185/4.7	150	212	300	425	640
FT32-21	1"	.157/4	110	155	219	310	440
FT32-32	1"	.126/3.2	65	91	129	183	275
FT32-4.5	1 1/2"	.689/17.5*	1209	1710	2418	3420	5733
FT32-10	1 1/2"	.591/15*	624	882	1247	1764	2734
FT32-14/21/32	1 1/2"	.531/13.5*	407	575	813	1150	1764
FT32-4.5	2"	1.122/28.5*	4289	6065	8577	12130	18522
FT32-10	2"	.807/20.5*	1559	2205	3118	4410	6950
FT32-14/21/32	2"	.657/16.7*	780	1103	1559	2705	3528

*Each orifice of Double Seated Trap.

(Reprinted by Permission from Spirax Sarco Inc.)

Figure 22.7

Antifreeze Solution (Glycol)

Glycol solutions in preheat coils are virtually freeze-proof. Glycol preheat coil systems avoid all of the disadvantages of freeze protection pumps and perform better than the other preheat systems. As Figure 22.15 shows, the installed cost of glycol preheat coil systems could also be less than the installed cost of preheat coil freeze protection pumps.

A 20% glycol solution will prevent coil bursting. At 0°F, a 30% glycol solution will turn to slush, but it will not freeze solid.

Designing Glycol Systems

The effect of glycol on system design must be taken into consideration. Heated glycol systems require an increase of the following, as compared to water.

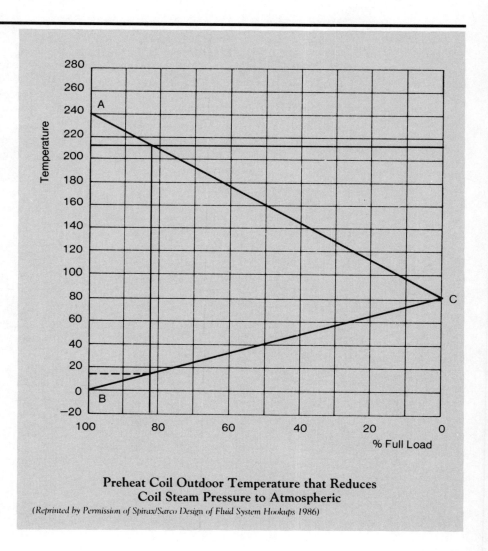

**Preheat Coil Outdoor Temperature that Reduces
Coil Steam Pressure to Atmospheric**

(Reprinted by Permission of Spirax/Sarco Design of Fluid System Hookups 1986)

Figure 22.8

302

30% Glycol Solution	@ 120°F	@ 40°F
• Gpm	6%	8%
• Heat Transfer Surface	12%	20%
• Pump Head	0	50%
• Expansion Tank Volume	15%	10%

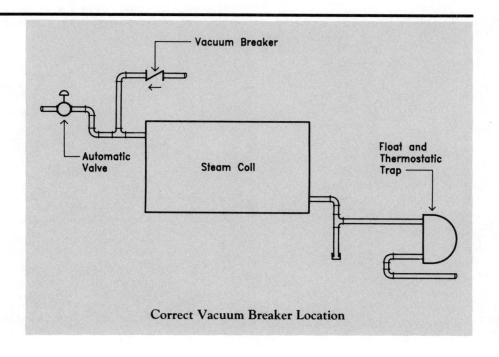

Correct Vacuum Breaker Location

Figure 22.9

Gpm: The gpm of fluid circulated to transfer heat is derived from the following formula:

gpm = btuh/60 m/h x 8.3 #/gal. x sh x sg x td

gpm = gallons per minute
btuh = btu per hour
m/h = minutes per hour
sh = specific heat
sg = specific gravity
td = temperature difference

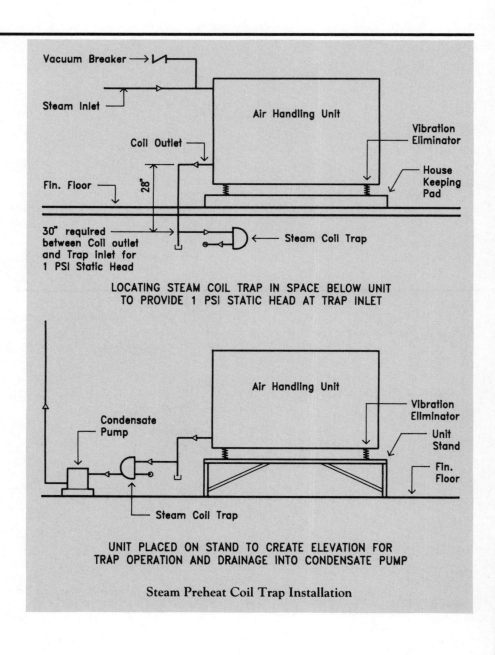

LOCATING STEAM COIL TRAP IN SPACE BELOW UNIT
TO PROVIDE 1 PSI STATIC HEAD AT TRAP INLET

UNIT PLACED ON STAND TO CREATE ELEVATION FOR
TRAP OPERATION AND DRAINAGE INTO CONDENSATE PUMP

Steam Preheat Coil Trap Installation

Figure 22.10

The specific heat of a 120°F 30% glycol solution is 0.92, as compared to 1.0 for water. The specific gravity is 1.025. Therefore, approximately 6% more fluid must be circulated, compared to water. When a 40°F 30% glycol solution is used in a cooling or "run around" heat recovery system, the 40°F glycol specific gravity is 1.045 and the specific heat is 0.88. Therefore, 8% *more gpm must be circulated.*

gpm (water) = btuh/60 x 8.3 x 1 x 1 x td = btuh/500 x td

gpm (glycol) = btuh/60 x 8.3 x .92 x 1.025 x td = btuh/470 x td

470/500 = 0.94 ....1 − 0.94 = 0.06 or 6% more gpm

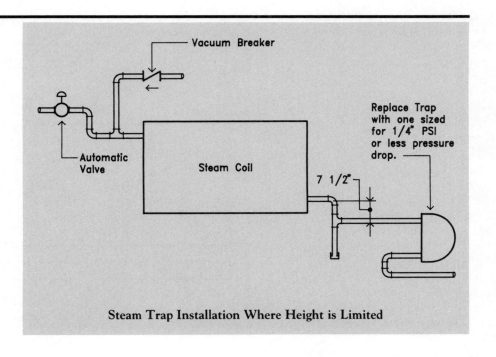

Steam Trap Installation Where Height is Limited

Figure 22.11

Heat Transfer Surface: The heat transfer of glycol is impeded by the change in the film coefficient of the surface in contact with the solution. This is primarily due to the change in viscosity. *Approximately 12% more heat exchange surface is needed for a 120°F 30% glycol solution compared to water. A 40°F glycol solution requires 20% more heat exchange surface compared to water.*

Pump Head: The friction loss in piping varies with changes in viscosity. The viscosity of 120°F water is 0.60 centipoises, and the viscosity of 120°F 30% glycol is 1.10 centipoises. *The friction loss of a 120°F 30% glycol piping system is 20% more than that of 120°F water.*

The viscosity of 40°F water is 1.5 centipoises and the viscosity of 40°F 30% glycol is 4 centipoises. *The friction loss in 40°F 30% glycol systems is 1.5 times that of water.*

Expansion Tank Volume: The glycol mixture has a higher expansion rate than that of water alone. Glycol solutions expand approximately 1.2 times the amount of water. *The expansion tank size for a 30% glycol mixture should be 15% larger than a water expansion tank, and 20% larger for 40% solutions.*

It is imperative that the correct glycol is used. Improper glycol, such as the type used in automobile radiators, can produce a sludge by reacting with the zinc on galvanized piping, or with the water treatment in the system. Or, it may react with pipe dope, cutting oils and dirt in the piping system.

Glycol manufacturers produce a special glycol for HVAC systems, that does not react with zinc, and the inhibitor is extremely durable. Glycol replacement may not be required for 10-15 years.

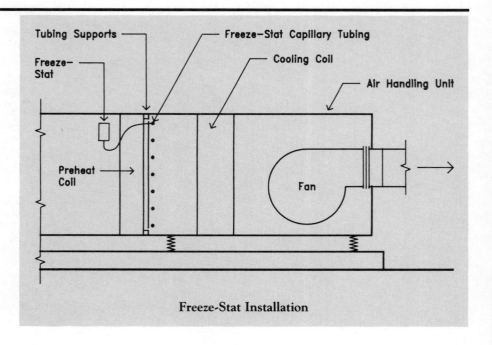

Freeze-Stat Installation

Figure 22.12

The following precautions should be taken when installing and operating a glycol system:

- Automatic vents should not be installed. Float-type vents will leak because of the difference in specific gravity between glycol and water.
- Chromate should not be used as a water treatment. Adding any other chemicals to a glycol system will only create problems. The correct glycol, as it is provided by the manufacturer, will be formulated with the necessary corrosion inhibitors.
- Pumps must be equipped with mechanical seals. Gland-type pumps leak glycol.
- Glycol systems must not be directly connected to city water with pressure-reducing valves. Most codes call for backflow preventers. A loss of glycol could allow more fresh water in and a dilution of the glycol solution could result. Refilling of the system should be under the direct control of the operator. Maintaining static pressure on the system should be accomplished by manually opening a cold water valve to the suction side of the glycol pump piping until the desired pressure is reached, or by an automatic glycol fill pump as shown in Figure 22.16.
- The piping system should be thoroughly flushed and cleaned with a heated trisodium phosphate solution, prior to filling it with glycol. Glycol can react with pipe dope, cutting oil, solder paste, and dirt inside the piping to create a sludge that will affect the heat exchange process.

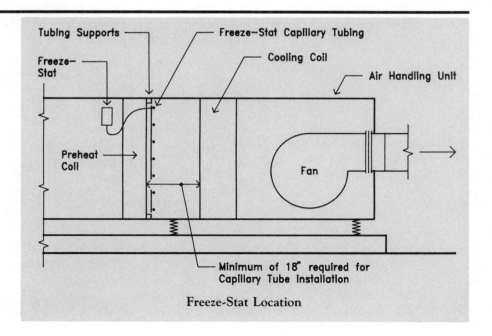

Freeze-Stat Location

Figure 22.13

- Galvanized pipe should not be used in the system. The zinc in galvanized pipe can react with the glycol inhibitor to create a sludge.
- The glycol inhibitor and glycol strength should be checked once a year. Inhibitors and glycol should be added as the test results indicate.

Face and Bypass Preheat Coils

Figure 22.17 shows a typical face and bypass steam preheat coil. Steam is continuously supplied to the top of the preheat coil header. The vertical tubes distribute the steam and condensate to the return header at the bottom of the coil. The automatic face and bypass dampers are operated by an actuator that is controlled by a supply air thermostat through a low limit control. The advantages of steam distribution coils are:

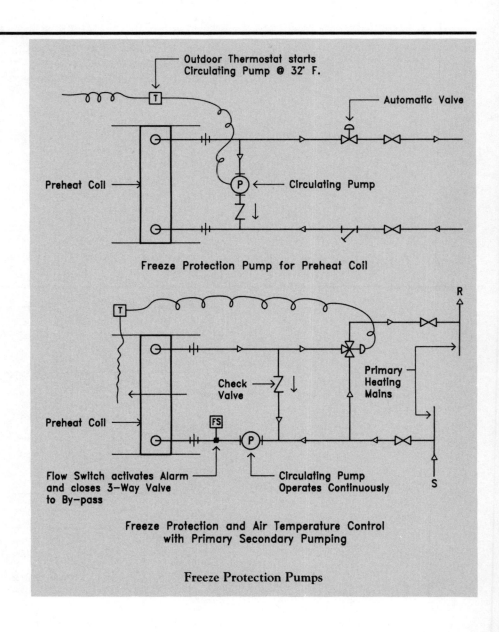

Freeze Protection Pumps

Figure 22.14

- The vertical steam tubes permit condensate to drain to the steam trap.
- The constant steam supply through the tubes induces condensate removal.

The disadvantages of face and bypass protection and steam distributing coils are:

- High installed cost
- High air pressure drop
- Damper control produces ineffective temperature control because of damper leakage and loose linkages.
- High maintenance of dampers and linkage
- Limited height for steam trap installation due to bottom steam return header

Freeze Protection Pumps Four 30 gpm preheat coils	
Item	**Installed Cost**
4 pumps $900 ea.	$ 3,600
Pipe, valves, fittings	2,000
Labor	2,800
Insulation	600
Starters	2,800
Electric wiring	1,200
Thermostat & wiring	200
Total	$13,200
Glycol Preheat System	
Heat exchanger package, converter, expansion tank, circulating pump, starter, automatic control valve and controls	$11,000
Electric wiring	500
Glycol 150 gal. x $5/gal.	750
Labor	700
Total	$12,950

Example of Comparative Installed Cost of Freeze Protection Pumps and Glycol Preheat Coil System

Figure 22.15

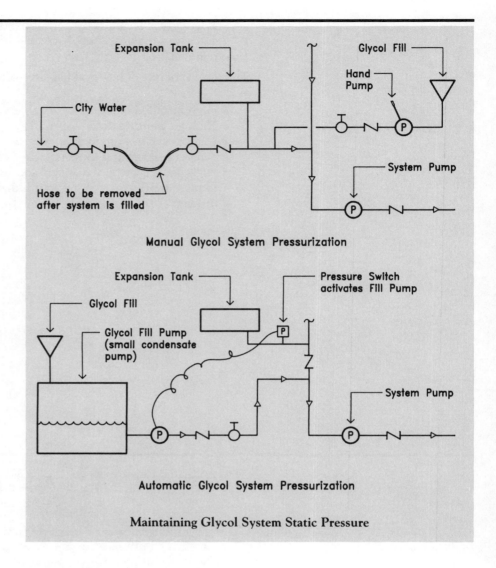

Manual Glycol System Pressurization

Automatic Glycol System Pressurization

Maintaining Glycol System Static Pressure

Figure 22.16

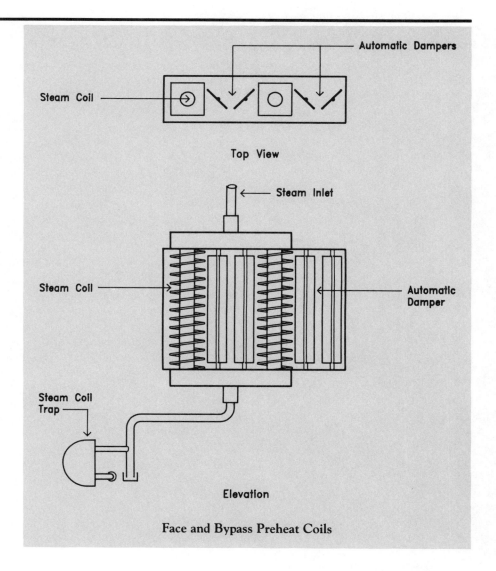

Top View

← Steam Inlet

Automatic Dampers

Steam Coil

Steam Coil

Automatic Damper

Steam Coil Trap

Elevation

Face and Bypass Preheat Coils

Figure 22.17

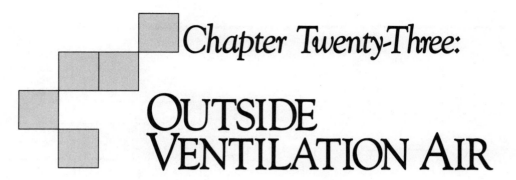

Chapter Twenty-Three:

OUTSIDE VENTILATION AIR

To satisfy building department codes and to avoid the "sick building syndrome" of poor ventilation, the required amount of outside ventilation air must be circulated throughout the building. Codes are being revised to require that a building must be served with a minimum of 15-20 cfm of outside air per person.

Let us examine a VAV (variable air volume) system with a maximum supply air volume of 20,000 cfm and a minimum outside air requirement of 2,000 cfm. The outside air volume represents 10% of the maximum supply air quantity. As the cooling requirements reduce by 50%, the supply air volume also reduces by 50%. In the process, the outside ventilation air to the building could also be reduced by 50%—from 2,000 cfm to an unacceptable 1,000 cfm. The outside air quantity delivered by the air handling unit can remain constant by the use of an air-measuring station (AMS) in the outside air duct.

Figure 23.1 shows the outdoor air control sequence used to ensure that correct ventilation air volume is delivered for a fixed outdoor air system (no economizer). Whenever the velocity controller reads a velocity that represents less than 2,000 cfm, it throttles the automatic return air damper until the velocity that corresponds to 2,000 cfm is reached. Economizer cycle systems are illustrated in Figure 23.2.

As the minimum outside air volume falls below 2,000 cfm, the return air automatic damper is throttled until the velocity controller reads 2,000 cfm. Minimum outside air control is disabled whenever the economizer cycle opens the maximum outside air damper. Minimum outside ventilation air volume is obtained when the maximum outside air damper moves to the "open" position.

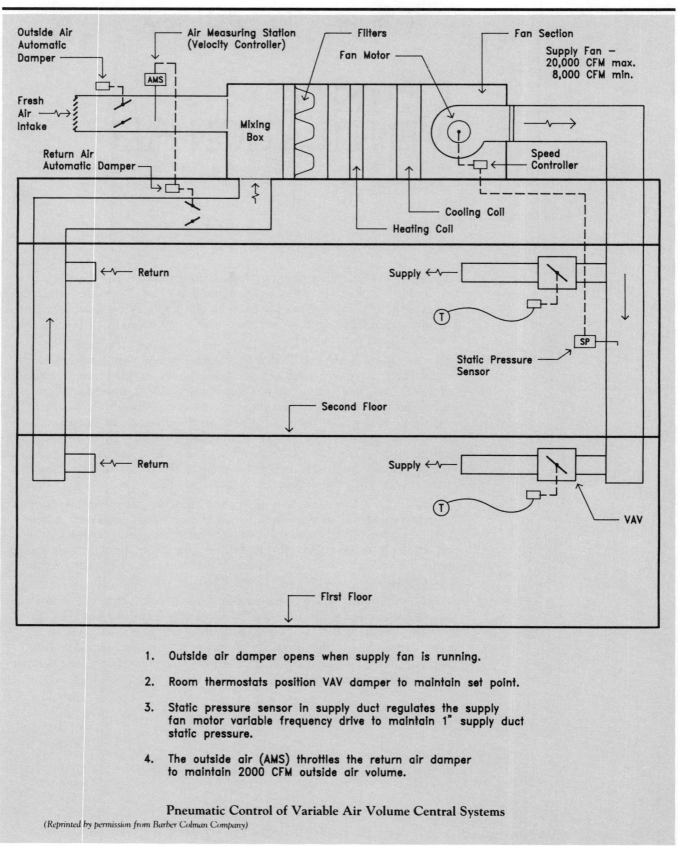

Pneumatic Control of Variable Air Volume Central Systems

(Reprinted by permission from Barber Colman Company)

Figure 23.1

314

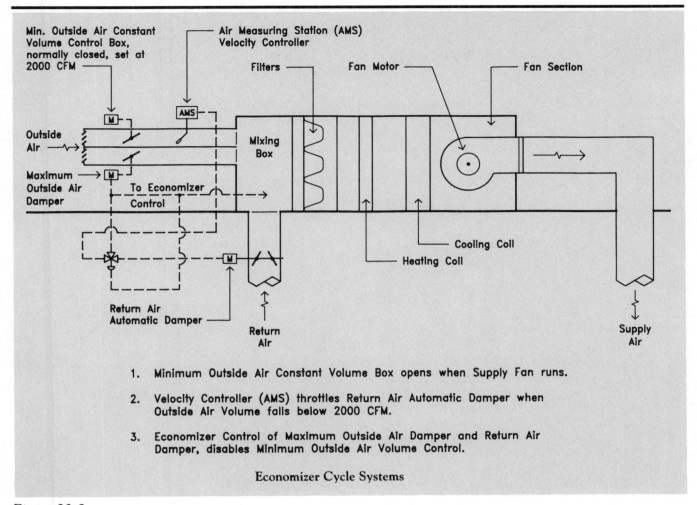

Min. Outside Air Constant Volume Control Box, normally closed, set at 2000 CFM

Air Measuring Station (AMS) Velocity Controller

Filters

Fan Motor

Fan Section

Outside Air

Maximum Outside Air Damper

To Economizer Control

Mixing Box

Cooling Coil

Heating Coil

Return Air Automatic Damper

Return Air

Supply Air

1. Minimum Outside Air Constant Volume Box opens when Supply Fan runs.

2. Velocity Controller (AMS) throttles Return Air Automatic Damper when Outside Air Volume falls below 2000 CFM.

3. Economizer Control of Maximum Outside Air Damper and Return Air Damper, disables Minimum Outside Air Volume Control.

Economizer Cycle Systems

Figure 23.2

315

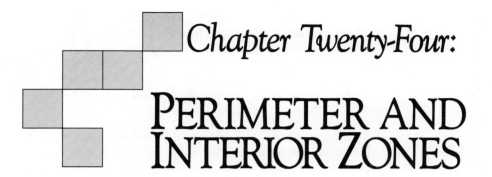

Chapter Twenty-Four:

PERIMETER AND INTERIOR ZONES

Perimeter and interior spaces should be on separate zones from one another. Interior spaces frequently require cooling year-round because of space heat gain from lights, electric equipment, and people, with little or no heat loss from the space. Perimeter spaces, on the other hand, require heat in cold weather because of heat loss through walls and windows.

As Figure 24.1 illustrates, the mass of cold air is introduced to the conditioned space at a temperature that is cooler than the desired room temperature. The heat generated within the space then warms the cold air mass to the required room temperature. In order for the space temperature to remain stable, the heat gain of the cold supply air must be equal to the heat accumulated within the conditioned space. Interior zones demand cool air supply, even during winter, to remove the internal heat generated in the space.

The problem is that if the perimeter air supply is taken from the same supply air duct that serves the interior space, the perimeter space will be overcooled. There is insufficient heat yielded within the perimeter areas to warm the cold supply air. This overcooling of the perimeter zone can be eliminated by:

- Reducing the volume of cool supply air to the perimeter zone (in a VAV system). The amount of this supply air reduction is limited by the occupant's ventilation requirements.
- Reheating the supply air to the perimeter zone.
- Adding heat to the perimeter zone by means of radiation heat. To be effective, the radiation heating system must also be available in mild weather.

Reducing the Supply Air Volume

Figure 24.2 shows the heat balance that is achieved when the space heat gain is matched to the warming of the cold supply air mass. The supply air volume to a space is determined by the maximum internal sensible heat gain of the space.

Supply Air Volume (cfm) = internal sensible heat gain (btuh)/1.08 (room air temperature − supply temperature)

Figure 24.3 shows how the supply air volume can be reduced to match the reduced space heat gain.

The interior space sensible heat gain can be reduced by computers being turned off, or fewer people in a room. Perimeter space heat gain is dramatically influenced by solar heat gain and outdoor air temperatures. During periods of peak space heat gain, the sunlight enters the perimeter room, warms the walls and roof, and raises the outdoor temperature to 90°F. During those periods when the sun is not creating maximum space heat gain, satisfactory space temperatures can be maintained by reducing the volume of cold air supplied to the room to match the heat gain of the room.

Reheat   Perimeter zones require reheat, when interior spaces do not. Fifty-five degree air is required to cool interior spaces, even in cold weather. Heat generated by lights, computers, copy machines, and people is continually added to the interior space during occupied hours. The interior space of a building does not have any contact with the outdoors, and cannot lose its heat to the outdoors. As Figure 24.1 shows, perimeter spaces would be overcooled by 55°F supply air, since these spaces have little internal heat gain, but do have transmission heat losses due to contact with the outdoors. The supply air to the perimeter spaces should be heated to or reset up to 68-70°F to maintain perimeter space room temperatures of 70-72°F. Even though the supply air automatic damper and fan may reduce the supply air volume, the outside air quantity should remain constant. Proper ventilation should be retained.

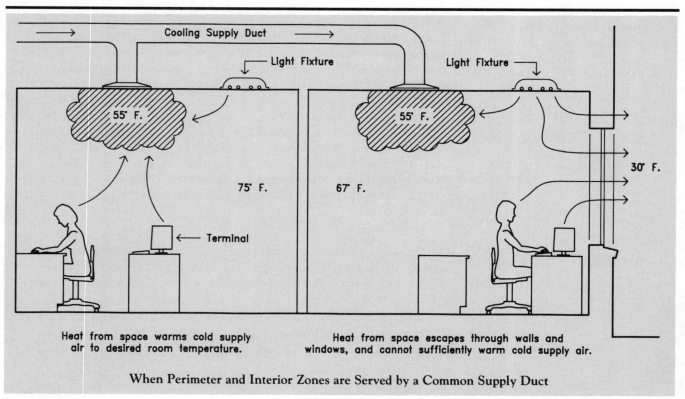

Heat from space warms cold supply air to desired room temperature.

Heat from space escapes through walls and windows, and cannot sufficiently warm cold supply air.

When Perimeter and Interior Zones are Served by a Common Supply Duct

Figure 24.1

Chapter 18 on VAV operational concerns explains that the total air supply to the perimeter zone should not be less than four air changes per hour. The four air changes of supply are required to prevent "stuffiness" complaints from occupants due to reduced air movement, and to preserve the outside air ventilation requirements.

Reheat is important to prevent overcooling of the perimeter zone. Whenever the space heat gain is not sufficient to warm the 55°F four air changes of supply air to the desired room temperature, the reheat coil in the supply air duct adds the additional heat to the supply air. See Figure 24.4.

Separate Perimeter Air Handling Unit

Reheating air to prevent overcooling wastes energy. One way to prevent overcooling while minimizing energy waste is to supply air to the perimeter zone from a separate air handling unit. The supply air temperature to that zone can be increased by reducing the amount of cool outside air drawn in by the unit, or by reducing the amount of chilled water or refrigerant that is delivered to the cooling coil. Furnishing warmer supply air to the perimeter zone without using any added energy improves system efficiency.

Resetting supply air to the perimeter spaces is only viable if the perimeter and interior zones are on different air handling units. The interior zone requires cooling at times when the perimeter zone needs heating. When

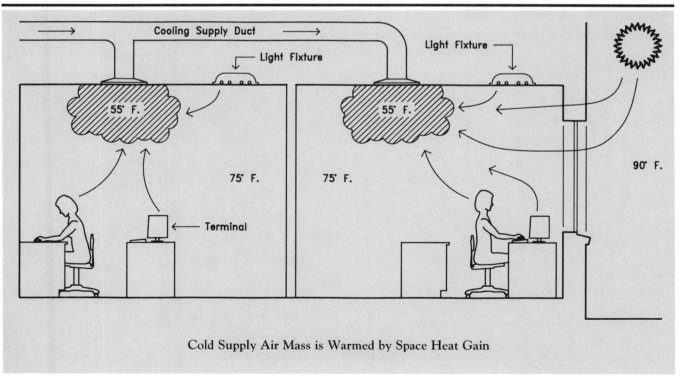

Cold Supply Air Mass is Warmed by Space Heat Gain

Figure 24.2

the perimeter zone is served by its own air handling unit, the 55°F supply air that the interior zone needs to satisfy the space heat gain can be provided without overcooling other zones. A separate air handling unit dedicated to the perimeter zone can supply warmer air to that zone when needed, with little or no reheat.

Perimeter Heating

The human body radiates heat to cold surfaces in the immediate surroundings, thereby making the body uncomfortably cold. For this reason, perimeter heating is required under the windows in cold climates. Heat under the windows warms the inside wall surface, and reduces the body's heat loss. See Figure 24.5.

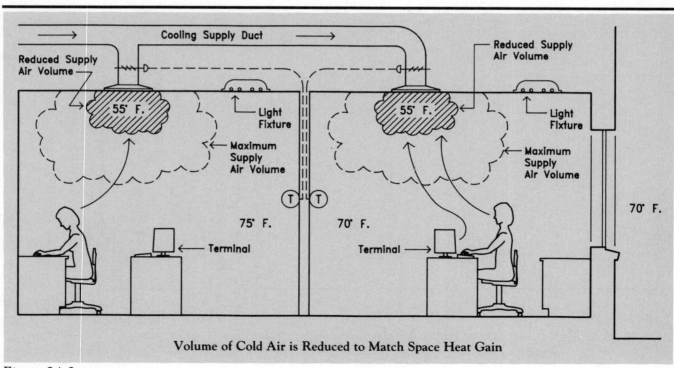

Volume of Cold Air is Reduced to Match Space Heat Gain

Figure 24.3

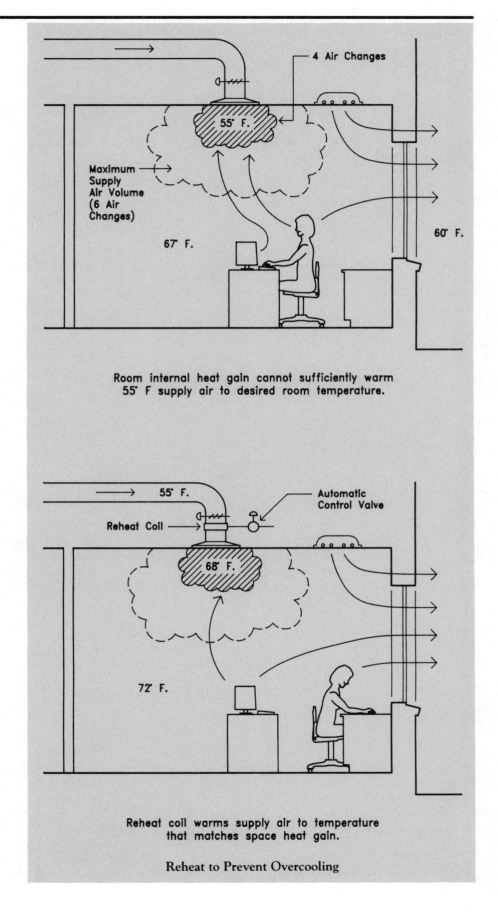

Room internal heat gain cannot sufficiently warm
55° F supply air to desired room temperature.

Reheat coil warms supply air to temperature
that matches space heat gain.

Reheat to Prevent Overcooling

Figure 24.4

321

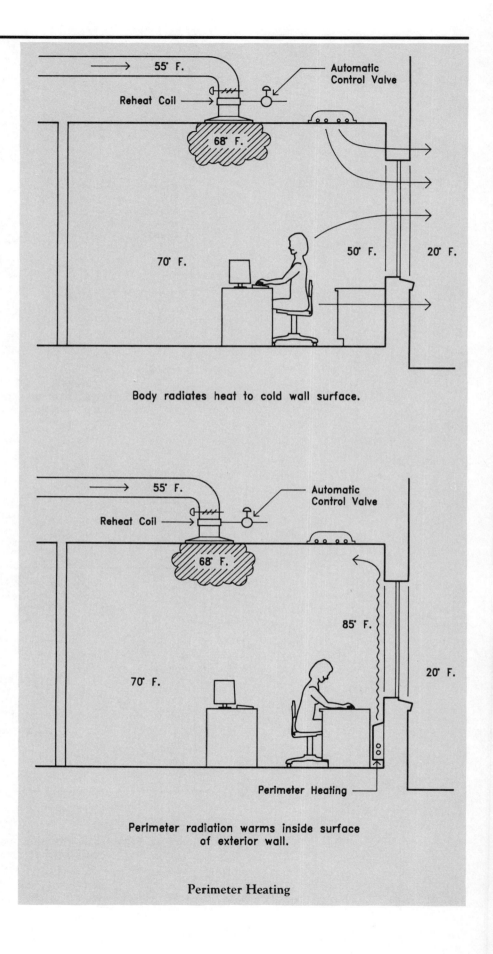

Body radiates heat to cold wall surface.

Perimeter radiation warms inside surface
of exterior wall.

Perimeter Heating

Figure 24.5

Chapter Twenty-Five:

EXPANSION TANKS AND SYSTEM PRESSURE CONTROL

The purpose of the expansion tank in a hydronic system is to provide water system pressure control. The expansion tank stores the excess water in the system that results from an increase in volume due to the rise in system water temperature. The expansion tank limits the system maximum pressure and maintains the minimum pressure necessary to vent air and prevent pump cavitation. Cavitation at the pump suction is avoided by maintaining the proper minimum system pressure above the vapor pressure of the water at the eye of the pump impeller.

The location of the expansion tank must be taken into consideration when selecting the proper expansion tank size and determining system performance. When an expansion tank is undersized or improperly located, the system pressure could increase to a point that will cause the pressure relief valve to discharge to waste the excess water volume. Most systems are equipped with an automatic fill valve set to maintain the minimum design pressure in the system by supplying water to make up for volume and pressure losses due to venting and leakage. It should be noted, however, that every water recharging cycle introduces additional oxygen that will promote corrosion.

Loss of pressurization may cause flashing within the system. The flashing of water to a gas at pressures below the water vapor pressure can damage pumps, heat exchangers, and piping. The connection of the expansion tank to the system is the point of no pressure change in the system, regardless of whether or not the pump is operating. In Figure 25.1, the pump is seen discharging away from the expansion tank and the boiler.

The full pump pressure, less piping friction loss, is added to all points downstream of the pump. The fill pressure need only be slightly above the system static head in order to avoid pump cavitation. Note that a 72-gallon expansion tank satisfies expansion requirements.

In Figure 25.2, the pump discharges into the boiler, expansion tank, and relief valve. The full pump pressure acts to decrease the system pressure below the fill pressure. The fill pressure must be above the pump pressure. Otherwise, a vacuum will be created in the pump suction line. A 433 gallon expansion tank is needed to prevent relief valve discharge.

As illustrated in Figure 25.3, when the pump operates, all points between the pump and the expansion tank (point of no pressure change) appear as a pressure increase. The pump in this figure is discharging into the system pressure relief valve. The pump pressure at this point must be below the relief valve setting, or the relief valve will discharge.

The relief valve set point is determined by the maximum pressure that the weakest point in the system can withstand. Low pressure hot water boilers are usually rated at 30 psi working pressure; therefore, low pressure relief valves are set at 30 psi. System design pressures should be 5 psi below the relief valve setting, or 25 psi.

All points between the expansion tank and the pump suction will show a pressure decrease below fill pressure. The fill pressure does not have to overcome system static pressure because the expansion tank is positioned at the top of the system, where there is no static pressure. When there is a considerable pressure drop due to long piping runs between the expansion

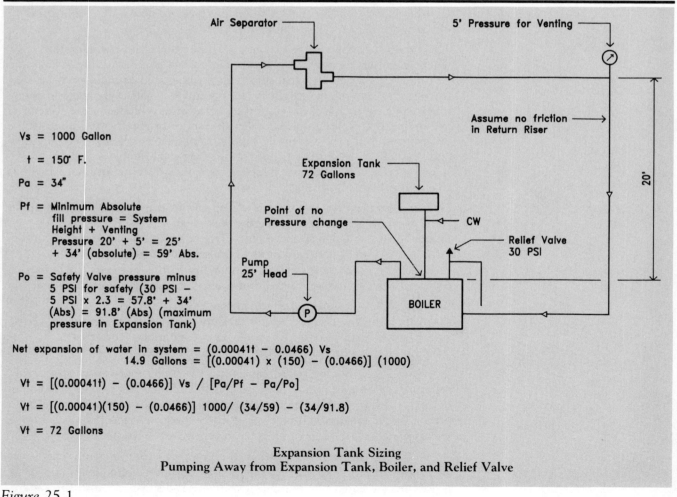

V_s = 1000 Gallon

t = 150° F.

P_a = 34"

P_f = Minimum Absolute fill pressure = System Height + Venting Pressure 20' + 5' = 25' + 34' (absolute) = 59' Abs.

P_o = Safety Valve pressure minus 5 PSI for safety (30 PSI − 5 PSI x 2.3 = 57.8' + 34' (Abs) = 91.8' (Abs) (maximum pressure in Expansion Tank)

Net expansion of water in system = $(0.00041t − 0.0466) V_s$
14.9 Gallons = $[(0.00041) \times (150) − (0.0466)] (1000)$

$V_t = [(0.00041t) − (0.0466)] V_s / [P_a/P_f − P_a/P_o]$

$V_t = [(0.00041)(150) − (0.0466)] 1000 / (34/59) − (34/91.8)$

V_t = 72 Gallons

Expansion Tank Sizing
Pumping Away from Expansion Tank, Boiler, and Relief Valve

Figure 25.1

tank and the pump, care must be taken to ensure that a negative pressure is not created in the pump suction line. Initial system fill pressurization must be set above any possible water vapor pressure in the system to avoid pump cavitation. The initial fill pressure should equal the system static height plus 5' for venting, plus the friction loss between the expansion tank and the pump. The expansion tank size in the example system is 122 gallons.

Figure 25.4 shows the expansion tank located at the system high point, and the pump discharging away from the boiler and relief valve. A 97-gallon expansion tank volume is required.

All points in the system between the pump discharge and the expansion tank appear as a pressure increase over fill pressure. All points in the

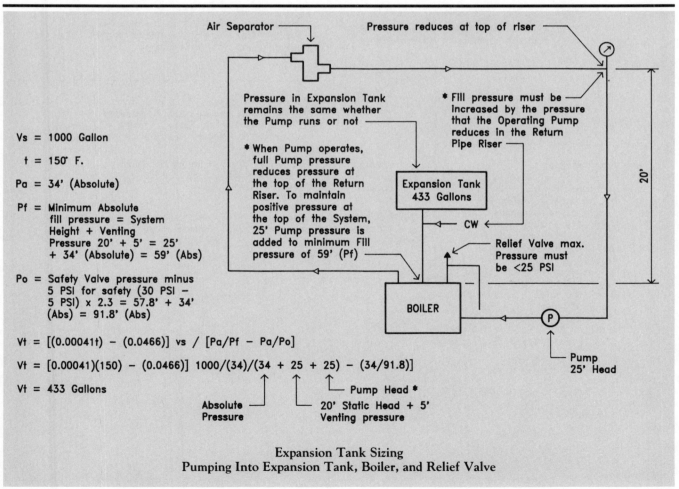

Air Separator

Pressure reduces at top of riser

Pressure in Expansion Tank remains the same whether the Pump runs or not

* When Pump operates, full Pump pressure reduces pressure at the top of the Return Riser. To maintain positive pressure at the top of the System, 25' Pump pressure is added to minimum Fill pressure of 59' (Pf)

* Fill pressure must be increased by the pressure that the Operating Pump reduces in the Return Pipe Riser

Expansion Tank 433 Gallons

CW

Relief Valve max. Pressure must be <25 PSI

BOILER

P

Pump 25' Head

20'

V_s = 1000 Gallon

t = 150° F.

P_a = 34' (Absolute)

P_f = Minimum Absolute fill pressure = System Height + Venting Pressure 20' + 5' = 25' + 34' (Absolute) = 59' (Abs)

P_o = Safety Valve pressure minus 5 PSI for safety (30 PSI − 5 PSI) x 2.3 = 57.8' + 34' (Abs) = 91.8' (Abs)

$V_t = [(0.00041t) - (0.0466)] \, v_s \, / \, [P_a/P_f - P_a/P_o]$

$V_t = [0.00041)(150) - (0.0466)] \, 1000/(34)/(34 + 25 + 25) - (34/91.8)]$

V_t = 433 Gallons

Absolute Pressure

Pump Head *
20' Static Head + 5' Venting pressure

Expansion Tank Sizing
Pumping Into Expansion Tank, Boiler, and Relief Valve

Figure 25.2

system between the expansion tank and the pump suction show a decrease in the fill pressure. As in Figure 25.3, the fill pressure must be set high enough to prevent pump cavitation due to the piping pressure loss between the expansion tank and the pump suction. The location of the expansion tank connection to the system determines whether the pump pressure is added to, or subtracted from, the system static pressure.

Common Operational Problems

Some common expansion tank operating problems are listed below. Explanations and solutions to these problems are provided in the paragraphs that follow.

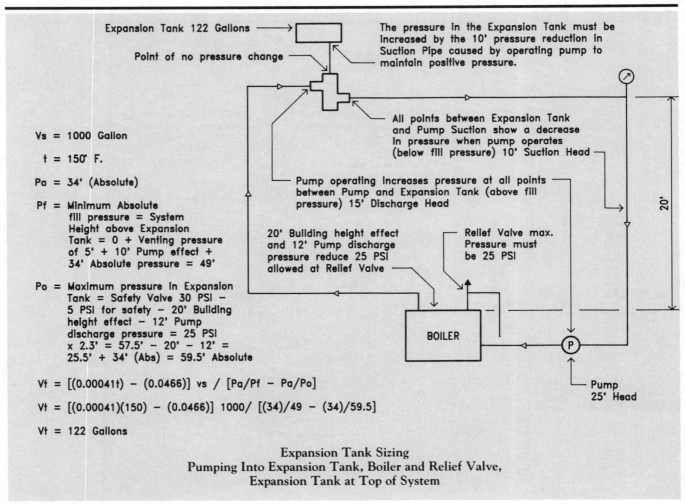

Expansion Tank 122 Gallons →

Point of no pressure change

The pressure in the Expansion Tank must be increased by the 10' pressure reduction in Suction Pipe caused by operating pump to maintain positive pressure.

All points between Expansion Tank and Pump Suction show a decrease in pressure when pump operates (below fill pressure) 10' Suction Head

Pump operating increases pressure at all points between Pump and Expansion Tank (above fill pressure) 15' Discharge Head

20' Building height effect and 12' Pump discharge pressure reduce 25 PSI allowed at Relief Valve

Relief Valve max. Pressure must be 25 PSI

BOILER

P

Pump 25' Head

20'

Vs = 1000 Gallon

t = 150° F.

Pa = 34' (Absolute)

Pf = Minimum Absolute fill pressure = System Height above Expansion Tank = 0 + Venting pressure of 5' + 10' Pump effect + 34' Absolute pressure = 49'

Po = Maximum pressure in Expansion Tank = Safety Valve 30 PSI – 5 PSI for safety – 20' Building height effect – 12' Pump discharge pressure = 25 PSI x 2.3' = 57.5' – 20' – 12' = 25.5' + 34' (Abs) = 59.5' Absolute

Vt = [(0.00041t) – (0.0466)] vs / [Pa/Pf – Pa/Po]

Vt = [(0.00041)(150) – (0.0466)] 1000/ [(34)/49 – (34)/59.5]

Vt = 122 Gallons

Expansion Tank Sizing
Pumping Into Expansion Tank, Boiler and Relief Valve,
Expansion Tank at Top of System

Figure 25.3

326

Relief Valve Discharge

If the expansion tank loses its air cushion, the water temperature increase expands the water volume. There is then no other place for the excess water to go other than out through the relief valve. To correct this condition, drain the expansion tank and reestablish the air cushion. As Figure 25.5 shows, a stop valve is needed between the expansion tank and the system. A drain valve at the bottom of the tank, and an air inlet valve at the top of the tank are required to service it. Plugs or caps should be installed in the open ends of all expansion tank drain and air inlet valves to ensure a tight expansion tank system. Combination valves are available to drain and recharge the expansion tank.

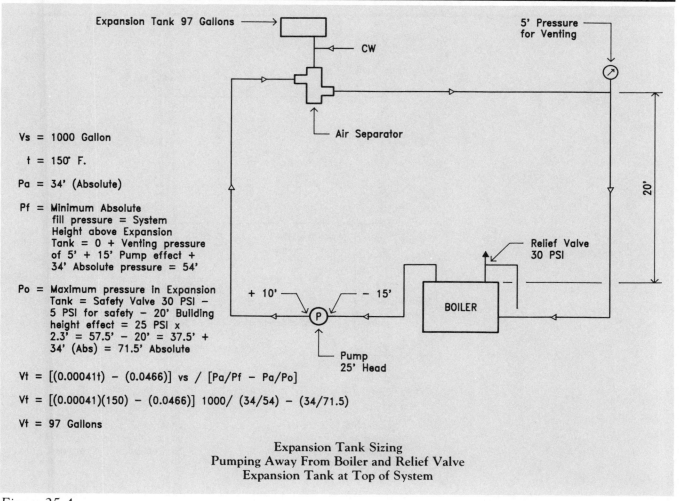

V_s = 1000 Gallon

t = 150° F.

P_a = 34' (Absolute)

P_f = Minimum Absolute fill pressure = System Height above Expansion Tank = 0 + Venting pressure of 5' + 15' Pump effect + 34' Absolute pressure = 54'

P_o = Maximum pressure in Expansion Tank = Safety Valve 30 PSI − 5 PSI for safety − 20' Building height effect = 25 PSI x 2.3' = 57.5' − 20' = 37.5' + 34' (Abs) = 71.5' Absolute

$V_t = [(0.00041t) − (0.0466)]$ vs / $[P_a/P_f − P_a/P_o]$

$V_t = [(0.00041)(150) − (0.0466)]$ 1000/ (34/54) − (34/71.5)

V_t = 97 Gallons

Expansion Tank Sizing
Pumping Away From Boiler and Relief Valve
Expansion Tank at Top of System

Figure 25.4

Expansion Tank Air Leaks

When gauge glasses are attached to the expansion tank, air leaks may develop at the top gauge glass connection, or at the gauge cocks. Figure 25.6 shows some sources of expansion tank air leaks. The solution to air leaks is to tighten the connections, replace the gauge glass assembly, or plug or cap the gauge glass tappings at the expansion tank. Then drain and recharge the tank.

Expansion Tank Too Small

Very often, additions are made to existing systems and increase the volume of system water. Since the correct expansion tank size is determined by the volume of water in the system, any changes to the system that increase the water volume should include proper expansion

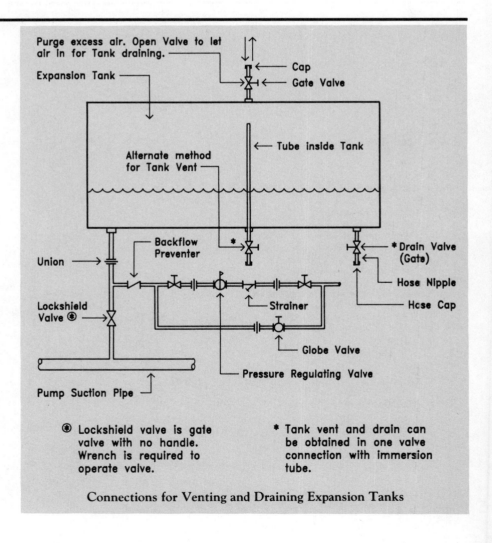

Purge excess air. Open Valve to let air in for Tank draining.

Expansion Tank

Cap
Gate Valve

Tube inside Tank

Alternate method for Tank Vent

Union

Backflow Preventer

Drain Valve (Gate)

Hose Nipple

Lockshield Valve ⊛

Strainer

Hose Cap

Globe Valve

Pressure Regulating Valve

Pump Suction Pipe

⊛ Lockshield valve is gate valve with no handle. Wrench is required to operate valve.

* Tank vent and drain can be obtained in one valve connection with immersion tube.

Connections for Venting and Draining Expansion Tanks

Figure 25.5

tank sizing. Figure 25.7 is an example of how expansion tanks are sized. Adding another properly sized expansion tank in parallel with the existing tank is one way to solve the problem.

Pump Discharges Into the Expansion Tank

As demonstrated in Figure 25.2, a much larger expansion tank is necessary when a pump discharges into the expansion tank, exchanger, and relief valve than if the expansion tank is connected to the pump suction. This condition can be rectified by relocating the expansion tank pipe connection to the suction side of the pump. This is usually a simpler modification than relocating the pump to the leaving side of the boiler or heat exchanger. The expansion tank connecting pipe is considerably smaller (1") than pump piping (2"-10").

Too Low a Pressure Regulating Valve Setting

As Figure 25.8 shows, the pressure regulating valve maintains the initial fill pressure at the expansion tank. The initial fill pressure must be equal to or greater than the elevation of the highest point in the system, plus

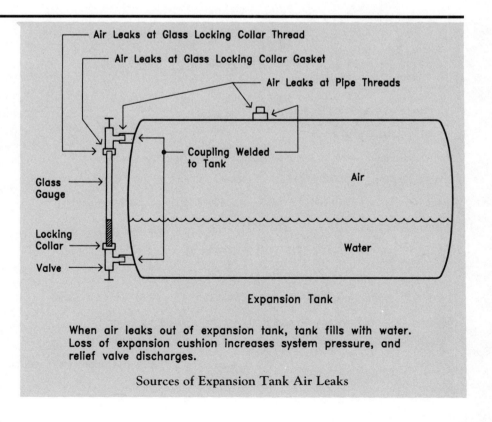

Figure 25.6

When air leaks out of expansion tank, tank fills with water. Loss of expansion cushion increases system pressure, and relief valve discharges.

Sources of Expansion Tank Air Leaks

another five feet of venting pressure, plus the pipe friction between the expansion tank and the pump inlet. The initial fill pressure is determined with the pump off.

It is important that the system pressurization connection be made at the expansion tank, and not at some other point in the system. The expansion tank is the point of no pressure change in the system.

Figure 25.9 shows what happens when the expansion tank is connected at the pump discharge side of the system, and the pressurization connection is

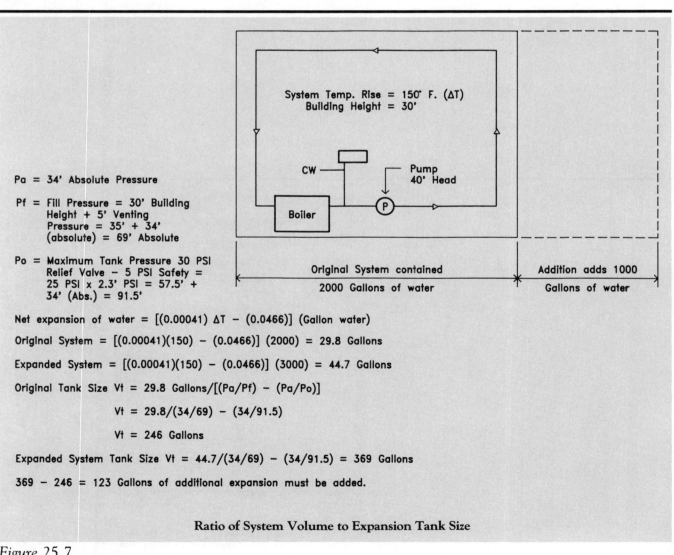

Pa = 34' Absolute Pressure

Pf = Fill Pressure = 30' Building Height + 5' Venting Pressure = 35' + 34' (absolute) = 69' Absolute

Po = Maximum Tank Pressure 30 PSI Relief Valve − 5 PSI Safety = 25 PSI x 2.3' PSI = 57.5' + 34' (Abs.) = 91.5'

System Temp. Rise = 150° F. (ΔT)
Building Height = 30'

CW

Pump 40' Head

Boiler

Original System contained 2000 Gallons of water

Addition adds 1000 Gallons of water

Net expansion of water = $[(0.00041) \Delta T - (0.0466)]$ (Gallon water)

Original System = $[(0.00041)(150) - (0.0466)] (2000) = 29.8$ Gallons

Expanded System = $[(0.00041)(150) - (0.0466)] (3000) = 44.7$ Gallons

Original Tank Size Vt = 29.8 Gallons/$[(Pa/Pf) - (Pa/Po)]$

$\qquad$ Vt = 29.8/(34/69) − (34/91.5)

$\qquad$ Vt = 246 Gallons

Expanded System Tank Size Vt = 44.7/(34/69) − (34/91.5) = 369 Gallons

369 − 246 = 123 Gallons of additional expansion must be added.

Ratio of System Volume to Expansion Tank Size

Figure 25.7

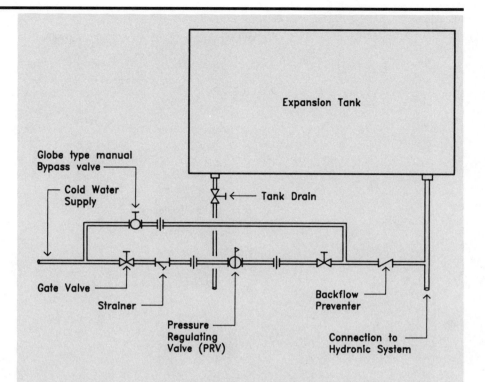

Globe type manual
Bypass valve

Cold Water
Supply

Expansion Tank

Tank Drain

Gate Valve

Strainer

Pressure
Regulating
Valve (PRV)

Backflow
Preventer

Connection to
Hydronic System

PRV must be set for Static Head of system. Vertical distance
between PRV and highest point of hydronic system + 5 ft.
pressure for venting + pipe friction loss between expansion
tank and pump suction (to maintain positive pressure at top
of pump suction riser.)

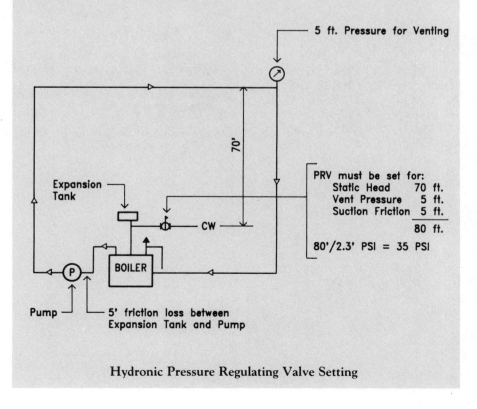

5 ft. Pressure for Venting

70'

Expansion
Tank

CW

PRV must be set for:
Static Head   70 ft.
Vent Pressure   5 ft.
Suction Friction 5 ft.
                 ――――――
                 80 ft.

80'/2.3' PSI = 35 PSI

BOILER

Pump

5' friction loss between
Expansion Tank and Pump

Hydronic Pressure Regulating Valve Setting

Figure 25.8

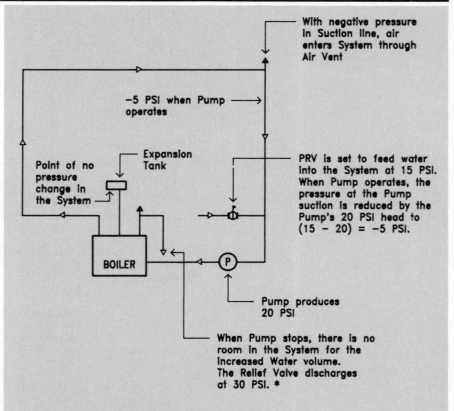

With negative pressure in Suction line, air enters System through Air Vent

−5 PSI when Pump operates

PRV is set to feed water into the System at 15 PSI. When Pump operates, the pressure at the Pump suction is reduced by the Pump's 20 PSI head to (15 − 20) = −5 PSI.

Point of no pressure change in the System

Expansion Tank

BOILER

P

Pump produces 20 PSI

When Pump stops, there is no room in the System for the increased Water volume. The Relief Valve discharges at 30 PSI. *

Correct PRV setting (at Expansion Tank) :

Static System Height 32'/2.3'	=	14.0 PSI
5' Venting Pressure 5'/2.3'	=	2.2
Pressure loss between Expansion Tank and Pump	=	20.0
Total PRV setting	=	* 36.2 PSI

* Relief Valve will discharge at 30 PSI

Incorrect location of system pressurization connection.

Correct location is at the expansion tank, the point of no system pressure change.

Expansion tank should be piped to pump suction, to avoid relief valve discharge.

Incorrect PRV Connection

Figure 25.9

made at the pump suction. The pressure regulating valve is set to maintain a 15 psi system pressure with the pump off. When the pump operates, the pressure is reduced from the expansion tank to the pump inlet. Since the expansion tank is on the discharge side of the pump, the 20 psi developed by the pump reduces the pump inlet pressure to −5 psi (15 psi − 20 psi). Not only can air be sucked into the system by the negative pressure, but water is continuously added to the system by the 15 psi pressure-regulating valve discharging into a −5 psi pump inlet pipe. There is no place for the excess water to hide, so the relief valve will discharge, especially when the pump is stopped.

The following requirements should be addressed when an expansion tank is to be installed:

- Note that the circulating pump suction pipe is usually the best location for the expansion tank system connection.
- Do not pump into boilers and relief valves when the expansion tank is connected to the boiler outlet. The safety relief valve may discharge.
- Be sure the system water pressure regulating valve is set for the correct system static height. The regulating valve must develop sufficient pressure to raise water to the top of the system, vent air from the system, and overcome the pipe friction loss between the expansion tank and the pump inlet.
- Determine the correct expansion tank size using the system water volume, initial fill pressure, static height, relief valve setting, and system temperature.
- Provide structural framing that will support the weight of the expansion tank filled with water. (A 900 gallon expansion tank filled with water weighs over 8,000 pounds.)
- Install a good backflow preventer in the domestic water connection used to pressurize the system.
- If glycol is used in the system, be sure to incorporate a method of preventing glycol from entering the potable water piping. Glycol fill systems can be manual or automatic. Manual fill systems must not have a direct city water connection to the glycol system. Even though a hose is used as the temporary pressurization connection, a backflow preventer must be installed in the potable water circuit. The potable water piping may be under a negative or low pressure because the potable water circuit was drained for servicing. If a direct connection to a pressurized glycol system is made, the glycol could flow from the higher pressure glycol system into the lower pressure potable water system. The backflow preventer is also used on non-glycol systems that might use chemical water treatment or cleaning agents. Figure 25.10 shows a manual glycol fill system.

Automatic glycol fill systems are connected to city water via a pump and receiver. Figure 25.11 is an illustration of an automatic glycol fill system.

The water or glycol mixture level in the receiver is maintained by a float valve in the receiver. The valve permits city water to enter the receiver when the water in the receiver drops below the set level. Whenever the pressure in the pump suction pipe drops below its set point, the pump is started to raise the system pressure to the desired level. The air gap between the glycol level and the city water supply prevents glycol (at the atmospheric pressure in the receiver) from entering the potable water supply. It should be noted that this method is not acceptable in many locales without the addition of a backflow preventer.

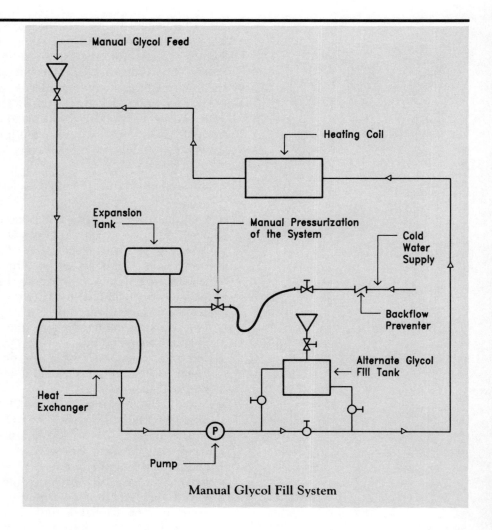

Manual Glycol Feed

Heating Coil

Expansion Tank

Manual Pressurization of the System

Cold Water Supply

Backflow Preventer

Heat Exchanger

Alternate Glycol Fill Tank

Pump

Manual Glycol Fill System

Figure 25.10

334

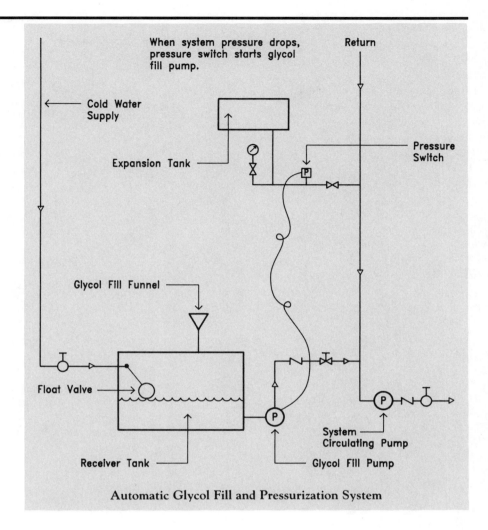

When system pressure drops, pressure switch starts glycol fill pump.

Cold Water Supply

Return

Expansion Tank

Pressure Switch

Glycol Fill Funnel

Float Valve

Receiver Tank

Glycol Fill Pump

System Circulating Pump

Automatic Glycol Fill and Pressurization System

Figure 25.11

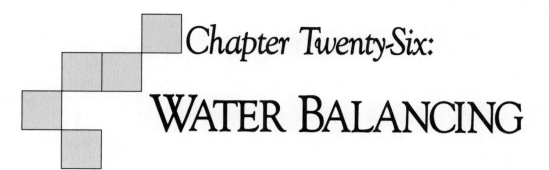

Chapter Twenty-Six:

WATER BALANCING

A hydronic system can only effectively contribute to the heat exchange process if the correct water quantity is delivered to each heat exchange unit in the system. The fact is, very few water systems are properly balanced. Balancing personnel are rarely given the installed equipment needed to establish the desired flow in each circuit.

Figure 26.1 shows that the water velocity in the unbalanced lower equivalent length circuit "A" will increase above the design flow, and the water velocity will decrease below the design flow level in the higher equivalent length circuit "B" until the pressure drops are equal in each circuit. The increase in velocity increases the flow above the design gpm of "A," and the reduced velocity decreases the flow in "B" below the design level.

To balance a system, the pressure loss of each circuit must be adjusted so that the required flow in each circuit is established. This is best accomplished using a balancing valve and a flow measuring device similar to the one in Figure 26.2, which are installed in each circuit to be balanced.

An unbalanced system wastes energy and deprives the high pressure loss circuits of the fluid they need. As Figure 26.3 demonstrates, a balanced system ensures that each terminal maintains the required flow.

In Figure 26.3a, balancing is done at the terminal units only, and all of the excess system head is absorbed by the balancing valves. Although the design gpm is delivered, the pump needs 13 hp to do it. In Figure 26.3b, proportional balancing is achieved by throttling the balancing valve in the circuit with the lowest pressure drop to produce the same proportional flow as the other circuits. The unbalanced flow in Figure 26.1 is 190 gpm for circuit "A" and 260 gpm for circuit "B." Circuit "A" has a proportional flow of 190/190 = 1. Circuit "B" has a proportional flow of 260/190 = 1.37. The balancing valve in circuit "B" is throttled until the proportional flow in circuit "B" is equal to the proportional flow of circuit "A." This will occur at 225 gpm with a proportional balance of 1.18 (225/190).

The balancing valve also enables the trimming of the pump impeller for the lowest horsepower input to deliver the desired gpm. In Figure 26.3c, after proportional balance is completed, the actual system curve is established and plotted as per Figure 26.4.

Figure 26.4 indicates that the design flow of 380 gpm can be delivered with a 9″ impeller in lieu of the installed 10-1/2″ impeller. After the impeller is trimmed to 9″, the pump will deliver 380 gpm, but at an input of only 8.8 bhp. Balancing at the terminal units requires 13 bhp, and proportional balancing only consumes 14 bhp. Impeller trimming reduces pump energy by 37%. "Read-out" type balancing valves, similar to the one depicted in Figure 26.2, are effective for balancing the hydronic system. A read-out meter is required to measure the pressure loss across the flow measuring device. A chart must be available to relate the pressure drop

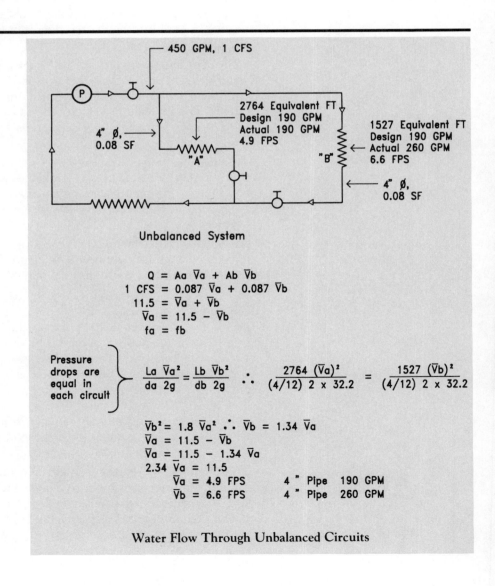

450 GPM, 1 CFS

2764 Equivalent FT
Design 190 GPM
Actual 190 GPM
4.9 FPS

1527 Equivalent FT
Design 190 GPM
Actual 260 GPM
6.6 FPS

4″ ∅,
0.08 SF

"A"

"B"

4″ ∅,
0.08 SF

Unbalanced System

$$Q = A_a \, \overline{V}_a + A_b \, \overline{V}_b$$
$$1 \text{ CFS} = 0.087 \, \overline{V}_a + 0.087 \, \overline{V}_b$$
$$11.5 = \overline{V}_a + \overline{V}_b$$
$$\overline{V}_a = 11.5 - \overline{V}_b$$
$$f_a = f_b$$

Pressure drops are equal in each circuit

$$\frac{L_a \, \overline{V}_a^2}{d_a \, 2g} = \frac{L_b \, \overline{V}_b^2}{d_b \, 2g} \quad \therefore \quad \frac{2764 \, (\overline{V}_a)^2}{(4/12) \, 2 \times 32.2} = \frac{1527 \, (\overline{V}_b)^2}{(4/12) \, 2 \times 32.2}$$

$$\overline{V}_b^2 = 1.8 \, \overline{V}_a^2 \quad \therefore \quad \overline{V}_b = 1.34 \, \overline{V}_a$$
$$\overline{V}_a = 11.5 - \overline{V}_b$$
$$\overline{V}_a = 11.5 - 1.34 \, \overline{V}_a$$
$$2.34 \, \overline{V}_a = 11.5$$
$$\overline{V}_a = 4.9 \text{ FPS} \qquad 4 \text{ '' Pipe } 190 \text{ GPM}$$
$$\overline{V}_b = 6.6 \text{ FPS} \qquad 4 \text{ '' Pipe } 260 \text{ GPM}$$

Water Flow Through Unbalanced Circuits

Figure 26.1

and gpm for the flow measuring device. Figure 26.5 shows the procedure to be followed to properly balance a system.

The following measures should be taken when a hydronic system is balanced.

1. Make sure all automatic and balancing valves are open when balancing is done.
2. Measure the flow through each riser.
   a. Calculate the ratio of actual flow to design flow. If the actual flow is 260 gpm and the design flow is 190 gpm, the proportional balance is 260/190 = 1.37.
   b. Do not balance risers yet.
3. Balance each unit on a branch.
   a. Measure the flow in each unit. (For example, in Figure 26.1, the flow in circuit "A" is 190 gpm and the flow in circuit "B" is 260 gpm.)

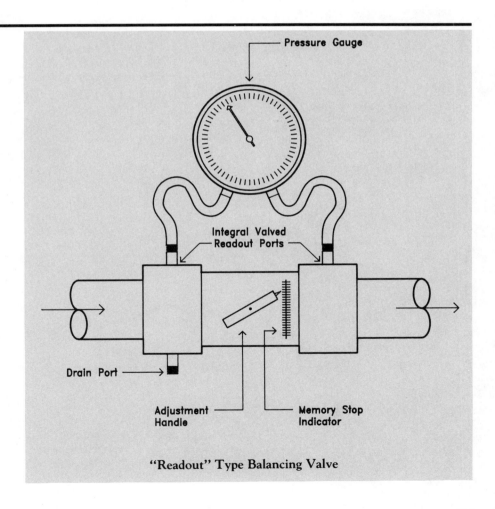

"Readout" Type Balancing Valve

Figure 26.2

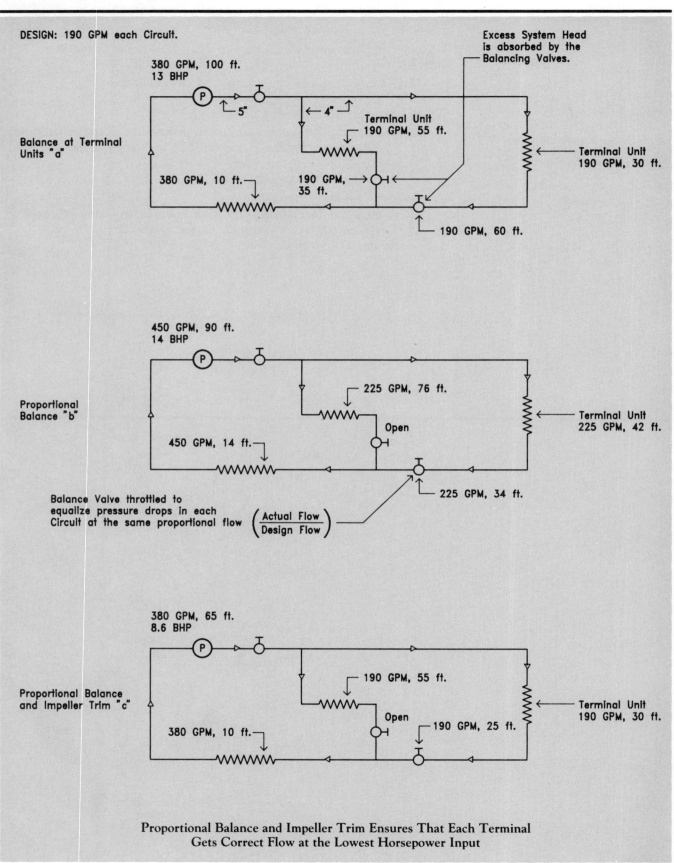

DESIGN: 190 GPM each Circuit.

380 GPM, 100 ft.
13 BHP

Excess System Head
is absorbed by the
Balancing Valves.

Balance at Terminal
Units "a"

← 5"   ← 4" →

Terminal Unit
190 GPM, 55 ft.

Terminal Unit
190 GPM, 30 ft.

380 GPM, 10 ft.

190 GPM,
35 ft.

190 GPM, 60 ft.

450 GPM, 90 ft.
14 BHP

Proportional
Balance "b"

225 GPM, 76 ft.

Open

Terminal Unit
225 GPM, 42 ft.

450 GPM, 14 ft.

225 GPM, 34 ft.

Balance Valve throttled to
equalize pressure drops in each
Circuit at the same proportional flow

$$\left(\frac{\text{Actual Flow}}{\text{Design Flow}} \right)$$

380 GPM, 65 ft.
8.6 BHP

Proportional Balance
and Impeller Trim "c"

190 GPM, 55 ft.

Open

Terminal Unit
190 GPM, 30 ft.

380 GPM, 10 ft.

190 GPM, 25 ft.

Proportional Balance and Impeller Trim Ensures That Each Terminal
Gets Correct Flow at the Lowest Horsepower Input

Figure 26.3

b. The unit with the lowest proportional flow, the last unit on the branch, should be made the "base" unit. In this way, one cannot forget which is the "base" unit.

c. Adjust the last unit on the branch to the same proportion as the unit with the lowest proportional flow. (Circuit "B" in Figure 26.1 has a proportion of 260/190 = 1.37. The flow in circuit "B" was balanced to equal the circuit "A" proportion at 225 gpm in each circuit.)

d. Balance the unit with the highest proportion to the proportion of the last unit on the branch. See "3c" above.

e. Keep checking the last unit to maintain the same flow ratio in the other units as in the last unit. The flow in the last unit will increase. (The flow in circuit "A" in Figure 26.1 increased from 190 gpm to 225 gpm.)

f. All units on a branch must have the same flow ratio of actual to design flow.

4. Balance the branches.

a. Balance all units on a branch per the instructions given in number 3 of this list.

b. Find the branch with the lowest flow proportion.

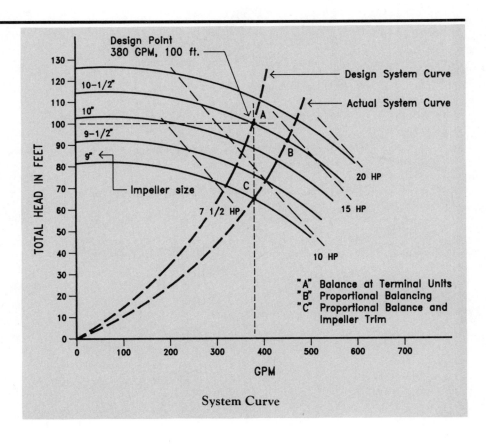

Figure 26.4

System Curve

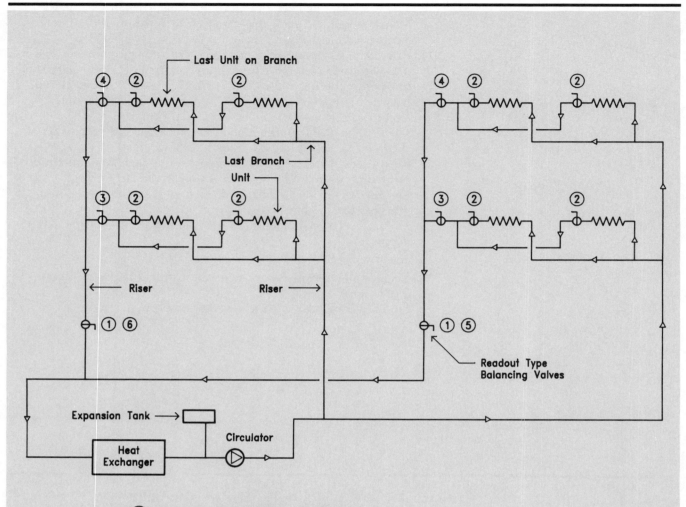

- ① Measure flow through each riser.

- ② Balance units on branch first.

- ③ Balance branches on riser with highest proportion of required flow.
  Proportion = Actual GPM/Design GPM

- ④ Balance branches with next highest proportional flow until all branches are balanced.

- ⑤ Balance risers with highest flow proportion.

- ⑥ Balance remaining risers in the order of their flow ratio.

HYDRONIC BALANCING

Hydronic Balancing

Figure 26.5

    c. Balance the last branch to the same flow ratio as the branch with the lowest flow proportion.

    d. Balance the larger flow branches first and the smaller flow branches later.

    e. All branches must have the same flow ratio of actual flow to design flow.

5. Balance the risers.

    a. Risers are balanced using a procedure similar to that used for branches.

Summary of Balancing Procedures

1. Measure flow through each riser.
2. Balance units first.
3. Balance branches on riser with highest proportion of required flow.
4. Balance the branches with the next highest proportional flow until all branches are balanced.
5. Balance risers with the highest flow proportion.
6. Balance the remaining risers in the order of their flow ratio.

Chapter Twenty-Seven:

KITCHEN EXHAUST SYSTEMS

Because of the life safety aspect of kitchen exhaust systems, it is particularly important to address certain concerns of this HVAC component.

Operational Problems

Operational problems with kitchen exhaust systems may be attributed to the following causes, which are discussed in the paragraphs that follow.

Grease Accumulation

Kitchen exhaust duct velocities should be greater than 2,000 feet per minute (fpm) to prevent grease from accumulating in the ducts. At velocities below 2,000 fpm, grease adheres to the inside walls of the exhaust duct, where it almost doubles the duct friction loss. The increased friction resistance reduces the fan air volume. Grease that is not trapped by the hood filters should be exhausted from the building, and not remain in the exhaust system.

Grease accumulation not only causes friction loss in the duct, but can also build up in the exhaust fan discharge cowl. (See Figure 27.1.) A more dangerous effect of grease build-up is the potential for fires.

Contaminated Air Re-entry

To disperse the range hood exhaust air away from the building and to prevent re-entry of the contaminated air into outside air intakes, a vertical high velocity discharge with a 2,500 fpm outlet velocity should be employed. Figure 27.2 shows the recommended exhaust fan discharge.

The system effect factor penalty for fan discharges without discharge ducts is presented in Chapter 19, "Ductwork." Exhaust fans should have 2-1/2 diameters of straight duct connected to the fan outlet in order to achieve the fan's catalogue performance. Figure 27.3 shows how installing a discharge duct on an existing horizontal discharge exhaust fan will improve the exhaust fan's performance.

Improper Fan Selection (Neglecting Fan Density)

Fan manufacturers rate their fan's performance on its ability to handle standard air. *Standard air* has a density of .075 pounds per cubic foot (lb/cf). Fans are constant volume machines, moving air at the same volume. When that volume of air is a different density than standard air volume, 0.075 lb/cf, the fan's performance is dramatically affected. The pressure and horsepower vary in direct proportion to the ratio of the gas density at the fan inlet compared to standard air density. The density ratio is:

$$\frac{\text{actual density}}{\text{standard air density}} = \frac{0.0375}{0.075} = 0.5$$

Figure 27.4 plots the fan's performance with both standard air (0.075 lb/cf) and air at 0.0375 lb/cf.

When the fan is moving 0.0375 lb/cf air (one half the density of standard air), it can develop only half the pressure that it could if it was handling standard air. The relationship between density (d), total pressure (TP), and brake horsepower (bhp) is:

$$\text{Density Ratio} = \frac{da}{ds} = \frac{TPa}{TPs} = \frac{bhpa}{bhps}$$

Since the system resistance of the lower density air is less than that of standard air, the system operates at the intersection of the 0.0375 lb/cf fan pressure curve and the 0.0375 duct system curve.

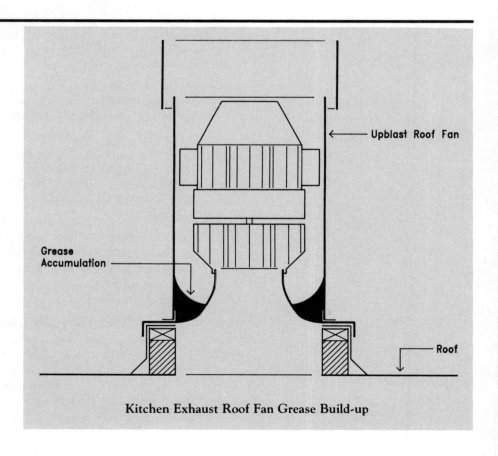

Kitchen Exhaust Roof Fan Grease Build-up

Figure 27.1

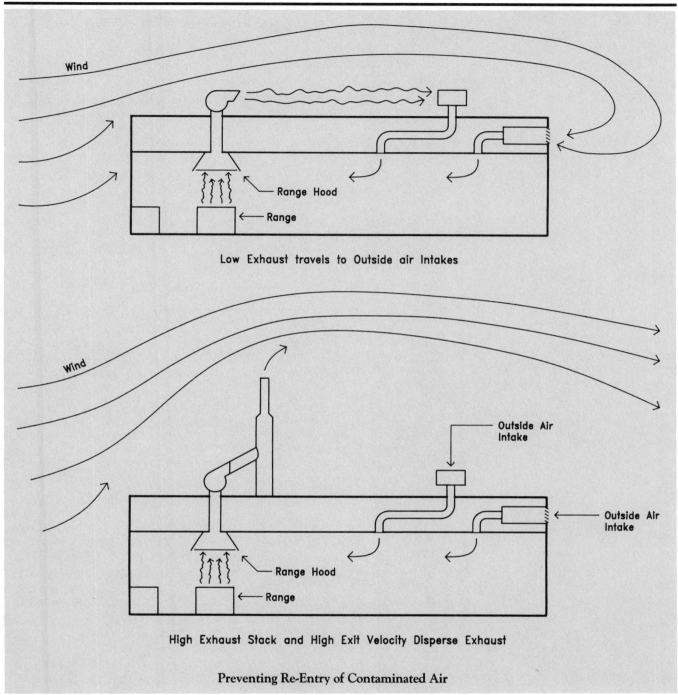

Low Exhaust travels to Outside air Intakes

High Exhaust Stack and High Exit Velocity Disperse Exhaust

Preventing Re-Entry of Contaminated Air

Figure 27.2

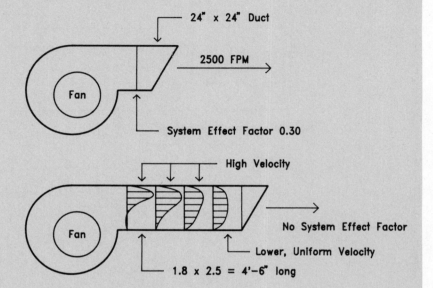

System Effect Factor:

$$\text{Duct Diameter} = \sqrt{\frac{4\ ab}{\pi}} = \sqrt{\frac{4 \times 2 \times 2}{\pi}} = 1.8'$$

Adding discharge duct 2-1/2 duct diameters long avoids System Effect Factor.

24" x 24" Duct

2500 FPM

Fan

System Effect Factor 0.30

High Velocity

Fan

No System Effect Factor

Lower, Uniform Velocity

1.8 x 2.5 = 4'-6" long

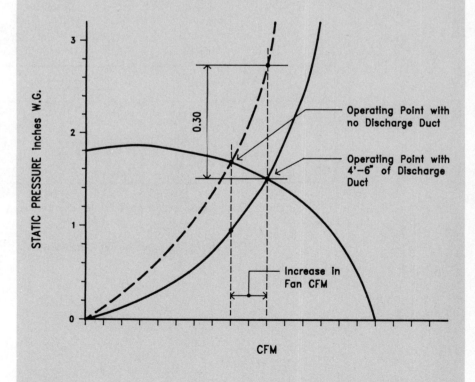

STATIC PRESSURE Inches W.G.

0.30

Operating Point with no Discharge Duct

Operating Point with 4'-6" of Discharge Duct

Increase in Fan CFM

CFM

Improving Exhaust Fan Performance by Adding Straight Duct at Fan Outlet

Figure 27.3

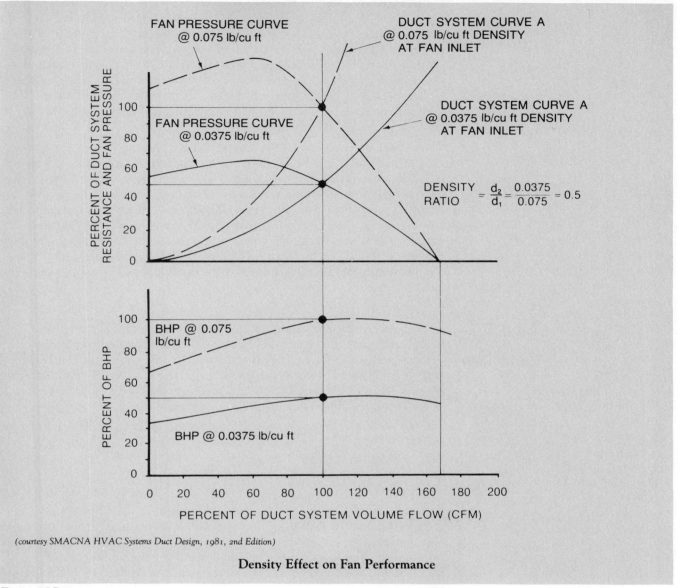

DENSITY RATIO $= \dfrac{d_2}{d_1} = \dfrac{0.0375}{0.075} = 0.5$

(courtesy SMACNA HVAC Systems Duct Design, 1981, 2nd Edition)

Density Effect on Fan Performance

Figure 27.4

Fan Brake Horsepower

The fan brake horsepower (bhp) is also directly proportional to the air density. Since the 0.0375 lb/cf air is half as dense as 0.075 lb/cf air, the bhp will be half that of the fan moving standard air. The lower fan bhp occurs only when the fan is handling a lower density air. If the fan is to take in standard air as well as low density air, the fan motor must be sized for the higher air density and higher bhp. Figure 27.5 is an example of what happens when air density is not taken into consideration in range hood exhaust system design and operation.

The system is to exhaust 10,000 cfm at a static pressure (sp) of 2″ wg at standard air conditions of 0.075 lb/cf. Because of the cooking heat from the range, the fan also handles air at 275°F and a density of 0.058 lb/cf. With the lower density air entering the fan, it can only develop 77% (0.058/0.075) of the pressure that could be achieved with standard air. The fan is now capable of exhausting only 8,000 cfm instead of 10,000 cfm. To make things worse, the 8,000 cfm entering the fan at 275°F and 0.058 lb/cf is only 0.058/0.075 x 8,000 or 6,187 cfm of 70°F standard air. Only 62% of the air that should be exhausted from the hood is actually passing across the hood face area.

Correct Hood Fan Selection

The correct method of sizing kitchen hood exhaust fans is explained below, and illustrated in Figure 27.6. To produce the desired capture velocity across the kitchen hood face area, 10,000 cfm of standard 70°F (0.075 lb/cf) air must enter the hood. Since the fan has 275°F air entering its inlet, it must exhaust 13,000 cfm of 275°F (0.058 lb/cf) air to permit 10,000 cfm of 70°F (0.075 lb/cf) air to enter the hood.

The actual cfm is inversely proportional to the air density. The actual cfm (acfm) =

> standard cfm (scfm) x ds/da acfm =
> 10,000 cfm x 0.075/0.058 = 13,000 cfm.

The 10,000 cfm of standard (0.075 lb/cf) air becomes 13,000 cfm when it is heated to 275°F (0.058 lb/cf) by the cooking surface. The lower air density reduces the duct friction to 0.76 of standard air friction. The duct friction is also increased by 1.9 because of the rough duct surface created by the grease on the duct. The fan must be capable of handling the 13,000 cfm of standard air with grease on the ducts until the air is heated to 275°F. The fan must also be able to handle 13,000 cfm of air at 275°F with grease on the ducts. The fan is a constant volume machine, so it must deliver the 13,000 cfm against the following static pressures:

> s.p. 10,000 cfm, clean duct = 2″
> grease on duct factor = 1.9
> air density correction factor for 275°F air = 0.76
> s.p.(70°F) = 2″ x 1.9 x (13,000/10,000)^2 = 6.42″
> s.p.(275°F) = 6.42 x 0.76 = 4.9″

The fan should be selected for 13,000 scfm (0.075 lb/cf) at 4.9 x 0.075/0.058 = 6.34″ s.p. This would be a Class II 30″ backward-curved (bc) single-inlet single-width (sisw) fan operating at 1,378 rpm and 16.6 bhp. A 20 hp motor must be used to handle the standard air load. When the fan is exhausting 275°F air, it will move 13,000 cfm at 4.9″ s.p. and 12.8 bhp. When starting up with 70°F air, the fan will exhaust 12,700 cfm at 6.4″ s.p. at 16.4 bhp.

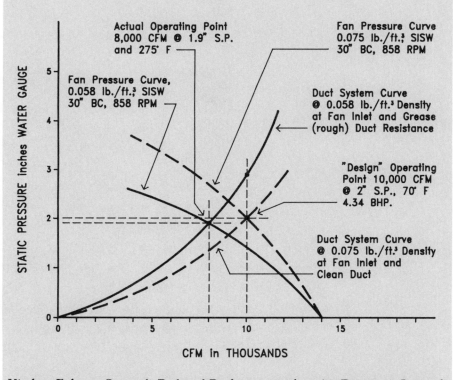

S.P. at 10,000 CFM of 70° F Air and Clean Duct = 2.0″

Air Density Friction Correction Factor =

$$Kt = \left(\frac{530}{Tact + 460} \right)^{0.825} = \left(\frac{530}{275 + 460} \right)^{0.825} = 0.76$$

Grease (rough) Duct Friction Factor = 1.9

Actual S.P. at 275° F and Grease Duct = 2″ x 0.76 x 1.9 = 2.9

$$Density\ Ratio = \frac{0.058}{0.075} = 0.77$$

8,000 CFM at 275° F at Fan Inlet is
8,000 x 0.77 = 6156 CFM at the Hood
6,156/8,000 = .62 ∴ 62%

Actual Operating Point
8,000 CFM @ 1.9″ S.P.
and 275° F

Fan Pressure Curve
0.075 lb./ft.³ SISW
30″ BC, 858 RPM

Fan Pressure Curve,
0.058 lb./ft.³ SISW
30″ BC, 858 RPM

Duct System Curve
@ 0.058 lb./ft.³ Density
at Fan Inlet and Grease
(rough) Duct Resistance

"Design" Operating
Point 10,000 CFM
@ 2″ S.P., 70° F
4.34 BHP.

Duct System Curve
@ 0.075 lb./ft.³ Density
at Fan Inlet and
Clean Duct

STATIC PRESSURE inches WATER GAUGE

CFM in THOUSANDS

Kitchen Exhaust System's Reduced Performance when Air Density is Ignored

Figure 27.5

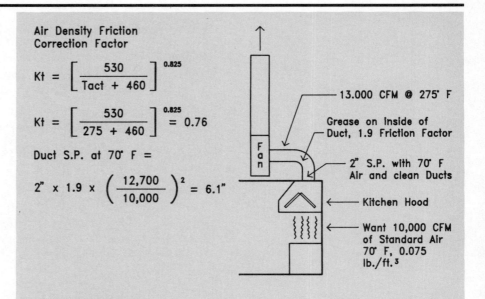

Air Density Friction
Correction Factor

$$Kt = \left[\frac{530}{Tact + 460} \right]^{0.825}$$

$$Kt = \left[\frac{530}{275 + 460} \right]^{0.825} = 0.76$$

Duct S.P. at 70° F =

$$2" \times 1.9 \times \left(\frac{12,700}{10,000} \right)^2 = 6.1"$$

13.000 CFM @ 275° F

Grease on Inside of
Duct, 1.9 Friction Factor

2" S.P. with 70° F
Air and clean Ducts

Kitchen Hood

Want 10,000 CFM
of Standard Air
70° F, 0.075
lb./ft.³

$$\text{Density Ratio} = \frac{\text{Actual Air 70° F } (0.058 \text{ lb./ft.}^3)}{\text{Standard Air 70° F } (0.075 \text{ lb./ft.}^3)} = 0.77$$

$$\text{Actual CFM} = \text{Standard CFM} \times ds/da = 10,000 \times \frac{0.75}{0.58} =$$
13,000 CFM @ 275° F

Actual S.P. due to 13,000 CFM of 275° F =

$$2" \times 1.9 \times 0.76 \times \left(\frac{13,000}{10,000} \right)^2 = 4.9"$$

Fan Tables are based on Standard Air, 70° F 0.075 lb./ft.³

Fan Table S.P. = 4.9/0.77 = 6.34" S.P.

Select fan for 13,000 CFM @ 6.34" S.P., 0.075 lb./ft.³ SISW,
      B.C. 30", 1,378 RPM, 16.6 BHP, 20 HP motor.

BHP @ 275° F = 16.6 × 0.77 = 12.8 BHP.

Fan handles 70° F air before cooking heats it to 275° F.

Fan must handle 70° F, 0.075 lb./ft.³ air at 1,378 RPM, 12,700 CFM @
      6.1" S.P., 16.4 BHP, 20 HP motor.

Correct Fan Selection for Kitchen Hood and Grease Duct with 275°F Air

Figure 27.6

352

How to Increase Kitchen Exhaust Performance

This section addresses the problem of improving kitchen exhaust performance in situations where the initial fan selection process did not properly consider the air density factor. Figure 27.7 shows that the exhaust fan selected in Figure 27.5 could be a 30″ backwardly inclined single inlet single width fan rated at 10,000 cfm, 2″ s.p., 858 rpm, and 4.34 bhp. The fan would have Class I construction with a maximum rpm of 1,273.

Figure 27.8 shows an alternate selection of a ventilating set with similar performance characteristics, except for a 5″ maximum s.p.

In order to produce 13,000 cfm at 6.34″ s.p., the fan in Figure 27.7 would have to be operated at 1,378 rpm and would draw 16.6 bhp. The fan would have to manage 105 rpm more than the 1,273 maximum rpm rating. The fan structure would have to be modified to Class 2 construction. Class 2 construction for a 30″ fan is 12-gauge scroll back in lieu of 14-gauge, 10-gauge housing side in lieu of 12-gauge, and a 1-15/16″ shaft in lieu of 1-11/16″. This modification would necessitate a fan replacement. The fan manufacturer should be consulted for the possibility of welding reinforcing angles to the scroll back and housing sides of the existing fan. It is not recommended that the ventilating set in Figure 27.8 operate that far beyond its maximum rating. The ventilating set should be replaced with a Class 2 fan, or as Figure 27.9 demonstrates, an additional ventilating set could be installed in series with the original fan.

It is a good idea to try to have each of the two fans in series operate at the same rpm in the Class 1 construction category. This will simplify maintenance of belts, drives, bearings and motors. The existing fan's rpm should be increased so that each of the two equal fans in series will deliver 13,000 cfm at half of the total s.p. of 6.34. When operating together, the existing and the new 30″ sisw bc Class 1 fans will operate at 1106 rpm and deliver 13,000 scfm at 3.17″ s.p. at 9.4 bhp, requiring a 15 hp motor. Having two fans in series also provides redundancy. Should one fan be out of service, the other fan will exhaust 11,500 cfm of 275°F air.

Hood Design

As Figure 27.10 indicates, the face area of a hood is the vertical distance from the bottom of the hood to the cooking surface. ASHRAE recommends a design face velocity of 75-100 fpm, with 60 fpm as the absolute minimum to produce the "capture" velocity to prevent contaminated air from entering the occupied space. Figure 27.11 points out how blanking off unused sides of the hood can reduce the exhaust requirements.

The lower exhaust cfm can dramatically reduce the operating cost. Not only is the exhaust fan motor power reduced, but the cost of transporting, heating, and cooling the make-up air is reduced. Safing off part of the exhaust hood can be a simple solution to an inadequately performing hood exhaust system. Figure 27.12 shows how reducing the exhaust requirements may make it possible for the existing hood exhaust fan to maintain a 60 fpm hood face velocity.

Makeup Air

All of the air that is exhausted from the kitchen, including that from range hoods and dishwashers, must be replaced by outside air. Figure 27.13 shows that outside air must be introduced into the kitchen at a blended temperature above 65°F, in order to avoid cold drafts that are objectionable to the staff.

Air entering the hood below 60°F may coagulate grease on the filters and create a high filter pressure drop. The kitchen exhaust fan can only take

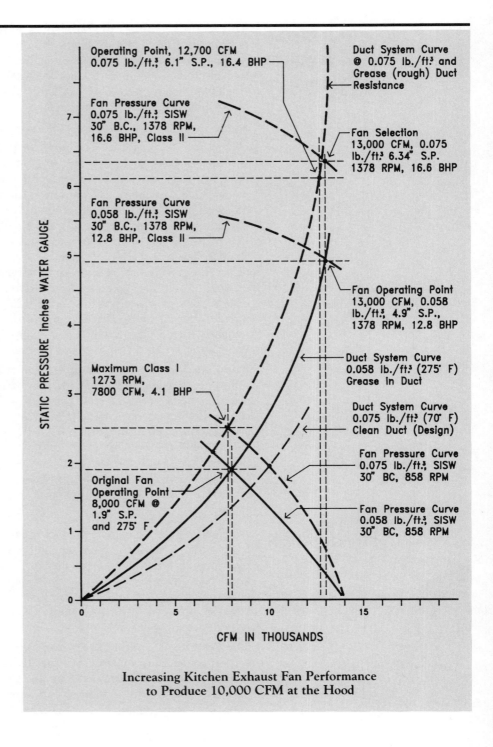

Operating Point, 12,700 CFM
0.075 lb./ft.³, 6.1" S.P., 16.4 BHP

Fan Pressure Curve
0.075 lb./ft.³ SISW
30" B.C., 1378 RPM,
16.6 BHP, Class II

Fan Pressure Curve
0.058 lb./ft.³ SISW
30" B.C., 1378 RPM,
12.8 BHP, Class II

Maximum Class I
1273 RPM,
7800 CFM, 4.1 BHP

Original Fan
Operating Point
8,000 CFM @
1.9" S.P.
and 275° F

Duct System Curve
@ 0.075 lb./ft.³ and
Grease (rough) Duct
Resistance

Fan Selection
13,000 CFM, 0.075
lb./ft.³ 6.34" S.P.
1378 RPM, 16.6 BHP

Fan Operating Point
13,000 CFM, 0.058
lb./ft.³, 4.9" S.P.,
1378 RPM, 12.8 BHP

Duct System Curve
0.058 lb./ft.³ (275° F)
Grease in Duct

Duct System Curve
0.075 lb./ft.³ (70° F)
Clean Duct (Design)

Fan Pressure Curve
0.075 lb./ft.³ SISW
30" BC, 858 RPM

Fan Pressure Curve
0.058 lb./ft.³ SISW
30" BC, 858 RPM

STATIC PRESSURE Inches WATER GAUGE

CFM IN THOUSANDS

Increasing Kitchen Exhaust Fan Performance
to Produce 10,000 CFM at the Hood

Figure 27.7

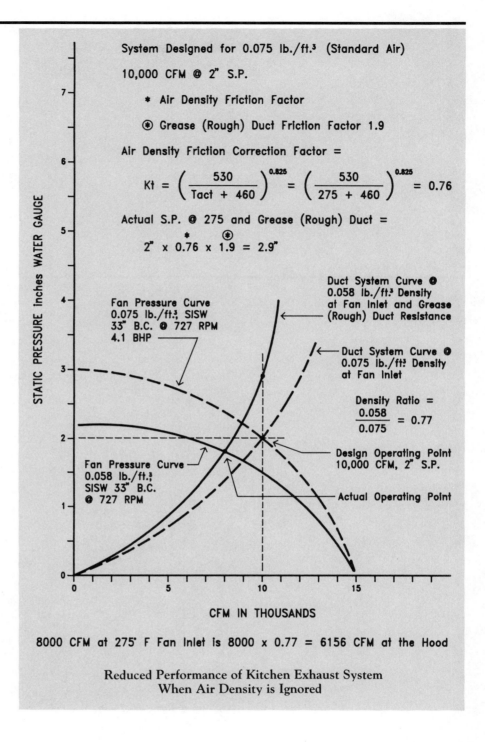

System Designed for 0.075 lb./ft.³ (Standard Air)

10,000 CFM @ 2" S.P.

＊ Air Density Friction Factor

⊕ Grease (Rough) Duct Friction Factor 1.9

Air Density Friction Correction Factor =

$$Kt = \left(\frac{530}{Tact + 460} \right)^{0.825} = \left(\frac{530}{275 + 460} \right)^{0.825} = 0.76$$

Actual S.P. @ 275 and Grease (Rough) Duct =

2" × 0.76 × 1.9 = 2.9"

Duct System Curve ⊕ 0.058 lb./ft.³ Density at Fan Inlet and Grease (Rough) Duct Resistance

Fan Pressure Curve 0.075 lb./ft.³ SISW 33" B.C. @ 727 RPM 4.1 BHP

Duct System Curve ⊕ 0.075 lb./ft.³ Density at Fan Inlet

Density Ratio = $\frac{0.058}{0.075}$ = 0.77

Design Operating Point 10,000 CFM, 2" S.P.

Fan Pressure Curve 0.058 lb./ft.³ SISW 33" B.C. @ 727 RPM

Actual Operating Point

CFM IN THOUSANDS

STATIC PRESSURE Inches WATER GAUGE

8000 CFM at 275° F Fan Inlet is 8000 × 0.77 = 6156 CFM at the Hood

**Reduced Performance of Kitchen Exhaust System
When Air Density is Ignored**

Figure 27.8

out of the kitchen the amount of outside air that enters the kitchen. If the outside air is not delivered directly to the kitchen, but must move through grease-lined filters, negative pressures below 0.02″ wg in the kitchen can create combustion problems with gas-fired cooking equipment.

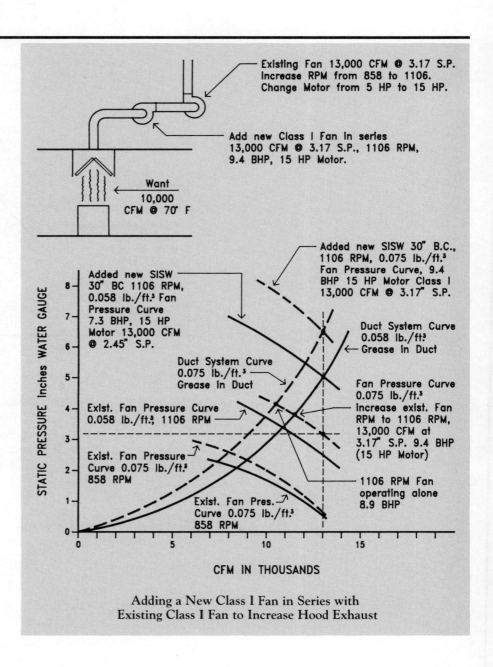

Adding a New Class I Fan in Series with
Existing Class I Fan to Increase Hood Exhaust

Figure 27.9

Figure 27.14 shows how outside air make-up discharged from the hood face close to the bottom of the hood entrains air from the cooking surface in a "roll out" pattern. Air discharged from the ceiling can be uncomfortably cold to the staff. Outside air injected into the hood does nothing to produce the hood face "capture" velocity, and if it is below 60°F, it can create grease coagulation on the filters.

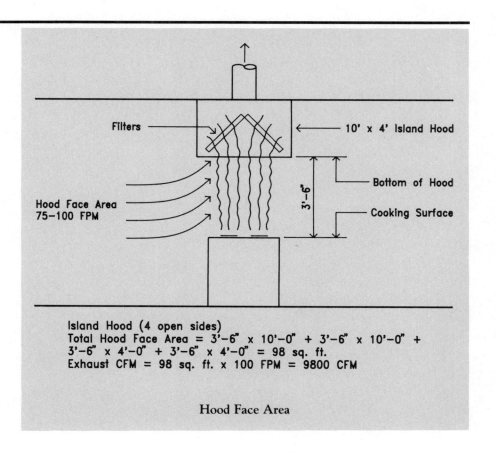

Island Hood (4 open sides)
Total Hood Face Area = 3'-6" x 10'-0" + 3'-6" x 10'-0" + 3'-6" x 4'-0" + 3'-6" x 4'-0" = 98 sq. ft.
Exhaust CFM = 98 sq. ft. x 100 FPM = 9800 CFM

Hood Face Area

Figure 27.10

357

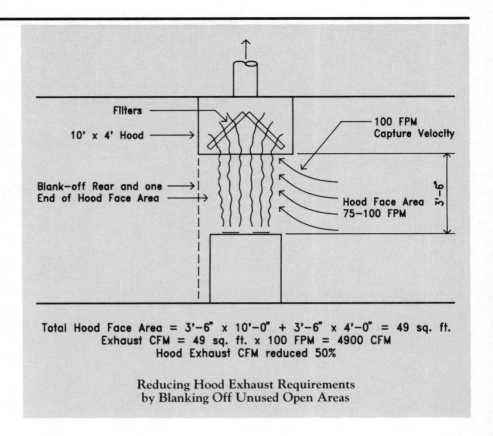

Total Hood Face Area = 3'-6" x 10'-0" + 3'-6" x 4'-0" = 49 sq. ft.
Exhaust CFM = 49 sq. ft. x 100 FPM = 4900 CFM
Hood Exhaust CFM reduced 50%

**Reducing Hood Exhaust Requirements
by Blanking Off Unused Open Areas**

Figure 27.11

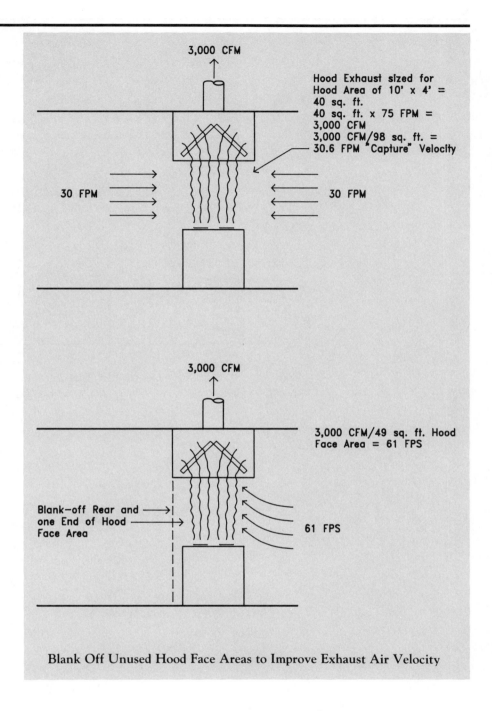

3,000 CFM

Hood Exhaust sized for
Hood Area of 10' x 4' =
40 sq. ft.
40 sq. ft. x 75 FPM =
3,000 CFM
3,000 CFM/98 sq. ft. =
30.6 FPM "Capture" Velocity

30 FPM   30 FPM

3,000 CFM

3,000 CFM/49 sq. ft. Hood
Face Area = 61 FPS

Blank—off Rear and
one End of Hood
Face Area

61 FPS

Blank Off Unused Hood Face Areas to Improve Exhaust Air Velocity

Figure 27.12

359

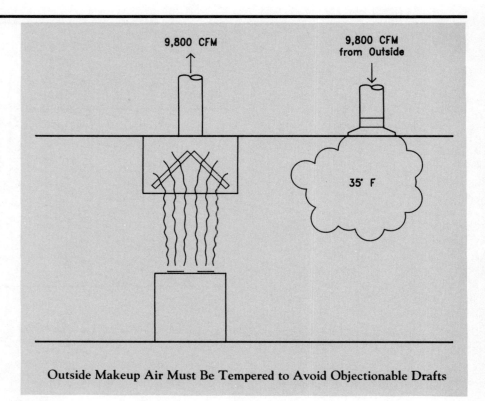

Outside Makeup Air Must Be Tempered to Avoid Objectionable Drafts

Figure 27.13

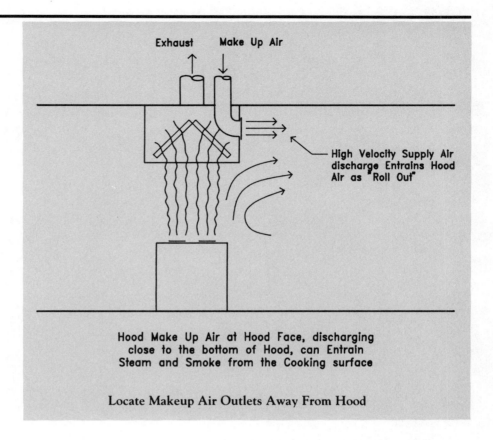

Exhaust   Make Up Air

High Velocity Supply Air
discharge Entrains Hood
Air as "Roll Out"

Hood Make Up Air at Hood Face, discharging
close to the bottom of Hood, can Entrain
Steam and Smoke from the Cooking surface

Locate Makeup Air Outlets Away From Hood

Figure 27.14

Chapter Twenty-Eight:

REFRIGERATION PIPING

Piping should be type "L" nitrogen-charged and sealed, and connected with wrought copper fittings. Pre-sealed piping keeps air, dirt, and moisture out of the pipe sections that are stored at the construction site before installation. Installed pipe sections must have their open pipe ends sealed to prevent dirt and moisture intrusion. One drop of moisture in the refrigerant system will create corrosion and operational problems, such as ice blockage at the expansion valve.

Minimizing Oxidation

Field-fabricated pipe joints should be made with dry nitrogen passing through the piping during the silver soldering process. This will prevent the formation of copper oxide scale that is produced on the pipe interior during soldering. The scale is difficult to remove and any scale retained in an operating refrigeration system could impede or stop refrigerant flow.

Pipe Joints

All copper pipe connections should be made with silver solder, not 50/50, 95/5, or other soft solders. Joints soldered with 50/50 solder cannot withstand the temperature and pressure developed by the hot refrigerant gas. Chase Brass and Copper Company's maximum pressure rating for 50/50 solder joints at 150°-200°F is 95 psi. Freon 22 hot gas lines can reach 300 psi and 200°F.

It is not practical to use soft solder on suction lines and silver solder on liquid and hot gas lines on the same job, as this can lead to accidental mixing of the two. All freon 22 piping should be silver-soldered.

Oil Return

Hot gas and suction risers must be sized for oil return to the compressor. The risers must be sized for the minimum refrigerant velocity at the minimum refrigerant flow. Minimum refrigerant flow occurs at low load, when the compressors are unloaded. Too low a refrigerant gas velocity will prevent oil from traveling up the riser with the refrigerant. Oil remaining in the piping does not supply adequate lubrication for the compressor and the result is early compressor failure.

There are two methods of piping hot gas and suction risers for oil return: the single riser and double riser techniques.

Example

The following examples show how to calculate the piping required for hot gas and suction risers for oil return. We will show the single suction riser method first, followed by the double riser method. Assume these conditions:

> 20 ton maximum load, 5 ton minimum load, F22, 40°F saturated gas temperature, 50°F refrigerant suction gas temperature, 105°F condensing temperature

Single Suction Riser Method (See Figure 28.1)

1. Determine the minimum refrigerant load that the riser will handle. For the 20-ton system, the cylinder unloaders reduce the capacity to 5 tons.

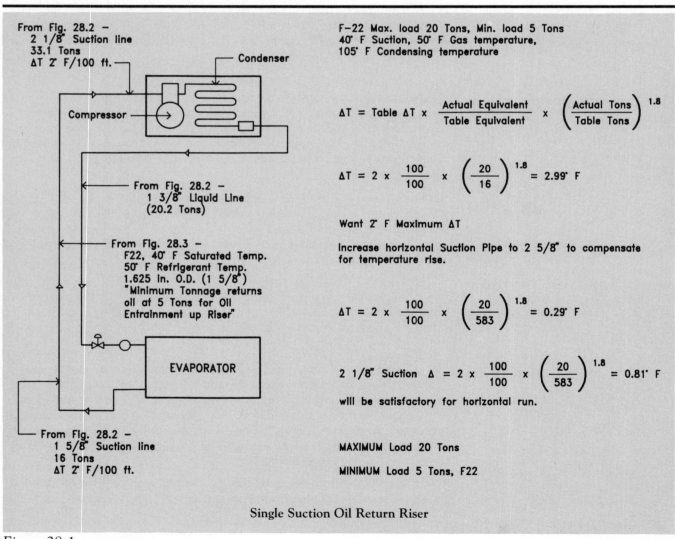

From Fig. 28.2 –
2 1/8" Suction line
33.1 Tons
ΔT 2° F/100 ft.

Condenser

Compressor

From Fig. 28.2 –
1 3/8" Liquid Line
(20.2 Tons)

From Fig. 28.3 –
F22, 40° F Saturated Temp.
50° F Refrigerant Temp.
1.625 in. O.D. (1 5/8")
"Minimum Tonnage returns oil at 5 Tons for Oil Entrainment up Riser"

EVAPORATOR

From Fig. 28.2 –
1 5/8" Suction line
16 Tons
ΔT 2° F/100 ft.

F-22 Max. load 20 Tons, Min. load 5 Tons
40° F Suction, 50° F Gas temperature,
105° F Condensing temperature

$$\Delta T = \text{Table } \Delta T \times \frac{\text{Actual Equivalent}}{\text{Table Equivalent}} \times \left(\frac{\text{Actual Tons}}{\text{Table Tons}}\right)^{1.8}$$

$$\Delta T = 2 \times \frac{100}{100} \times \left(\frac{20}{16}\right)^{1.8} = 2.99° \text{ F}$$

Want 2° F Maximum ΔT

Increase horizontal Suction Pipe to 2 5/8" to compensate for temperature rise.

$$\Delta T = 2 \times \frac{100}{100} \times \left(\frac{20}{583}\right)^{1.8} = 0.29° \text{ F}$$

$$2 \ 1/8" \text{ Suction } \Delta = 2 \times \frac{100}{100} \times \left(\frac{20}{583}\right)^{1.8} = 0.81° \text{ F}$$

will be satisfactory for horizontal run.

MAXIMUM Load 20 Tons

MINIMUM Load 5 Tons, F22

Single Suction Oil Return Riser

Figure 28.1

2. From Figure 28.2, select the suction line pipe size for the maximum load of 20 tons, 40°F suction temperature, 2 degree delta "T." A 2-1/8″ pipe will handle 33.1 tons.

3. Select the largest pipe that will return oil up the suction riser at the minimum load of 5 tons. From Figure 28.3, F22, 40°F saturated suction temperature, 50°F suction gas temperature, a 1.625″ o.d. pipe will move oil at a 3.81-ton load, which is lower than the 5-ton minimum load. If the refrigerant velocity in the 1.625″ pipe is high enough to move oil at a 3.81-ton load, then the velocity will be even higher at a 5-ton load. The oil return requirement is satisfactory.

4. At the maximum load of 20 tons, the 1.625″ o.d. suction riser will increase the pressure drop due to friction to an equivalent of 3°F/100′. Systems are designed for 2°F/100′. To compensate for the increased pressure drop of the smaller suction riser, the horizontal and vertical suction pipes should have a lower equivalent pressure loss than 2°F/100′. A 2.125″ o.d. suction pipe carries a 20-ton refrigerant load at an 0.80°F/100′ pressure drop. This more than offsets the greater pressure drop of the 1.625″ pipe. The single suction oil return riser will usually satisfy oil transport velocities and pressure drops at capacity reductions to 25% of maximum load.

Double Suction Riser Method (Figure 28.4)

When multiple compressors with capacity control operate in parallel, a vertical suction line sized to transport oil will have an excessive pressure drop at maximum load. Multiple compressors connected in parallel can unload to a system capacity of 12.5% or lower. These systems require double gas risers.

In our example, riser "A" is sized at 1.625″ (1.78 square inches) to return oil at the minimum 5-ton load.

1. The area of "A" + "B" should be the same or greater than the area of a single pipe with a satisfactory pressure drop. A 2.125″ o.d. (3.094 square inches) suction pipe transports 33.1 tons of F22 at 40°F saturated suction temperature at a 2°F/100′ equivalent pressure drop. 3.094 − 1.78 = 1.314 square inches is the minimum area of riser "B."

   The pipe with the closest area that is larger than 1.314 square inches is the 1.625″ o.d. pipe, with an area of 1.78 square inches.

Table 2 Refrigerant Suction, Discharge and Liquid Line Capacities in Tons for Refrigerant 22
(Single or High Stage Applications)

[Tons of Refrigeration Resulting in a Line Friction Drop (ΔP in psi) per 100 ft Equivalent Pipe Length as Shown, with Corresponding Change (ΔT) in Saturation Temp.]

Line Size Type L Copper, OD	Suction Lines ΔT = 2 deg F					Discharge Lines ΔT = 1 deg F ΔP = 3.05		Line Size Type L Copper, OD	Liquid Lines[a]	
	Saturated Suction Temp, F					Saturated Suction Temp, F			Velocity = 100 fpm	ΔT = 1 deg F ΔP = 3.05 psi
	−40 ΔP = 0.79	−20 ΔP = 1.15	0 ΔP = 1.6	20 ΔP = 2.22	40 ΔP = 2.91	−40	40			
1/2	—	—	—	0.40	0.6	0.75	0.85	1/2	2.3	3.6
5/8	—	0.32	0.51	0.76	1.1	1.4	1.6	5/8	3.7	6.7
7/8	0.52	0.86	1.3	2.0	2.9	3.7	4.2	7/8	7.8	18.2
1 1/8	1.1	1.7	2.7	4.0	5.8	7.5	8.5	1 1/8	13.2	37.0
1 3/8	1.9	3.1	4.7	7.0	10.1	13.1	14.8	1 3/8	20.2	64.7
1 5/8	3.0	4.8	7.5	11.1	16.0	20.7	23.4	1 5/8	28.5	102.5
2 1/8	6.2	10.0	15.6	23.1	33.1	42.8	48.5	2 1/8	49.6	213.0
2 5/8	10.9	17.8	27.5	40.8	58.3	75.4	85.4	2 5/8	76.5	376.9
3 1/8	17.5	28.4	44.0	65.0	92.9	120.2	136.2	3 1/8	109.2	601.5
3 5/8	26.0	42.3	65.4	96.6	137.8	178.4	202.1	3 5/8	147.8	895.7
4 1/8	36.8	59.6	92.2	136.3	194.3	251.1	284.4	4 1/8	192.1	1263.2

Steel IPS	SCH								Steel IPS	SCH		
1/2	40	—	0.38	0.58	0.85	1.2	1.5	1.7	1/2	80	3.8	5.7
3/4	40	0.50	0.8	1.2	1.8	2.5	3.3	3.7	3/4	80	6.9	12.8
1	40	0.95	1.5	2.3	3.4	4.8	6.1	6.9	1	80	11.5	25.2
1 1/4	40	2.0	3.2	4.8	7.0	9.9	12.6	14.3	1 1/4	80	20.6	54.1
1 1/2	40	3.0	4.7	7.2	10.5	14.8	19.0	21.5	1 1/2	80	28.3	82.6
2	40	5.7	9.1	13.9	20.2	28.5	36.6	41.4	2	40	53.8	192.0
2 1/2	40	9.2	14.6	22.1	32.2	45.4	58.1	65.9	2 1/2	40	76.7	305.8
3	40	16.2	25.7	39.0	56.8	80.1	102.8	116.4	3	40	118.5	540.3
4	40	33.1	52.5	79.5	115.9	163.2	209.5	237.3	4	40	204.2	1101.2

NOTES:

(1) For other ΔT's and Equivalent Lengths, L_e, Line Capacity (Tons) =

$$\text{Table Tons} \cdot \left(\frac{\text{Table } L_e}{\text{Actual } L_e} \cdot \frac{\text{Actual } \Delta T \text{ Loss Desired}}{\text{Table } \Delta T \text{ Loss}} \right)^{0.55}$$

(2) For other Tons and Equivalent Lengths in a given pipe size

$$\Delta T = \text{Table } \Delta T \cdot \frac{\text{Actual } L_e}{\text{Table } L_e} \cdot \left(\frac{\text{Actual Tons}}{\text{Table Tons}} \right)^{1.8}$$

(3) Values are based on 105 F condensing temperature. For capacities at other condensing temperatures, multiply table tons by the following factors:

Condensing Temp, F	Suction Lines	Hot Gas Lines
80	1.11	0.79
90	1.07	0.88
100	1.03	0.95
110	0.97	1.04
120	0.90	1.10
130	0.86	1.18
140	0.89	1.26

(Reprinted by permission from 1986 ASHRAE Handbook – Refrigeration)

Figure 28.2

Table 11 Minimum Tonnage for Oil Entrainment up Suction Risers Type L Copper Tubing

Refriger-ant	Sat. Temp, F	Suction Gas Temp, F	0.500	0.625	0.750	0.875	1.123	1.375	1.625	2.125	2.625	3.125	3.625	4.125
			Pipe OD, in.											
			0.146	0.233	0.348	0.484	0.825	1.256	1.780	3.094	4.770	6.812	9.213	11.970
12	−40.0	−30.0	0.045	0.061	0.133	0.201	0.391	0.662	1.02	2.04	3.51	5.48	7.99	11.1
		−10.0	0.044	0.078	0.130	0.196	0.381	0.645	0.997	1.99	3.42	5.34	7.78	10.8
		10.0	0.044	0.080	0.132	0.199	0.388	0.655	1.01	2.02	3.47	5.42	7.91	11.0
	−20.0	−10.0	0.059	0.106	0.175	0.264	0.513	0.868	1.34	2.68	4.60	7.19	10.5	14.5
		10.0	0.058	0.103	0.171	0.258	0.503	0.850	1.31	2.62	4.51	7.04	10.3	14.2
		30.0	0.059	0.105	0.173	0.262	0.510	0.863	1.33	2.66	4.57	7.14	10.4	14.4
	0.0	10.0	0.077	0.139	0.229	0.345	0.673	1.14	1.76	3.51	6.03	9.42	13.7	19.1
		30.0	0.075	0.134	0.221	0.334	0.650	1.10	1.70	3.39	5.82	9.09	13.3	18.4
		50.0	0.075	0.135	0.223	0.337	0.657	1.11	1.72	3.43	5.89	9.19	13.4	18.6
	20.0	30.0	0.094	0.169	0.279	0.421	0.820	1.39	2.14	4.28	7.35	11.5	16.7	23.2
		50.0	0.095	0.170	0.280	0.423	0.825	1.39	2.16	4.30	7.39	11.5	16.8	23.4
		70.0	0.095	0.170	0.281	0.425	0.828	1.40	2.17	4.32	7.42	11.6	16.9	23.4
	40.0	50.0	0.121	0.217	0.358	0.541	1.05	1.78	2.76	5.50	9.45	14.8	21.5	29.8
		70.0	0.117	0.210	0.347	0.524	1.02	1.73	2.67	5.33	9.16	14.3	20.8	28.9
		90.0	0.117	0.211	0.348	0.526	1.02	1.73	2.68	5.34	9.18	14.3	20.9	29.0
22	−40.0	−30.0	0.067	0.119	0.197	0.298	0.580	0.981	1.52	3.03	5.20	8.12	11.8	16.4
		−10.0	0.065	0.117	0.194	0.292	0.570	0.963	1.49	2.97	5.11	7.97	11.6	16.1
		10.0	0.066	0.118	0.195	0.295	0.575	0.972	1.50	3.00	5.15	8.04	11.7	16.3
	−20.0	−10.0	0.087	0.156	0.258	0.389	0.758	1.28	1.98	3.96	6.80	10.6	15.5	21.5
		10.0	0.085	0.153	0.253	0.362	0.744	1.26	1.95	3.88	6.67	10.4	15.2	21.1
		30.0	0.086	0.154	0.254	0.383	0.747	1.26	1.95	3.90	6.69	10.4	15.2	21.1
	0.0	10.0	0.111	0.199	0.328	0.496	0.986	1.63	2.53	5.04	8.66	13.5	19.7	27.4
		30.0	0.108	0.194	0.320	0.484	0.942	1.59	2.46	4.92	8.45	13.2	19.2	26.7
		50.0	0.109	0.195	0.322	0.486	0.946	1.60	2.47	4.94	8.48	13.2	19.3	26.8
	20.0	30.0	0.136	0.244	0.403	0.608	1.18	2.00	3.10	6.18	10.6	16.6	24.2	33.5
		50.0	0.135	0.242	0.399	0.603	1.17	1.99	3.07	6.13	10.5	16.4	24.0	33.3
		70.0	0.135	0.242	0.400	0.605	1.18	1.99	3.08	6.15	10.6	16.5	24.0	33.3
	40.0	50.0	0.167	0.300	0.495	0.748	1.46	2.46	3.81	7.60	13.1	20.4	29.7	41.3
		70.0	0.165	0.296	0.488	0.737	1.44	2.43	3.75	7.49	12.9	20.1	29.3	40.7
		90.0	0.165	0.296	0.488	0.738	1.44	2.43	3.76	7.50	12.9	20.1	29.3	40.7
502	−40.0	−30.0	0.051	0.092	0.152	0.230	0.447	0.756	1.17	2.33	4.01	6.26	9.13	12.7
		−10.0	0.053	0.095	0.157	0.237	0.461	0.779	1.21	2.41	4.13	6.45	9.41	13.1
		10.0	0.055	0.098	0.163	0.246	0.476	0.809	1.25	2.50	4.29	6.39	9.76	13.5
	−20.0	−10.0	0.068	0.122	0.201	0.303	0.591	0.999	1.54	3.08	5.30	8.27	12.1	16.7
		10.0	0.070	0.125	0.207	0.312	0.608	1.03	1.59	3.17	5.45	8.51	12.4	17.2
		30.0	0.072	0.129	0.213	0.322	0.627	1.06	1.64	3.27	5.62	8.78	12.8	17.8
	0.0	10.0	0.087	0.157	0.259	0.391	0.761	1.29	1.99	3.97	6.82	10.6	15.5	21.5
		30.0	0.089	0.160	0.264	0.399	0.777	1.31	2.03	4.05	6.96	10.9	15.9	22.0
		50.0	0.092	0.165	0.273	0.412	0.802	1.36	2.10	4.19	7.19	11.2	16.4	22.7
	20.0	30.0	0.110	0.197	0.325	0.491	0.957	1.62	2.50	4.99	8.58	13.4	19.5	27.1
		50.0	0.112	0.201	0.331	0.501	0.975	1.65	2.55	5.09	8.74	13.6	19.9	27.6
		70.0	0.115	0.207	0.342	0.516	1.01	1.70	2.63	5.25	9.02	14.1	20.5	28.5
	40.0	50.0	0.136	0.243	0.401	0.606	1.18	2.00	3.09	6.16	10.6	16.5	24.1	33.4
		70.0	0.138	0.247	0.408	0.616	1.20	2.03	3.14	6.28	10.8	16.8	24.5	34.0
		90.0	0.142	0.254	0.420	0.634	1.23	2.09	3.23	6.44	11.1	17.3	25.2	35.0

NOTE: The tonnage is based on 90 F liquid temperature and superheat as indicated by the listed temperature. For other liquid line temperatures, use correction factors in the table below.

Refrigerant	Liquid Temperature, F								
	50	60	70	80	100	110	120	130	140
12	1.17	1.13	1.09	1.04	0.96	0.91	0.87	0.81	0.76
22	1.17	1.14	1.10	1.06	0.98	0.94	0.89	0.85	0.80
502	1.24	1.18	1.12	1.06	0.94	0.87	0.81	0.74	0.67

(Reprinted by permission from 1986 ASHRAE Handbook – Refrigeration)

Figure 28.3

2. The combined area of "A" + "B" should not be greater than a single pipe that will return oil under maximum load.

   A 2.625" o.d. (4.77 square inches) suction pipe will return oil up a riser at a 13.1-ton minimum load. This is less than the 20-ton maximum load, and will function properly. The area of "A" (1.78 square inches) plus the area of "B" (1.78 square inches) is 3.56 square inches, which is less than 4.77 square inches of the 2.625" o.d. pipe that will ensure oil return.

3. During part load, the gas velocity will not return oil through both risers. The trap fills with oil until "B" is sealed off. Gas travels up "A" only, with enough velocity to return the oil. As Figures 28.1 and 28.4 show, the single oil return riser is easier to design and install than double gas risers.

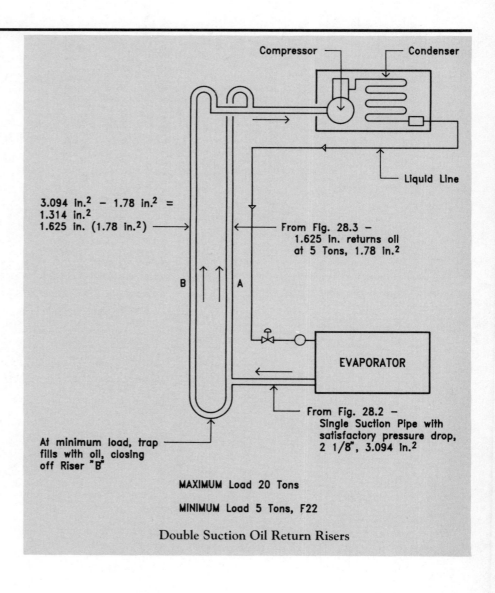

Double Suction Oil Return Risers

Figure 28.4

Hot Gas Risers

Minimum hot gas loads usually occur at a saturated condensing gas temperature of 90°F and a discharge gas temperature of 120°F. Like suction risers, hot gas risers must also be sized for minimum gas velocities to ensure oil return. See Figure 28.5, ASHRAE's chart for "Minimum Tonnage for Oil Entrainment up Hot Gas Risers."

As with suction risers, single suction hot gas risers usually suffice for oil return at reduced capacities down to 25%. When multiple compressors operate in parallel, double hot gas risers or single risers with an oil separator are needed in order to avoid excessive pressure loss.

Single Hot Gas Method (Figure 28.6)

1. The compressor unloaders reduce the refrigerant capacity to 25% of the maximum load, for a minimum hot gas load of 5 tons.
2. Select the maximum hot gas pipe size from Figure 28.5. Using a 1°F temperature differential results in a 1-5/8" pipe that will handle 23.4 tons.
3. The largest pipe size that will return oil up the hot gas riser at the minimum load of 5 tons is found in Figure 28.5.

   Using F22, 90°F saturated gas temperature and 120°F gas temperature, the 1.375" pipe is found to return oil at 3.56 tons. It will therefore return oil when 5 tons of cooling capacity is circulating through the pipe. From Figure 28.5, the 1.78" pipe will return oil at a minimum cooling capacity of 5.5 tons. Oil will not be entrained when only 5 tons of cooling capacity move through the 1.78" pipe. A 1.375" pipe must be installed to ensure oil return to the compressor.

   If excessive pressure drop would result because of the smaller pipe riser, the horizontal pipe should be one pipe size larger than the pipe size for maximum flow. The following formula will determine if the 2°F maximum temperature rise is exceeded.

1-3/8" pipe delta "T" = Table delta "T" x actual equivalent length per 100'/table equivalent length per 100' x (actual tons/table tons):

1-3/8" delta "T" = 1 x 1/1 x (20/14.8) = 1.72°F. Two degrees is the maximum corresponding temperature drop used in refrigerant hot gas piping practice. Since 1.72°F is less than 2°F, no horizontal pipe size increase over the 1-5/8" hot gas pipe is required.

Double Hot Gas Risers (Figure 28.7)

Double hot gas risers are usually necessary to return oil to the compressor whenever the minimum load is less than 25%. As in the double suction riser system, at low loads, the trap at the bottom of the riser fills with oil when the gas velocity is below the level necessary to move the oil. The oil seal will not permit refrigerant to flow up the "B" riser, diverting the refrigerant gas up the smaller "A" riser at a velocity that will entrain the oil in the gas.

1. The hot gas riser pipe size that gives a satisfactory pressure drop at the full load of 20 tons is found to be 1-5/8" (in Figure 28.2). Riser "A" (in Figure 28.7) must handle the minimum load of 5 tons. The 1.375" pipe (in Figure 28.7) will handle oil at the 5 ton minimum load.
2. The area of "A" − "B" should be equal to or more than the area of the 1-5/8" hot gas pipe that gives a satisfactory pressure drop. The 1.375" pipe has an area 1.256 square inches. The 1-5/8" pipe has an area of 1.78 square inches. The difference between 1.78 and 1.256 is the 0.52 square inch area of riser "B." The closest pipe size to a 0.52 area is the 1-1/8" pipe with an area of 0.825 square inches.

Table 12 Minimum Tonnage for Oil Entrainment up Hot-Gas Risers Type L Copper Tubing

Refriger- ant	Sat. Temp, F	Discharge Gas Temp, F	0.500	0.625	0.750	0.875	1.123	1.375	1.625	2.125	2.625	3.125	3.625	4.125
			0.146	0.233	0.348	0.484	0.825	1.256	1.780	3.094	4.770	6.812	9.213	11.970
12	80.0	110.0	0.161	0.289	0.478	0.721	1.41	2.38	3.67	7.33	12.6	19.7	28.7	39.8
		140.0	0.150	0.270	0.443	0.672	1.31	2.21	3.42	6.83	11.7	16.3	26.7	37.1
		170.0	0.143	0.256	0.423	0.638	1.24	2.10	3.25	6.49	11.1	17.4	25.4	35.2
	90.0	120.0	0.167	0.299	0.494	0.745	1.45	2.46	3.80	7.58	13.0	20.3	29.6	41.1
		150.0	0.155	0.278	0.459	0.694	1.35	2.29	3.53	7.05	12.1	18.9	27.6	38.3
		180.0	0.147	0.264	0.436	0.639	1.28	2.17	3.36	6.70	11.5	18.0	26.2	36.3
	100.0	130.0	0.171	0.307	0.506	0.765	1.49	2.52	3.89	7.77	13.4	20.8	30.4	42.2
		160.0	0.159	0.285	0.470	0.710	1.38	2.34	3.62	7.22	12.4	19.4	28.2	39.2
		190.0	0.151	0.271	0.448	0.677	1.32	2.23	3.43	6.88	11.8	18.4	28.9	37.3
	110.0	140.0	0.174	0.312	0.515	0.778	1.52	2.56	3.96	7.91	13.6	21.2	30.9	42.9
		170.0	0.162	0.290	0.479	0.724	1.41	2.38	3.69	7.36	12.6	19.7	28.8	39.9
		200.0	0.153	0.274	0.452	0.683	1.33	2.25	3.49	6.95	11.9	18.6	27.2	37.7
	120.0	150.0	0.175	0.314	0.518	0.782	1.52	2.58	3.96	7.95	13.7	21.3	31.1	43.2
		180.0	0.162	0.291	0.480	0.725	1.41	2.39	3.69	7.37	12.7	19.8	28.8	40.0
		210.0	0.153	0.274	0.452	0.682	1.33	2.25	3.47	6.93	11.9	18.6	27.1	37.6
22	80.0	110.0	0.235	0.421	0.695	1.05	2.03	3.46	5.35	10.7	18.3	28.6	41.8	57.9
		140.0	0.223	0.399	0.659	0.996	1.94	3.28	5.07	10.1	17.4	27.1	39.6	54.9
		170.0	0.215	0.385	0.635	0.960	1.87	3.16	4.89	9.76	16.8	26.2	38.2	52.9
	90.0	120.0	0.242	0.433	0.716	1.06	2.11	3.56	5.50	11.0	18.9	29.5	43.0	59.6
		150.0	0.226	0.406	0.671	1.01	1.97	3.34	5.16	10.3	17.7	27.6	40.3	55.9
		180.0	0.216	0.387	0.540	0.956	1.88	3.18	4.92	9.82	16.9	26.3	38.4	53.3
	100.0	130.0	0.247	0.442	0.730	1.10	2.15	3.83	5.62	11.2	19.3	30.1	43.9	60.8
		160.0	0.231	0.414	0.884	1.03	2.01	3.40	5.26	10.5	18.0	28.2	41.1	57.0
		190.0	0.220	0.394	0.650	0.982	1.91	3.24	3.00	9.96	17.2	26.8	39.1	54.2
	110.0	140.0	0.251	0.451	0.744	1.12	2.19	3.70	5.73	11.4	19.6	30.6	44.7	62.0
		170.0	0.235	0.421	0.693	1.05	2.05	3.46	3.35	10.7	18.3	28.6	41.8	57.9
		200.0	0.222	0.399	0.658	0.994	1.94	3.28	5.06	10.1	17.4	27.1	39.5	54.8
	120.0	150.0	0.257	0.460	0.760	1.15	2.24	3.78	5.85	11.7	20.0	31.3	45.7	63.3
		180.0	0.239	0.428	0.707	1.07	2.08	3.51	5.44	10.8	18.6	29.1	42.4	58.9
		210.0	0.225	0.404	0.666	1.01	1.96	3.31	5.12	10.2	17.6	27.4	40.0	55.5
502	80.0	110.0	0.192	0.344	0.567	0.857	1.67	2.82	4.36	8.71	15.0	23.4	34.1	47.3
		140.0	0.180	0.323	0.534	0.806	1.57	2.66	4.11	8.20	14.1	22.0	32.1	44.5
		170.0	0.173	0.310	0.512	0.773	1.50	2.54	3.94	7.85	13.5	21.1	30.7	42.8
	90.0	120.0	0.194	0.348	0.574	0.867	1.69	2.85	4.41	8.81	15.1	23.6	34.5	47.8
		150.0	0.182	0.326	0.538	0.813	1.58	2.68	4.14	8.26	14.2	22.2	32.3	44.8
		180.0	0.169	0.303	0.501	0.756	1.47	2.49	3.85	7.69	13.2	20.6	30.1	41.7
	100.0	130.0	0.194	0.348	0.575	0.869	1.69	2.86	4.42	8.83	15.2	23.7	34.5	47.9
		160.0	0.182	0.326	0.539	0.813	1.58	2.68	4.14	8.27	14.2	22.2	32.3	44.9
		190.0	0.170	0.304	0.503	0.739	1.48	2.50	3.87	7.71	13.3	20.7	30.2	41.9
	110.0	140.0	0.170	0.305	0.504	0.761	1.48	2.51	3.87	7.73	13.3	20.7	30.2	42.0
		170.0	0.162	0.291	0.481	0.726	1.41	2.39	3.70	7.38	12.7	19.8	28.9	40.1
		200.0	0.152	0.273	0.450	0.680	1.33	2.24	3.46	6.92	11.9	18.5	27.0	37.5
	120.0	150.0	0.170	0.305	0.503	0.760	1.48	2.50	3.87	7.73	13.3	20.7	30.2	41.9
		180.0	0.153	0.275	0.453	0.683	1.33	2.26	3.49	6.96	12.0	18.7	27.2	37.8
		210.0	0.149	0.267	0.440	0.665	1.30	2.19	3.39	6.76	11.6	18.1	26.4	36.7

NOTE: The tonnage is based on a saturated suction temperature of 20 F with 15 deg F superheat at the indicated saturated condensing temperature with 15 deg F subcooling. For other saturated suction temperatures with 15 deg F superheat, use the following correction factors:

Saturation suction temperature, F	−40	−20	0	40
Correction Factor	0.88	0.95	0.96	1.04

(Reprinted by permission from 1986 ASHRAE Handbook – Refrigeration)

Figure 28.5

3. The area of "A" and "B" should not be greater than the area of a single riser pipe that will return oil at the maximum load. Oil must be returned up the hot gas riser during the maximum load, too. From Figure 28.5 we find that a 2.625″ hot gas riser with an area of 4.77 square inches will return oil at 18.9 tons of cooling. Since 20 tons of cooling will produce a higher gas velocity than 18.9 tons, the 2.625″

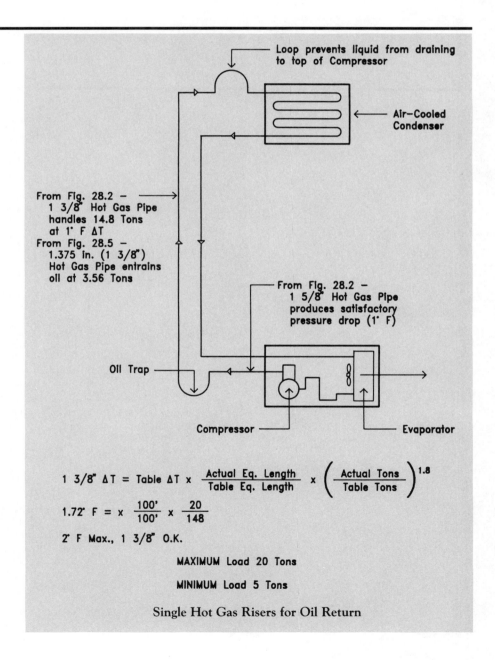

Loop prevents liquid from draining to top of Compressor

Air-Cooled Condenser

From Fig. 28.2 –
1 3/8″ Hot Gas Pipe handles 14.8 Tons at 1° F ΔT

From Fig. 28.5 –
1.375 in. (1 3/8″) Hot Gas Pipe entrains oil at 3.56 Tons

From Fig. 28.2 –
1 5/8″ Hot Gas Pipe produces satisfactory pressure drop (1° F)

Oil Trap

Compressor

Evaporator

$$1\ 3/8″\ \Delta T = \text{Table } \Delta T \times \frac{\text{Actual Eq. Length}}{\text{Table Eq. Length}} \times \left(\frac{\text{Actual Tons}}{\text{Table Tons}}\right)^{1.8}$$

$$1.72°\ F = \times \frac{100'}{100'} \times \frac{20}{148}$$

2° F Max., 1 3/8″ O.K.

MAXIMUM Load 20 Tons

MINIMUM Load 5 Tons

Single Hot Gas Risers for Oil Return

Figure 28.6

hot gas pipe will return oil at 20 tons of cooling. The 1.256 square inch area of "A," plus the 0.825 square inch area of "B" is 2.605 square inches. Since 2.605 square inches is less than the 4.77 square inch area of the single 2.625 inch hot gas pipe, oil entrainment up the hot gas risers is ensured at maximum load.

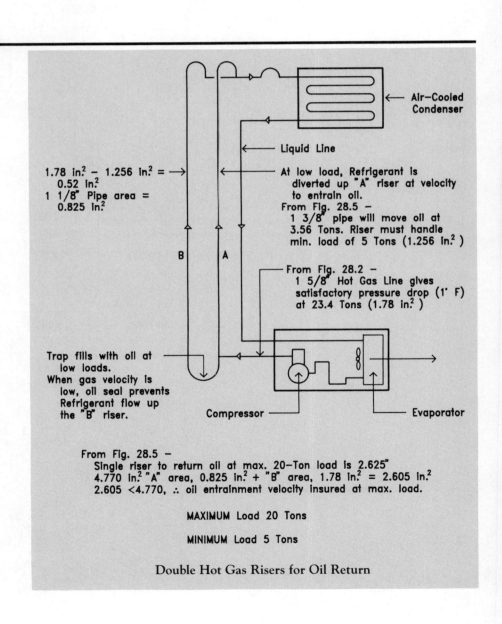

1.78 in.^{2} − 1.256 in.^{2} = →
0.52 in.^{2}
1 1/8" Pipe area =
0.825 in.^{2}

At low load, Refrigerant is
diverted up "A" riser at velocity
to entrain oil.
From Fig. 28.5 −
1 3/8" pipe will move oil at
3.56 Tons. Riser must handle
min. load of 5 Tons (1.256 in.^{2})

Air−Cooled
Condenser

Liquid Line

B   A

From Fig. 28.2 −
1 5/8" Hot Gas Line gives
satisfactory pressure drop (1° F)
at 23.4 Tons (1.78 in.^{2})

Trap fills with oil at
low loads.
When gas velocity is
low, oil seal prevents
Refrigerant flow up
the "B" riser.

Compressor   Evaporator

From Fig. 28.5 −
Single riser to return oil at max. 20−Ton load is 2.625"
4.770 in.^{2} "A" area, 0.825 in.^{2} + "B" area, 1.78 in.^{2} = 2.605 in.^{2}
2.605 <4.770, ∴ oil entrainment velocity insured at max. load.

MAXIMUM Load 20 Tons

MINIMUM Load 5 Tons

Double Hot Gas Risers for Oil Return

Figure 28.7

Chapter Twenty-Nine:

COOLING TOWERS

Cooling towers are the most common means of cooling the condenser water for water-cooled refrigeration units. The cooling tower theory of operation is given in Chapter 2, along with basic design and installation considerations.

Cooling towers are open to the atmosphere, and as such, present a challenge to the HVAC system operator. The principles by which cooling towers operate force them to endure a harsh environment.

In open systems, there is direct contact between the water and the outside air, providing the water with the opportunity to pick up contaminants from the air. Atmospheric gases are absorbed and concentrated by the water in the following manner: Evaporation permits pure water to leave the system, and the result is a concentration of aggressive fluid that attacks the condenser water containment apparatus.

Warm condenser water from the refrigerant condenser is pumped to the cooling tower, where it is sprayed over a fill material. Outside air is passed across the wet fill to transfer heat from the water to the air. The cooled water collects in an open basin, from which it is piped to a pump that returns the water back to the condenser, where it picks up heat. The warmed water is then returned to the tower. See Figure 29.1.

Maintenance Requirements

Figure 29.2 identifies the components of the cooling tower that need periodic attention. The basic maintenance requirements for each are outlined in the following section.

The open condenser water system requires the following maintenance. First, on a **weekly** basis:

- The cold water basin:
  — Inspect, clean, and repair any cracks or defects if necessary.
- The drive shaft:
  — Check for unusual vibration or noise.
  — Check alignment of the drive shaft couplings.
  — Check the oil in the gear reducer; also check oil fill and drain lines for corrosion.
  — Check oil seals and replace any that leak.
- The fans:
  — Check fan blades and hub for wear, corrosion and algae build-up.

— Clean and give the hub a protective coating.
— Clean screen of leaves, bugs, and debris.

Monthly:

- The drift eliminators:
  — Check to see if they are deteriorated, sagging, or have accumulated scum, algae, etc.
  — Check eliminator blades. Missing or improperly installed eliminator blades will allow water droplets to "drift" out of the tower.
  — Repair or replace any broken or missing eliminator slats or panels.
  — Check inlet louvers for obstructions of scale, mud, oil, and algae.
- The headers:
  — Check for corrosion and clogging from leaves, debris, and algae growth.
- The spray trees:
  — Check for deterioration. Replace with schedule 80 pvc pipe or other recommended by the manufacturer.

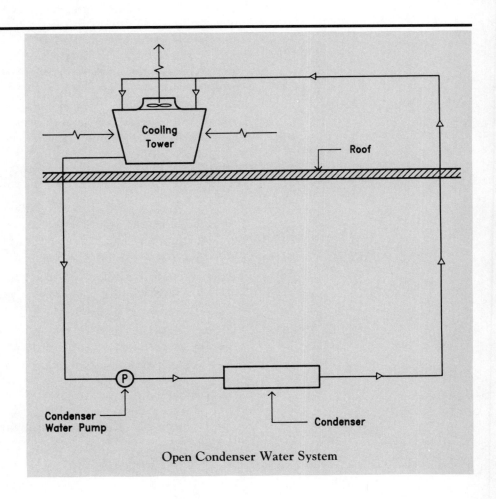

Open Condenser Water System

Figure 29.1

374

Quarterly:
- The distribution:
  — Check for clogged or broken nozzles. Clean and clear clogged nozzles and replace broken ones.

Semi-annually:
- The fill:
  — Check tower fill to be sure it is evenly spaced, level, and free from algae and deterioration. Sagging, broken, or eroded fill should be repaired or replaced.
- The drive shaft:
  — Change the oil in the gear reducer.

Annually:
  — Conduct a thorough inspection of all tower components, including fill, hardware, safety devices, railings, stairs, and walkways. The

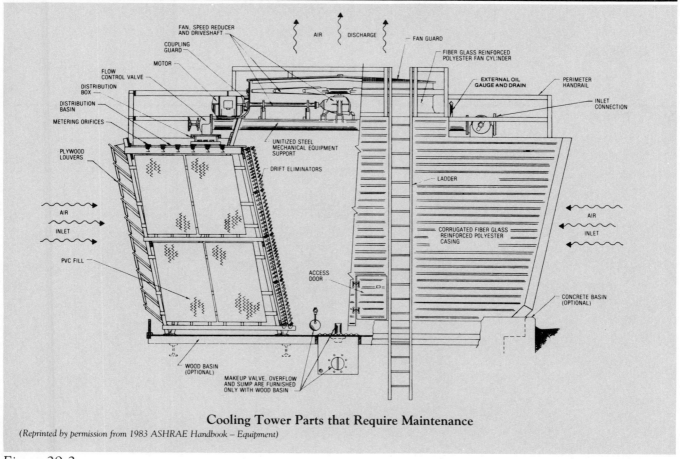

Cooling Tower Parts that Require Maintenance

(Reprinted by permission from 1983 ASHRAE Handbook – Equipment)

Figure 29.2

structural integrity of the tower may be undermined by chemical and biological attack.

— If vibration is present, check columns, braces, ties, fill supports, drift eliminator supports, deck joists, and louvers.

— Check for "iron rot" (corrosion) at structural connections.

Water Treatment

It is essential that the open cooling tower water be treated to avoid scale, corrosion, algae, slime, fungi, bacteria, mud, and dirt. Weekly or monthly water analysis and treatment is usually advised.

Operation in Freezing Weather

In areas where freezing occurs, cooling towers may present additional operating problems. Mechanical cooling may be necessary in late fall and early spring, when freezing weather is possible. The ice potential decreases when the airflow through the tower is reduced, thereby decreasing the volume of cold air in contact with the water.

Two-speed fans are the minimum requirement to permit tower airside control. Variable speed drives will produce excellent airside control for both ice control and energy reduction. The ice potential increases when the volume of water flowing over the fill is reduced. Modulation of bypass water that allows a small quantity of water to flow over the fill should be avoided. Figure 29.3 shows how the total water flow should be bypassed directly into the sump.

During cold weather start-up, the cold water in the basin may be near freezing. In this case, the total water flow should be returned from the condenser directly to the basin without going over the fill. The water

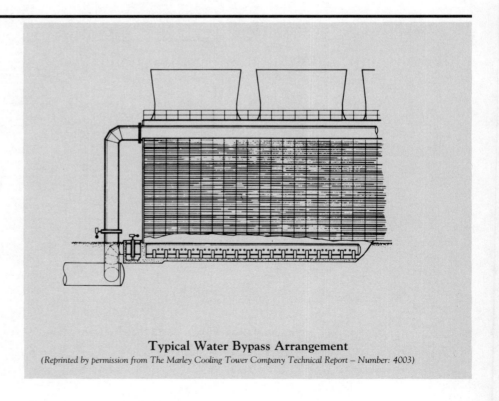

Typical Water Bypass Arrangement
(Reprinted by permission from The Marley Cooling Tower Company Technical Report – Number: 4003)

Figure 29.3

should bypass the fill until the basin water temperature reaches 80°F, at which point the total water flow should be diverted to circulate over the fill. The circulation should be diverted through the water distribution system such that the greatest volume of water flows over the fill near the tower intakes. This ensures a rapid air temperature rise and puts the warmest water in contact with the ice formation.

Electric, steam, or glycol heating coils should be installed in the tower basin to prevent freezing of the tower water. It should be noted that this additional equipment increases not only the system installed cost, but also the cost of operation and maintenance. If the cooling tower is not equipped with basin heaters, the tower basin must be drained of water to prevent its freezing. As soon as cooling is needed, the tower must be refilled and the water treatment replenished. This situation might occur several times a season.

As Figure 29.7 shows, an indoor tower basin for storing the cooled water would avoid the necessity for basin heaters and drain down and refilling the basin. Indoor space is required for the indoor basin.

Winterization

The need for year-round cooling tower operation is increasing because of high internal cooling loads that cannot be satisfied by cooling with outside air. These applications include computer room cooling by either chilled water air handling units or water-cooled self-contained air conditioning units. In cases where provisions for 100% outside air intake and exhaust are not practical, mechanical cooling may have to be operated.

When water-cooled mechanical cooling is operated during possible freezing weather, the cooling tower basin condenser water, drain, overflow, and makeup water piping that are exposed to freezing temperatures must be protected to avoid freezing the condenser water.

Cooling towers can be winterized using the following methods. Each is described in the paragraphs that follow.

Adding Heat to the Tower Basin and Piping

Figure 29.4 demonstrates how adding steam, electric, or glycol heating cables to the tower basin, and wrapping electric heating elements around the condenser water, tower drain, and overflow piping (as well as to the makeup water piping that is exposed to freezing conditions) permits winter tower operation. The electric pipe heating (heat tracing) coils and piping must be insulated and weatherproofed to preserve the generated heat.

The tower basin should be protected from freezing when condenser water is not circulating, that is, when the tower pump is not operating. Freezing is a serious problem only during the shutdown period. The basin heater should be sized for the basin heat loss and the water surface loss of 133 btu/sf. It is recommended that the basin be insulated at the same time the pipes are insulated. The tower drain piping must be heated because it contains dormant water and is especially susceptible to freezing. A small quantity of water moves through the overflow piping when the tower is operating. This water bleeds off suspended solids that accumulate in the basin. There is no water flow in the overflow piping when the tower is not in operation. The overflow pipe must be heated to prevent freezing.

The makeup water pipe does not have water flowing through it at all times. Water flows through this pipe only when the basin water level drops. It is essential that the makeup water pipe be protected from freezing. One method of preventing freezing of the cold water makeup pipe is to connect it indoors to the pump suction pipe located in the mechanical equipment room, as in Figure 29.5. A float switch in the tower basin will activate the makeup water automatic valve to inject cold water into the pump suction pipe.

Indoor Cooling Towers

Installing cooling towers inside a building offers a degree of freeze protection. Figure 29.6 shows an indoor cooling tower installation.

The advantages of installing a cooling tower indoors are:

- The tower is virtually freezeproof.
- The tower exterior is isolated from weather deterioration.
- Noise impact to neighbors is reduced.
- This arrangement does not detract from the appearance of the building's exterior.

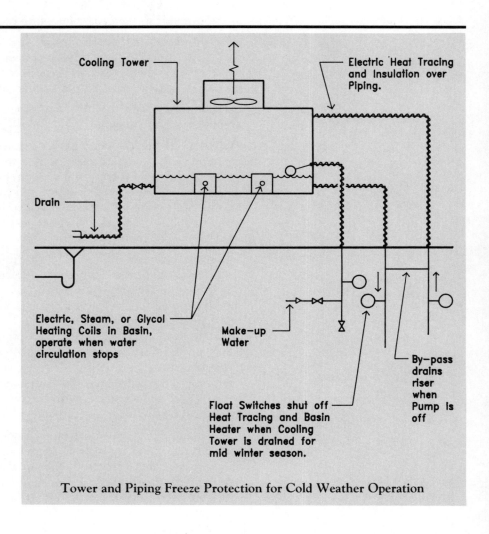

Tower and Piping Freeze Protection for Cold Weather Operation

Figure 29.4

378

The disadvantages of indoor cooling towers include the following:
- They occupy valuable floor space.
- There is an increased initial cost for the required centrifugal fan, air intake and exhaust ductwork, exterior wall louvers, and waterproof floor (floors with a membrane to prevent water penetration).
- The operating expense is increased for greater fan energy.
- Additional noise impact inside the building

Indoor Storage Tank

The "dry basin" system offers cold weather protection for smaller systems because the water that is normally pumped through the cooling tower and then collects in the basin, now drains directly into an indoor tank. Figure 29.7 shows an indoor storage tank condenser water system.

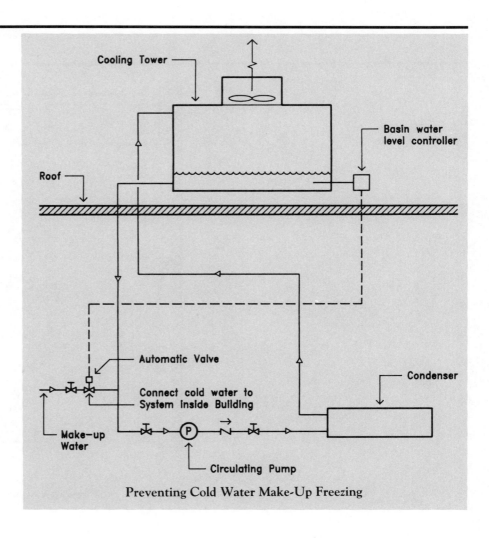

Preventing Cold Water Make-Up Freezing

Figure 29.5

The cooled condenser water collects in the indoor tank. The circulating pump takes its suction from this tank and pumps the condenser water back through the condenser and up to the tower. The indoor tank is usually three to five times the circulation rate, since it should contain the entire system capacity.

Tower Equalizing Lines

Uneven basin water levels in multiple tower installations can be avoided and corrected by installing water level equalizing lines between the towers. The uneven water levels are caused by the different equivalent lengths of the piping from the common pipe connection point to each of the cooling tower basins. The tower basin with the greater equivalent pipe length will fill up and overflow, while the tower basin with the shortest equivalent pipe length will have a lower water level, and water will be continuously added. Figure 29.8 is an illustration of cooling tower piping that will result in uneven basin water levels.

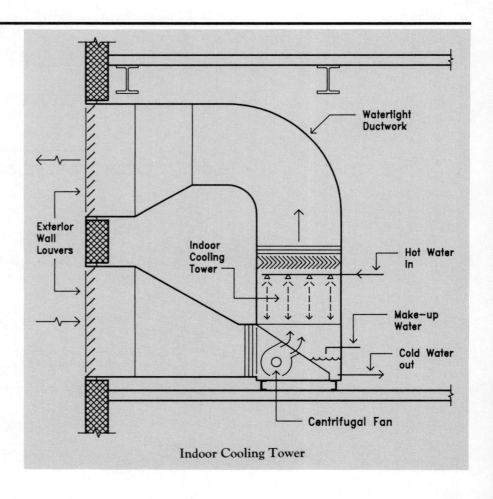

Indoor Cooling Tower

Figure 29.6

Equalizing lines can be added between existing cooling tower basins to restore an operating system's basin water level as per Figure 29.9. Equalizing lines should be sized so that 20% of the cooling tower gpm will flow through the equalizing pipe at a 2″ total head loss. As Figure 29.9 shows, the equalizing pipe is usually one or two sizes smaller than the tower condenser water pipe.

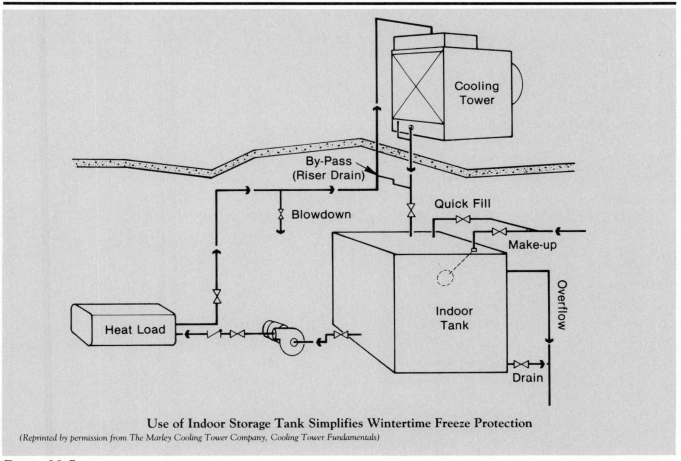

Use of Indoor Storage Tank Simplifies Wintertime Freeze Protection

(Reprinted by permission from The Marley Cooling Tower Company, Cooling Tower Fundamentals)

Figure 29.7

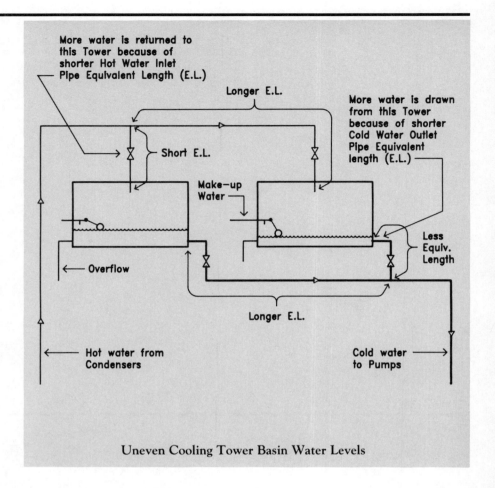

Uneven Cooling Tower Basin Water Levels

Figure 29.8

382

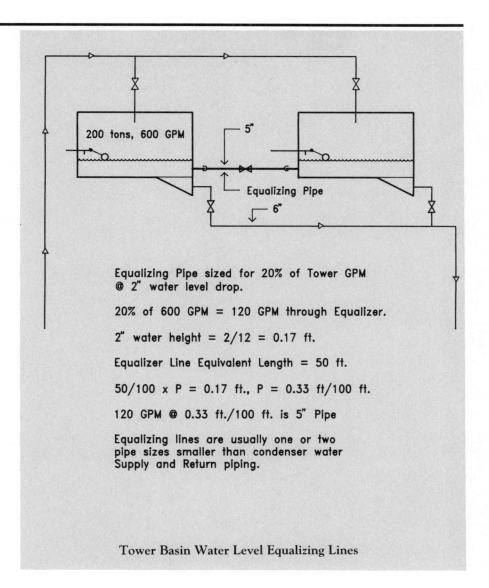

200 tons, 600 GPM

5"

Equalizing Pipe

6"

Equalizing Pipe sized for 20% of Tower GPM @ 2" water level drop.

20% of 600 GPM = 120 GPM through Equalizer.

2" water height = 2/12 = 0.17 ft.

Equalizer Line Equivalent Length = 50 ft.

50/100 x P = 0.17 ft., P = 0.33 ft/100 ft.

120 GPM @ 0.33 ft./100 ft. is 5" Pipe

Equalizing lines are usually one or two pipe sizes smaller than condenser water Supply and Return piping.

Tower Basin Water Level Equalizing Lines

Figure 29.9

Chapter Thirty:

AIR-COOLED CONDENSERS

Most maintenance and operational problems in air-cooled condensers are caused by exposure to the elements. Most air-cooled condensers are located out of doors where there is an inexhaustible supply of outside air to cool the hot refrigerant gas, and the same ambient space to discharge the hot condenser air. Unfortunately, whenever mechanical equipment is located outdoors, it suffers from the abuse of nature and man. The sun, wind, salt air, products of combustion, and other chemicals we pump into the air take their toll on the air-cooled condenser.

Most air-cooled condensers are constructed with baked enamel sheet metal casings, aluminum plate fins on copper coils, metal access panels, and aluminum fan blades. These materials installed as standard factory items cannot tolerate the harsh environment of most urban areas for any prolonged period of time.

The major problems with air-cooled condensers are:
- Dirt accumulation on the condenser coils, which reduces heat transfer
- Head pressure control (High pressure can be caused by dirty coils.)
- Incorrect refrigerant charge
- Physical deterioration

Dirt Accumulation on the Condenser Coils

Air-cooled condenser coils should be cleaned every six months to maintain proper heat transfer in order to keep refrigerant head pressures from climbing above the design point. Portable high pressure detergent cleaning units are available and very effective in removing dirt and debris that is drawn into the condenser coil.

Head Pressure Control

Air-cooled condensers are designed to condense the hot refrigerant gas by passing 95°F outside air over the condenser coils. The refrigerant transfers its latent heat to the outside air to permit the refrigerant to be condensed to a liquid. An R-22 air-cooled system should operate at 280 psi with 95°F air on the condenser and 72°F suction pressure (42°F coil temperature).

As the outside air temperature drops, the condenser's heat transfer capacity is increased, and more refrigerant is condensed than would be the case when 95°F air is used for condensing. The greater the temperature difference between the outside air and the refrigerant, the greater the air-cooled condenser capacity.

Excessive condensing lowers the head pressure. The low head pressure causes erratic operation of the expansion valve, and lowers the suction pressure. The minimum condensing temperature that will permit expansion valve operation with full compressor capacity is 90°F, or 169 psig for R-22. That means that with a 20°F temperature difference between the outside air and the refrigerant, 70°F is the minimum outdoor temperature at which the compressor can function at full capacity. The low suction pressure can also produce icing of the evaporator coil or stopping of the compressor by the low suction pressure control.

A good air-cooled refrigeration system operation requires that the head pressure be maintained above 170 psig.

Condensing pressure control can be accomplished by:
- Cycling the condenser fan motor
- Throttling the air passing through the condenser coil
- Condenser fan motor speed control

Cycling the Condenser Fan Motor

If the air-cooled condenser is under 10 tons, it may have only one condenser fan. Fan cycling of the only condenser fan is not recommended because the head pressure will rapidly fluctuate with every cycle of the fan. High head pressure during cool weather will start the fan. The head pressure will drop rapidly with the fan on, and the head pressure control will stop the condenser fan again, etc. Such rapid head pressure fluctuation is not a desirable condition for the expansion valve.

Single fan air-cooled condensers that were selected with a 20°F temperature difference (TD) without unloaders should not be operated under 70°F ambient.

When a condenser has two fans, one fan can be cycled to produce a stable head pressure. Two fan units selected at 20°F TD, operating at full capacity, can perform well with ambient temperatures of 30°F. Units above 20 tons usually have 3-6 condenser fans. In general, two of the fans are cycled together as a unit. At full compressor capacity and 20°F TD, they can operate down to 30-52°F ambient, depending on the unit size.

Fan cycling of air-cooled condensers should not be utilized with ambient temperatures below 30°F. Stable head pressure above 170 psig cannot be maintained with air below 30°F entering the condenser. Condensers with multiple fans can be controlled by cycling all but one fan by an outdoor thermostat. One of the two fans will be shut off by the outdoor thermostat at 40°F. With 3 fan units, one fan will be shut off at 50°F, and the third fan will be shut off at 25°F. Four fan units would be controlled by stopping one fan at 60°F, the second at 40°F and the third at 25°F. In each case, the last fan's speed is controlled by the head pressure. Solid state controls are available to modulate the condenser fan motor variable speed to maintain 100°F condensing temperature at ambient temperatures down to −20°F.

Throttling the Air Passing Through the Condenser Coil

An automatic damper located at the discharge of one of the condenser fans (with the actuator connected to the hot gas line entering the condenser) is used to restrict the airflow through the condenser. As the air volume through the condenser is reduced, the head pressure increases. Operational problems with this type of head pressure control develop when ice and snow collect on the damper and linkages.

Flooding the Condenser Coil

By flooding the condenser coil with liquid refrigerant, the effective condenser surface area is reduced. Condenser control systems of this type allow refrigeration systems to operate at $-30°F$ ambient. Figure 30.1 shows the arrangement of a flooded condenser head pressure control system.

A head pressure control valve is installed to accept liquid from the condenser or hot gas from the compressor. As the condensing temperature

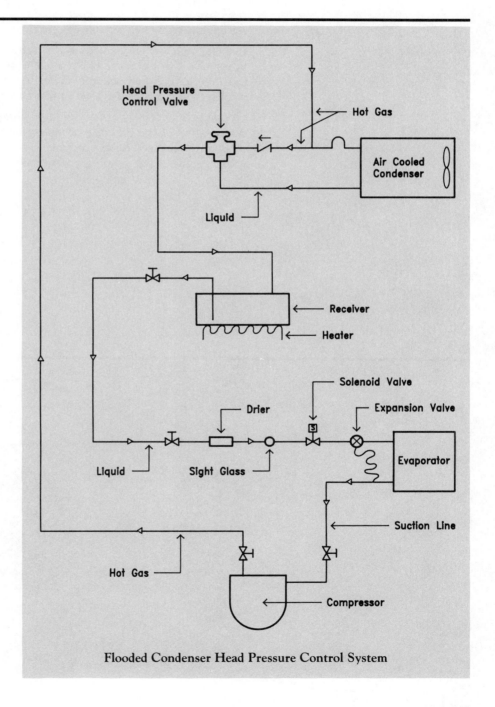

Flooded Condenser Head Pressure Control System

Figure 30.1

drops, the control valve throttles to restrict liquid flow from the condenser. This causes the liquid to back up into the condenser, reducing the effective condenser surface. The head pressure control valve bypasses hot gas to warm the liquid and ensure good expansion valve operation. A rise in condensing pressure causes the control valve to modulate more liquid and less hot gas to the receiver and to the expansion valve. The receiver remains warm, and the control valve establishes proper pressure for expansion valve operation.

Operational problems with flooded condensers relate to the following:

Correct Refrigerant Charge

More refrigerant is required for a flooded condenser system than for conventional air-cooled condensers. Incorrect refrigerant charge is responsible for many air-cooled system operational problems. The refrigerant charge is even more critical with the flooded condenser system.

If the system is charged with excess refrigerant in cold weather, the excess refrigerant will lie in the bottom of the condenser as sub-cooled liquid. The space that the liquid occupies is needed for condenser heat exchange in hot weather when excessive head pressures develop.

Flooded condenser systems should be charged, depending on the ambient temperature at the time the refrigerant is introduced to the system. Figure 30.2 shows refrigerant receivers with one and two sight glasses and their refrigerant levels depending on outdoor temperatures.

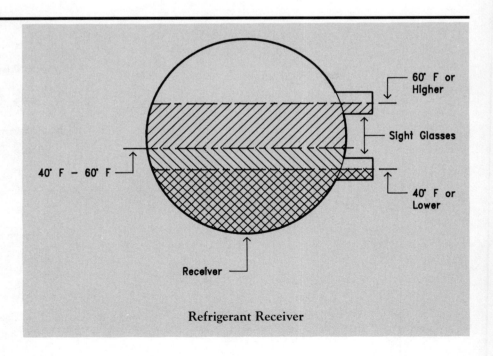

Figure 30.2

If the outdoor temperature is below 40°F the system should be charged until the receiver sight glass is one fourth full. When the outdoor temperature is 40-60°F, the system should be charged until the receiver is one half full. At ambient temperatures above 60°F, the system should be charged until the receiver is 75% full. It is important that the system is checked during all anticipated outdoor temperature fluctuations. It is extremely difficult to establish a refrigerant charge that will satisfy 98°F ambient, with 10°F air on the condenser coils.

Electric Heater Operation

Electric heaters are installed on the flooded condenser receiver to keep the refrigerant warm during the off cycle. Warm refrigerant should enter the expansion valve for proper expansion valve operation. Electric heat failure due to heater or control malfunction, blown fuses, etc., will lead to refrigerant problems in cool weather.

Correct Head Pressure Control Valve Setting

The lowest refrigerant temperature that will permit proper expansion valve operation is 90°F. This corresponds to 170 psig for R-22. The lowest head pressure produces the lowest kw/ton. Monitoring the refrigeration system operation during the coldest outdoor temperatures that the system will experience will verify that the 170 psig head pressure control valve setting is appropriate. If expansion valve problems occur, raise the head pressure in 5°F increments until stable operation persists.

Condenser Fan Speed Motor Control

Condenser fan speed motor control provides air delivery to the condenser in direct proportion to the actual heat rejection needs of the system. As the ambient temperature drops, the head pressure is reduced. The change in head pressure is sensed by the transducer, and the fan speed and air delivery are reduced to raise the head pressure to the desired set point.

Note that the hot gas connection to the air-cooled condenser should loop upward in an inverted trap before the gas enters the condenser. This is to prevent liquid refrigerant from draining into the hot gas line and damaging the compressor valves when the compressor is off.

Special care must be taken to prevent pulsating and vibrating hot gas lines from transmitting noise to the building. A material like polyethelene insulation should be placed between the hot gas pipe and any attachments to the building.

Advantages of Air-Cooled Condensers

No Water Freezing Problems

Perhaps the most significant advantage of air-cooled condensers is that they do not use water. This fact is tied in to all the other advantages of air-cooled condensers. Since no water is used, there is no water to freeze during the "winter" operating season.

No Seasonal Preparation

An air-cooled system can be put into operation by simply placing the control switch in the cooling mode. Water-cooled systems that utilize cooling towers have to be drained or winterized when the winter season approaches. Cooling tower systems must have the condenser water system refilled with fresh water and chemicals for water treatment when the summer cooling season is at hand, unless they were winterized. When the system is drained at the end of the cooling season, the cooling tower basin should be cleaned and painted.

No Water Treatment

Since air-cooled condensers use no water, water treatment is eliminated. (See Chapters 29 and 32 for more on water treatment.) Water treatment is not only an operating expense, but is a critical concern for system protection.

Possible Lower Installed Cost

For 6,000-60,000 S.F. offices with air- or water-cooled chillers with fan coil units, *Means Mechanical Cost Data 1989* indicates that the installed cost of packaged air-cooled systems is slightly lower (less than 10%) than water-cooled systems. Self-contained air-cooled unit systems of 125 tons show a 35% lower installed cost than water-cooled systems with cooling towers.

Possible Lower Yearly Operating Cost

Analysis of a 100,000 S.F. building with 375 tons of cooling shows that the yearly electric consumption of the air-cooled chiller is 10-15% less than that of a water-cooled chiller. This is primarily due to the energy use of the cooling tower fans and the condenser water pumps. Air-cooled units approach water-cooled unit efficiency at 80°F ambient. See Chapter 2 for more on air-cooled vs. water-cooled systems. In addition to the yearly reduced power consumption of air-cooled units, the operating cost of condenser water treatment is eliminated.

No Condenser Water Pump

Air-cooled systems do not use cooling towers and condenser water. The initial operating and maintenance costs of condenser water pumps are completely eliminated with air-cooled systems.

No Winterization Equipment Installation and Maintenance

Air-cooled systems do not use water; therefore, there is no water to freeze during the winter operating season. Most internal cooling loads require mechanical refrigeration during some part of the spring and fall when those systems that are equipped with economizer controls (using outside air for cooling) are functioning. There are many days when the outdoor temperature rises above 55°F and mechanical cooling is needed to maintain space comfort. When outdoor freezing temperatures are imminent, the condenser water system is protected by electric heaters in the tower basin, and by electric heating elements (heat tracing) wrapped around the condenser water, make up water, and tower drain and overflow piping.

Air-cooled systems avoid the installation, maintenance, and energy use of the condenser water heating system. Heating element failure due to blown fuses and burnout are, however, a reality.

No Water Carryover (Drift) and Fog

Drift is the water droplets that are entrained in the air passing through the tower and discharged from the tower. Efficient eliminators limit the drift to 0.2% of the water circulation rate. A 300 ton system circulates 600 gpm. Drift should be limited to 1.2 gpm. Most towers, especially the older ones, will produce drift three to five times that amount. Drift contains the minerals of the makeup water and water treatment chemicals, and can be corrosive enough to destroy automobile finishes. Cooling towers should not be located where drift can come into contact with autos, people, or adjacent building windows.

Fog develops when cooling towers are operated in cool weather. Fog is created when the ambient air surrounding the tower cannot absorb all of the moisture from the saturated warm cooling tower discharge air. The plume emitted from a cooling tower operating in cool weather can be extremely objectionable. Large cooling towers located near interstate highways can produce enough fog that traffic may be stopped in cold weather. Heating the tower exhaust can reduce the plume size, but this is not always an effective method. Figure 30.3 shows a heating coil installed at the tower discharge, utilizing warm condenser water to heat the tower exhaust.

The tower exhaust reheat coil adds a year-round operating expense for fan energy, as well as substantial coil maintenance.

No Water Consumption

Last, but not least, and perhaps the most meaningful of the air-cooled system advantages, is the fact that air-cooled systems do not use water. Modern society has placed water on the endangered species list. Our supply of potable water is rapidly disappearing.

Cooling towers do consume water by evaporation, drift and blowdown or bleed off. One percent of the circulating water is consumed by

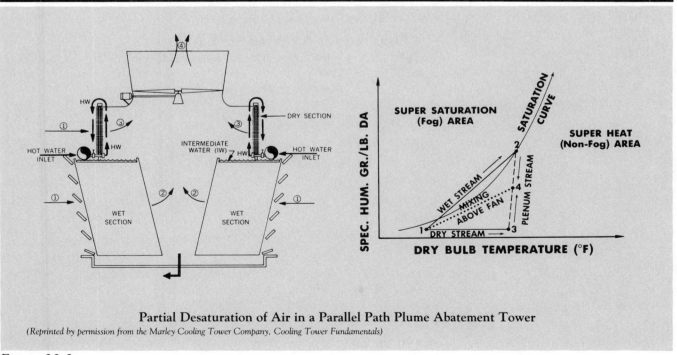

Partial Desaturation of Air in a Parallel Path Plume Abatement Tower

(Reprinted by permission from the Marley Cooling Tower Company, Cooling Tower Fundamentals)

Figure 30.3

391

evaporation. Seventy-five percent of the cooling tower's heat transfer is latent, and twenty-five percent is sensible cooling of the water. Water cooling is accomplished when the water gives up its latent heat through evaporation. Drift in most older towers wastes another one percent of the circulating water.

In practice, bleedoff wastes one percent of the total circulating water. Bleedoff also limits the concentration of impurities. Dissolved solids are created by the evaporation process, and airborne dust and gases are introduced into the system. Bleedoff and drift help remove these impurities.

The total quantity of makeup water to replace the wasted water is:

1% evaporation + 1% drift + 1% bleedoff = 3%.

A 100 ton system circulating 300 gpm wastes 3%, or **9 gpm.** Most operating cooling towers waste about 5% of the circulating water or about 15 gpm per 100 tons.

Disadvantages of Air-Cooled Condensers

Higher Installed Cost with Remote Air-Cooled Condensers, Especially in Larger Capacities

The installed cost of 60-200 ton air-cooled chilled water plants with remote air-cooled condensers is 40-50% higher than water-cooling chillers with cooling towers. Self-contained packaged air-cooled chiller installed costs for these sizes is comparable to the water-cooled chiller installation with cooling towers. The installed cost of air-cooled centrifugal chillers is more than that of water-cooled chillers because:

- The standard water-cooled chiller must be modified for air-cooled application.
- Very large air-cooled condensers cost three to five times as much as a cooling tower.
- Very large air-cooled condenser field erection costs are greater than cooling tower field erection costs.
- Industrial type purge units are required to maintain head pressure.
- Considerably more refrigerant is needed.

Not Suitable for Very Large Chillers

The high premium for air-cooled centrifugal chiller installation is addressed above.

Chapter Thirty-One:

AIR-COOLED CHILLERS

Because no seasonal preparation such as drain down, refill, or water treatment are necessary, an air-cooled chiller system is easier to operate than a water-cooled chiller system. However, air-cooled chillers require that provisions be made for head pressure control. Head pressure is the refrigerant pressure at the discharge side of the compressor. The hot refrigerant gas is cooled and condensed by the condenser. The temperature of the saturated gas corresponds to a saturated gas pressure (condensing pressure and temperature, usually 105-120°F). Controlling the head (discharge) pressure controls the head (discharge) temperature. As discussed in Chapter 2 and again in the following section, condenser fan control is usually employed to regulate the condensing temperature. The following paragraphs address some of the most significant operational concerns for air-cooled chillers.

Refrigerant Charging

When adding refrigerant to a new air-cooled system, it is important to take the ambient temperature into consideration. If the refrigerant charge is established in cool 40-60°F weather, a clear sight glass and normal suction and head pressures indicate a correct refrigerant level. Problems relating to high suction and head pressures can develop in 90°F weather. The refrigerant charge completed in 40-60°F ambient temperatures is too great for proper hot weather operation. Refrigerant must be evacuated from the system to create the required system pressures.

The opposite condition can occur when refrigerant is added during a 90°F ambient system start-up. The refrigerant quantity may be satisfactory for warm weather, but insufficient for 40-60°F ambient conditions. Refrigerant may have to be added in cool weather. The cool outside air passing over the air-cooled condenser coil changes more refrigerant gas into a liquid. A pound of refrigerant gas occupies much more coil space than a pound of refrigerant liquid. During hot weather, the gas is not condensed as rapidly as it was in cool weather. A smaller refrigerant charge fills the condenser coil, and head pressures are normal. The smaller charge will not properly fill the condenser during cold weather because the cold air will condense the gas and lower the head pressure and suction pressure. It is important to check the refrigerant cycle operation during the hottest and coolest operating conditions.

Controlling Condenser Pressure

Condensing pressure control is essential for good operation. Condenser fan cycling, condenser fan speed control, and head pressure regulating valves are all effective means of attaining pressure control. Without condenser control, during cool weather the condenser fans will blow the full volume of cool outside air over the condenser coil and lower the head pressure, possibly below 168 psig for R22 (22 grade refrigerant).

Condensing pressures must be above 168 psig in order to effectively operate the expansion valve. When the head pressure is lowered, the suction pressure is also reduced. R22 suction pressures below 57 psig are at 32°F. At refrigerant suction gas temperatures below 32°F, condensed water on DX (direct expansion) coils and water in the DX-to-water chiller can freeze. Safety depends on a functioning low temperature control that will stop the compressor.

Fan speed control regulates the air delivery to the condenser coil in direct proportion to the heat rejection needs. This arrangement maintains optimum system pressures as the ambient temperature varies. As the ambient temperature drops, the head pressure will also drop. As the head pressure drops, the pressure change is sensed by the transducer and the fan speed is adjusted to raise the head pressure and temperature to the set point. See Figure 31.1.

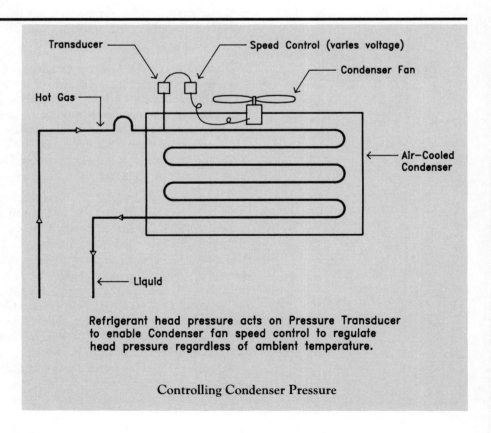

Refrigerant head pressure acts on Pressure Transducer to enable Condenser fan speed control to regulate head pressure regardless of ambient temperature.

Controlling Condenser Pressure

Figure 31.1

Condenser Cleaning

Debris on the condenser fins restricts airflow through the condenser coil, and reduces operating efficiency by raising the compressor head pressure. Routine maintenance procedures include coil inspection and removal of any impedant to the airflow through the coil.

Contaminated Refrigerant

Improper refrigerant system evacuation allows moisture to remain in the refrigerant circuit. The moisture will attack the compressor and affect expansion valve operation. (See Chapter 28.)

Crankcase Heaters

Crankcase heaters are electric heaters attached to the compressor body to warm the refrigerant and oil when the compressor is not operating. The heater vaporizes liquid refrigerant in the compressor to prevent the compressor from trying to compress the liquid. If the crankcase heater is not installed or is inoperative, compressor flooding and seizure could occur.

Phase Failure Relays

Loss or a reduction of voltage on any of the phase legs of a three-phase chiller could result in a compressor burn-out. Phase failure relays installed on each of the phase legs will instantly disengage the compressor starter when any phase is overpowered.

Insufficient Chilled or Condenser Water Flow

Each chiller has a minimum and maximum flow rate for the cooler and condenser. Too low a cooler flow rate results in low chilled water temperature cut-out or cooler freeze-up. A high chilled water flow results in a high chiller pressure drop or tube erosion. Low condenser water flow rates, below 3 ft./sec. contribute to condenser fouling and excessive tube cleaning.

Cooler and condenser flow switches must be set for the minimum allowable water flow, by creating the minimum flow to test the switches.

Water Treatment

The major chiller problem is attributed to improper water treatment. Coolers and condensers can be completely disabled in one season's operation due to mismanaged water treatment.

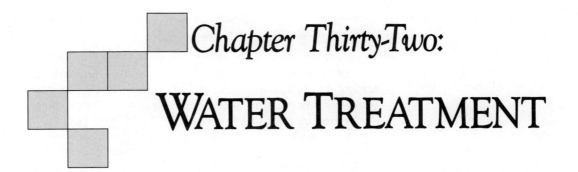

Chapter Thirty-Two:

WATER TREATMENT

Inadequate water treatment is a major factor in both pipe failure and ineffective heat transfer in chillers and condensers. Steam, hot water heating, and condenser water piping systems are especially vulnerable to pipe failures if water treatment is not properly maintained.

Water problems in heating and cooling systems are corrosion, scale formation, biological growth, and suspended solid matter. Even a small buildup of scale on a heat exchange surface will reduce heating and cooling capacity. Chiller and condenser heat exchangers are designed to operate with a maximum fouling factor of 0.0005. This is equivalent to a layer of calcium carbonate scale 0.006″ thick on the tubes. An increase scale thickness to 0.060″ will result in a 53% increase in energy to overcome the scale.

Scale Control

Scale control is aimed at minimizing the formation and accumulation of calcium carbonate. This is done by controlling the pH and solid content of the water.

Corrosion

Corrosion is the destruction of a metal by chemical or electrochemical reaction with its environment. A flow of electricity between certain areas of the metal is the electrochemical action that causes the eating away of a metal.

Corrosion control minimizes the damage to the metal by removing oxygen from the water, and by altering the water composition by adding corrosion inhibitors and pH control chemicals. Controlling pH alone will not minimize corrosion. An inhibitor is also required.

Biological Growth

Biological growth causes problems in cooling towers by blocking water distribution to nozzles and piping. The use of microbiocidal materials will minimize slime buildup.

Suspended Solids

Suspended solids can settle out at heat exchanger tubes, reduce heat transfer, and interfere with water flow. Treatment with polyelectrolytes and blow-down will prevent the most harmful effects of suspended solids. New piping should be thoroughly flushed to remove as much of the suspended material as possible before placing the system in operation. Strainers at control valves and coils will remove larger suspended solids.

Use of Chemicals

Careful application of chemicals to water in piping systems is essential for system preservation. Too much can create as many problems as too little. While pitting may develop if concentrations are low, high concentrations of chlorine can accelerate corrosion. A pH level below 7 can cause sodium nitrite to break down. Nitrite and nitrate concentrations must be checked frequently to prevent bacterial oxidation and conversion of nitrites to nitrate, which has no corrosion-inhibiting properties.

The services of a reliable water treatment company are essential to the preservation of HVAC piping systems. The following requirements should be addressed for an operational water treatment system:

- Domestic water supply is required for mixing chemicals.
- Electric power is needed for the chemical injection pump.
- A drain is necessary for bleed-off waste.
- Space must be furnished for chemical tanks and drums.
- Lack of turbulence, due to laminar flow, is a result of small flow in large condenser water pipes. Turbulence is essential to permit water treatment chemicals to come into contact with the metal surfaces they are to protect.

Precautions

Figures 32.1 and 32.2 show how a condenser water piping system of a tenant's year-round air-conditioning system can be destroyed by connecting it to the building's central condenser water system.

The building condenser water system should receive regular water treatment service. During the normal summer air-conditioning season, the water treatment is effective because water velocities in the condenser water system fluctuate between five and ten feet per second. During the winter season, however, the main condenser water pump is inoperative, and condenser water velocities are established by the tenant's small condenser water pumps. Water velocities drop below two feet per second, at which point laminar flow and the accompanying ineffective water treatment take their toll. Two-and-one-half inch black steel condenser water pipe interiors have been reduced to less than one inch within one winter season of operation.

Copper pipe should be used for condenser water service when a year-round air-conditioning system is connected to a central condenser water system.

It is important to size the tenant's condenser water pump for the central system's cooling tower "lift," as well as for the central system's piping friction loss that will be shared by the tenant's piping.

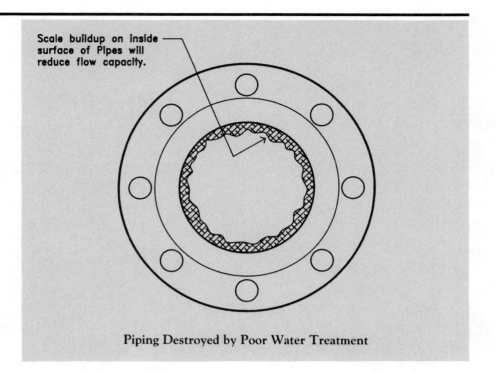

Piping Destroyed by Poor Water Treatment

Figure 32.1

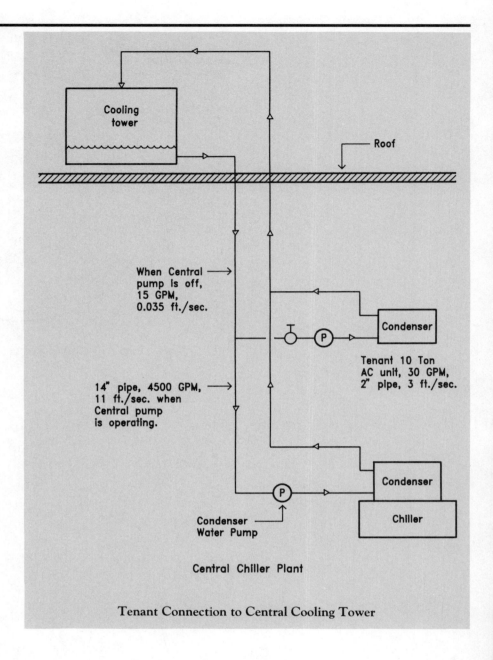

Figure 32.2

Tenant Connection to Central Cooling Tower

Chapter Thirty-Three:

BOILERS

The various boiler types have been introduced and compared in Chapter 8. This chapter will explore some of the most common operational problems that arise with boilers, along with preventive strategies and solutions.

Condensation

Boiler water temperatures must never be lower than the flue gas dew point temperature, except in the case of specially constructed condensing boilers. (For more information on condensing type boilers, see "Pulse Boilers" in Chapter 8.) Water is contained in the products of combustion from the burning of all types of fuel, since the chemical combination of hydrogen in the fuel and oxygen in the air produce water vapor (H_2O). If the boiler or flue pipe surfaces are below the flue gas dew point temperature, water will condense on those surfaces and create rust, scaling, and possible boiler failure.

Flue gas condensation problems develop when any of the following occur:

- **Vastly oversized boilers that cycle frequently under low-fire conditions.** When the low-fire rate is too low, the boiler extracts heat from the flue gases, lowering their temperature to the point where condensation takes place. Boilers should not be greatly oversized because of such condensation problems and because of the lower seasonal efficiency that results from oversizing.
- **The fuel contains sulphur.** As the sulphur content in the fuel increases, so will the flue gas dew point temperature. The flue gas dew point temperature is increased 15°F for every 0.5% increase in the sulphur content of the fuel. The sulphur in the fuel produces sulphur dioxide and sulphur trioxide, both of which will attack the boiler steel surfaces.

  Figure 33.1 shows the dew point temperature of various sulphur-free fuels, the minimum boiler water temperature to prevent condensation within the boiler, and the minimum stack temperature to prevent condensation in the breeching. If the fuel contains sulphur, the minimum temperatures must be increased.
- **The boiler water temperature is reduced as the outdoor temperature increases.** This practice can result in a boiler water temperature below 170°F. Because of the possible sulphur content in the fuel, boiler water temperatures should not be less than 170°F.

Continuous Flow-Through Hot Water Boilers

Hot water boilers should be installed so that there is continuous water circulation through the boiler under all operating conditions. Piping should be arranged so that automatic valves do not bypass water around the boiler. If the boiler continues to operate with no circulating water, stratification takes place, further increasing temperature, and subjects the boiler to thermal stress. The minimum water circulation through the boiler is 1 gpm for every 33,000 btu/h boiler input. Figure 33.2 shows the correct method of piping a hot water boiler to ensure that minimum boiler water circulation is provided.

Thermal Shock

Thermal shock usually occurs in a hot water boiler when large volumes of cold return water are introduced into the boiler to replace the hot water leaving the boiler. Thermal shock can also take place when large quantities of hot water enter the boiler to replace the cold water leaving the boiler. The second case might occur when a standby boiler is placed on line to take over for a malfunctioning or otherwise inactive boiler. It is not uncommon for large quantities of cold return water to be circulated into a hot water boiler, especially during start-up. The stress generated by the hot expanded and rapidly cooled metal can severely damage the boiler. Figure 33.3 shows how to pipe a hot water boiler to avoid thermal shock.

Fuel Oil Tank Installation

Anti-floatation pads

Storage tanks installed below the ground are subject to the buoyancy force of water below the ground. If sluggish ground water drainage persists after a rain, or the ground water table is high, the accumulation of water below the ground and the buoyant force can lift the tank out of the ground. The probable rupture of fuel oil piping and oil spillage that is caused by the tank movement could be a more serious problem than the task of reinstalling the tank and repairing the grade above the tank.

The buoyancy force is equal to the weight of displaced ground water. The underground storage tank must be secured to sufficient weight to overcome the buoyant force. The weight is in the form of a concrete pad poured at the bottom of the excavation. To calculate the required amount of

Sulphur-Free Fuel Dew Point Temperatures and Minimum Boiler and Stack Temperatures			
Type of fuel	Average Dew Point Temperature Degrees F	Minimum Boiler Water Temperature Degrees F	Minimum Gross Stack Temperature Degrees F
Heavy Oil	130	160	275
Light Oil	111	140	275
Natural Gas	127	150	275
Propane Gas	119	140	275

Figure 33.1

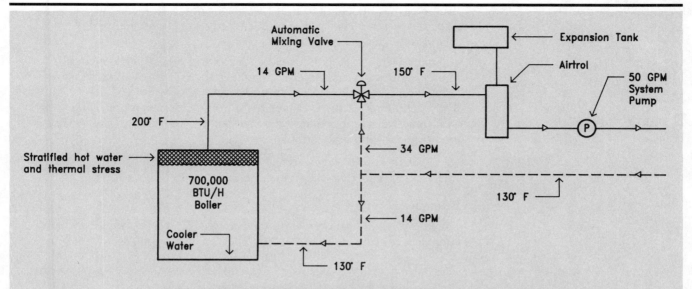

Incorrect Boiler Piping and Possibly Little or No Flow Through Boiler
When Automatic Mixing Valve is in Bypass, Hot Water

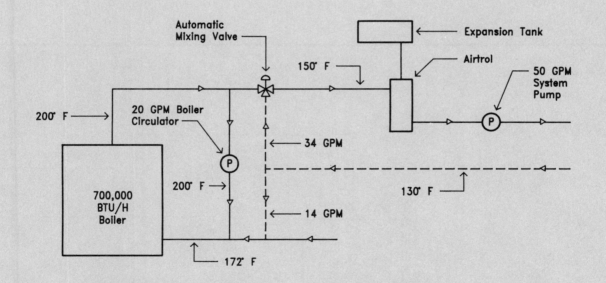

Correct Boiler Piping – Continuous Minimum Circulation Through Boiler – 1 GPM/35,000 BTUH

Figure 33.2

concrete, it is necessary to find the weight of the displaced water. The displaced water is the storage tank capacity in gallons, multiplied by 8.3 pounds of water per gallon.

Concrete weighs 150 pounds per cubic foot in air. When concrete is immersed in water, its effective weight is the weight of concrete in air (150 lb/cf) less the weight of an equal volume of water (62.5 lb/cf, or 87.5 lb/cf). The tank contributes weight to hold it in the ground. The additional weight of concrete required to hold the tank in the ground is found by the following formula:

$$\frac{\text{weight of displaced water } - \text{ weight of tank}}{\text{effective weight of concrete}} = \text{cf of concrete}$$

Example:

An empty 10,000 gallon fuel oil tank weighs 12,000 pounds. How many cubic feet of concrete is required for the anti-floatation pad?

10,000 gallons x 8.3 lb/gal =	83,000 lbs. displaced tank weight
tank weight	− 12,000 lbs.
net weight of displaced water	71,000 lbs.

$$\frac{71,000 \text{ lbs. of displaced water}}{87.5 \text{ lb/cf effective weight of concrete}} = 811 \text{ cf}$$

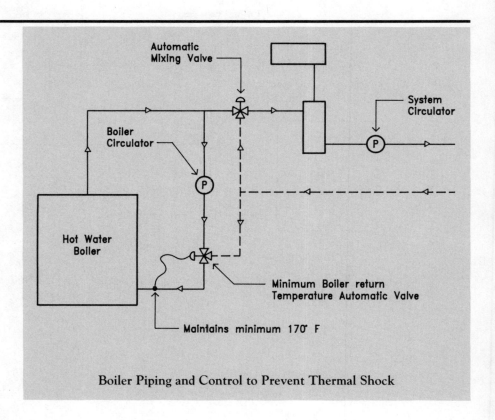

Boiler Piping and Control to Prevent Thermal Shock

Figure 33.3

Storage Tanks

To minimize ground contamination, many states are requiring that underground tanks and piping must be protected and monitored for leakage. Some of the regulations for new underground storage systems demand the following:

- Corrosion-resistant tanks and pipes
- Secondary containment
- A leak monitoring system
- Overfill prevention equipment
- Fill port labels
- Underground piping access ports

Figure 33.4 shows an underground tank with the above components, which are discussed in the following paragraphs.

Corrosion-Resistant Tanks and Pipes

Underground tanks can be constructed of fiberglass-reinforced plastic. Steel tanks may also be protected with fiberglass-reinforced coatings or by cathodic protection using sacrificial anodes or impressed current. Underground pipes can be constructed of stainless steel or fiberglass-reinforced plastic, or they can be protected cathodically or by galvanic coatings.

Secondary Containment

Secondary containment is a barrier constructed under the tank to contain a leak long enough to be detected by the leak monitoring system. A

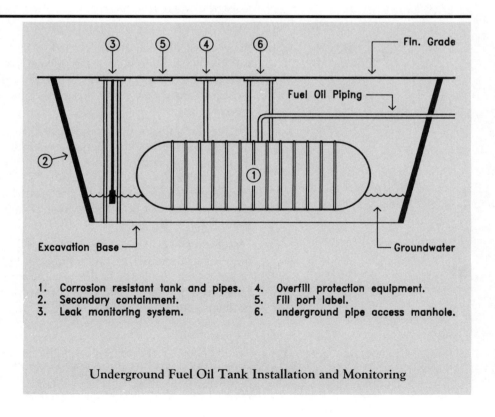

1. Corrosion resistant tank and pipes.
2. Secondary containment.
3. Leak monitoring system.
4. Overfill protection equipment.
5. Fill port label.
6. underground pipe access manhole.

Underground Fuel Oil Tank Installation and Monitoring

Figure 33.4

double-walled tank provides containment. When oil enters the space enclosed by the outer tank shell, the leak detector probe between the shells indicates the presence of oil.

Leak Monitoring System
A leak detector or well must be installed between the tank and the secondary barrier. The well could be a slotted or screened 4″ plastic pipe that can be checked weekly for oil odors. An electronic detector will energize a signal in the more elaborate leak detector system.

Overfill Prevention Equipment
Equipment must be installed to warn the operator of an imminent overfill or automatically shut off the fill system.

Fill Port Labels
A permanent label must be installed at underground fill ports, showing the date of installation, tank capacity, dimensions, and manufacturer.

Underground Piping Access Ports
Access ports must be installed to test the pipe for tightness without involving extensive excavation.

Hot Water Boiler System Water Treatment

Trouble-free operation of hot water heating systems requires periodic analysis and treatment of the water by a water testing company. Water treatment is almost always necessary for all hot water heating systems, as it prevents the accumulation of scale and sludge deposits that impair heat transfer in the boiler. In this way, water treatment prevents overheated metal, which would eventually fail, and cracking and metal embrittlement. A water meter should be installed on the water makeup line to determine the amount of raw water entering the system. This information will not only serve as an indication of leaks, but will also assist the water treatment firm in their program.

Standby Losses

Every time a burner starts and stops, seasonal efficiency declines. Relatively cool combustion air is passed through the boiler before ignition and fuel are sequenced. The air purge clears out any gases in the flue passages before new combustion takes place. This is called the *pre-purge cycle*.

The cool combustion air takes heat from the boiler surfaces, and sends the heat up the chimney. When the operating control stops the oil flow, the burner's combustion air fan continues to force air through the boiler, to clear out the flue gas. This is called the *post-purge cycle*. More heat is transferred from the boiler to the purged air, to be vented out of the chimney. The hot boiler also gives up a portion of its heat to the boiler room from jacket losses.

Matching the boiler size to the load reduces the standby losses. Multiple boilers can increase seasonal efficiency by providing boiler capacity to better match the load. Boiler tests performed by Brookhaven National Laboratory indicate that boilers operating at 25% of the design load, were producing at 25% seasonal efficiency.

Combustion Air

Oxygen must be provided to support combustion. Stoichiometric combustion for oil requires approximately 280 cf of air per gallon of oil. It is not uncommon to have 50% excess air provided. Many codes mandate 20-30 cfm of combustion air per gallon, per hour of oil burner capacity. At that rate, a 100 gallon per hour burner will need 3,000 cfm of outside air for combustion. It is essential that provisions are made for providing the required volume of combustion air. In cold climates, special attention must be given to the method used to introduce that volume of air into the boiler room. As Figure 33.5 shows, 3,000 cfm of 10°F air can ravage a piping system near the combustion air louver. Combustion air should be preheated to at least 40°F to prevent damage to the facility, and to 60°F if the combustion air supply will come into contact with personnel.

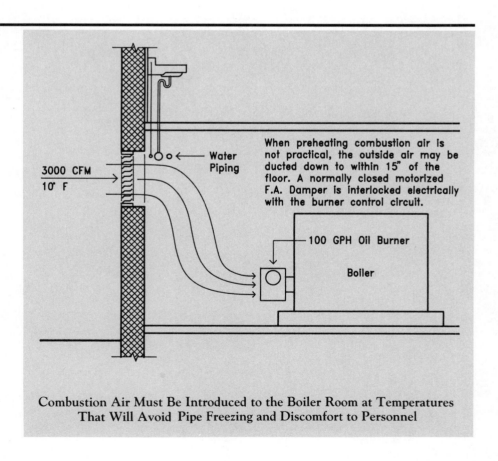

Combustion Air Must Be Introduced to the Boiler Room at Temperatures That Will Avoid Pipe Freezing and Discomfort to Personnel

Figure 33.5

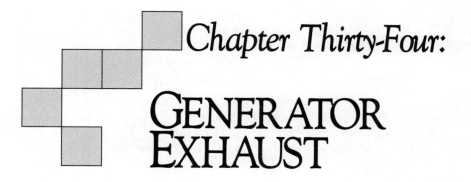

Chapter Thirty-Four:

GENERATOR EXHAUST

Engine performance is diminished and operational problems develop when engine back pressure is ignored. It is not unusual to find exhaust mufflers and piping installed with pipe sizes the same as the engine exhaust connection. One does not think twice about running a 1" heating pipe to a reheat coil with a 1/2" connection. But when it comes to sizing generator exhaust systems, we often overlook their impact on generator performance.

Diesel generator exhaust should discharge vertically with a "flapper" type exit rain cap valve. Figure 34.1 shows that a vertical discharge disperses the hot exhaust gas into the atmosphere more effectively than a horizontal discharge.

It is important that generator engine exhaust is not discharged where it can be drawn into ventilation fresh air intakes or building openings. Recirculation of engine exhaust is best avoided by a vertical discharge located as high above the roof as possible. The 900°F exhaust can severely harm anyone with whom it comes into contact. Figure 34.2 shows the non-recommended engine exhaust arrangement with a horizontal discharge less than six feet above the floor.

Generator Exhaust Piping

Generator exhaust pipes should not always be the same size as the muffler. The maximum allowable engine back pressure is 27" of water pressure. The silencer pressure drop should be about 4"-6" water gauge (wg). The exhaust piping, including the flexible connector and vent cap pressure drop, must not exceed the remaining 21"-23" wg pressure.

Excessive exhaust back pressure seriously affects engine horsepower output, durability, and fuel consumption. Poor combustion and higher operating temperatures develop when the discharge of cylinder gases is impeded. High back pressures are caused by:

- Too small an exhaust pipe
- Too many turns in the piping
- Piping that is too long
- An exhaust silencer that is too small (high pressure drop)

Figure 34.3 gives an example of how to size a generator exhaust piping system.

Engine Exhaust Pipe Expansion

Steel exhaust pipe expands 0.0076″ per foot of pipe for each 100°F rise of exhaust temperature. A steel exhaust pipe will expand 0.65″ for each 10′ of pipe that is heated from 100°F to 950°F. For the example in Figure 34.3, the 150′ of 12″ exhaust pipe is heated from 65°F to 865°F.

0.0076 in-ft-100°F x 150′ x 8 (100°F rise) = 9″

The 9″ expansion of the 12″ exhaust pipe must be properly accommodated or it could destroy the structure. Anchors, guides, and expansion joints must be provided for pipe expansion of that magnitude. Figure 34.4 shows an engine exhaust system with provisions for pipe expansion.

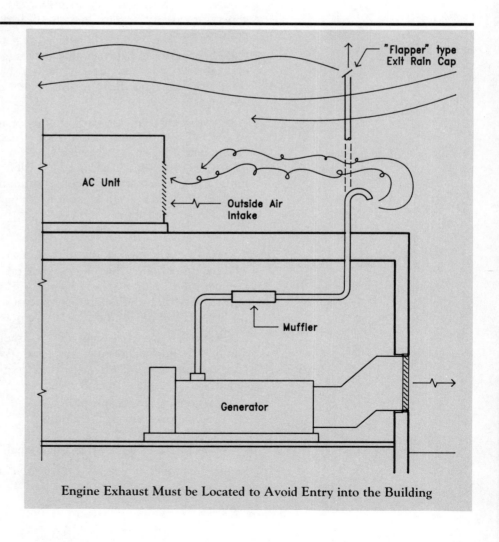

Engine Exhaust Must be Located to Avoid Entry into the Building

Figure 34.1

410

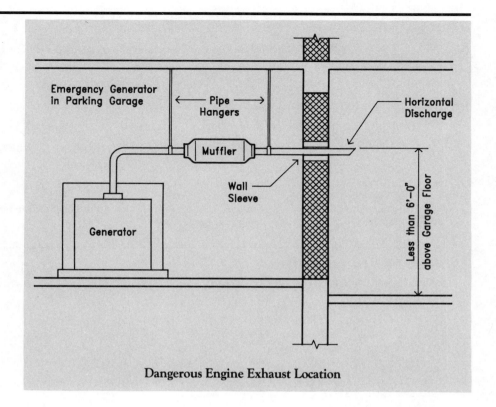

Figure 34.2

Emergency Generator
in Parking Garage

←— Pipe —→
Hangers

Horizontal
Discharge

Muffler

Wall
Sleeve

Generator

Less than 6'-0" above Garage Floor

Dangerous Engine Exhaust Location

411

Determining Engine Exhaust System Pressure Drop

Example: 800 kw diesel generator, 7150 acfm (865°F), 850°F Exhaust temperature (To), 10″ engine exhaust connection

Calculate muffler pressure drop.

70°F cfm (scfm) = acfm x (460 + 70) / (460 + To)
2860 cfm (70°F) = 7150 x (460 + 70) / (460 + 865)
Silencer pressure drop charts are plotted at 70°F. Assume the silencer will be the engine exhaust outlet size of 10″.
Area of 10″ silencer is 3.14 x (10/12) / 4 = 0.545 square feet.
V = Q / A V = 2860 / 0.545 V = 5244 fpm
Using the silencer pressure drop chart in figure 36.2, plot the 5255 fpm velocity to the intersection of the diagonal muffler line and read 6.5″ of wg as the muffler pressure drop.

Calculate piping pressure drop.

150 lineal feet of 10″ exhaust pipe, Four 10″ elbows, one exhaust cap (forced open by exhaust pressure to keep rain water from entering)
Equivalent length of elbows: L = 15 x Dia / 12
15 x 10 / 12 = 12.5 Four elbows x 12.5′ = 50′
10″ exhaust cap equivalent length = 39′
Total equivalent length = 150 + 50 + 39 = 239′

The engine backpressure must not exceed 27 in wg and is calculated by:
P(psi) = L x S x Q / 5184 x D
P = Backpressure (psi)
psi = 0.0361 x inches wg
L = Equivalent length of pipe (feet)
Q = Exhaust gas flow (cfm)
D = Inside pipe diameter (inches)
S = Specific weight of exhaust gas (lb-ft)
S (lb-ft) = 41.1 / Exhaust Temperature + 460°F
S = 41.1 / 865 + 460 = 0.03 lb-ft

The constant 41.1 is based on the weight of combustion air and fuel burned at the rated load and SAE conditions.

P = 239 ft x 0.03 x (7150) / 5184 x (10) = 0.707 psi
0.707 psi / 0.036 psi-in = 19.64″ wg
Total backpressure = (muffler) 6.5″ + (piping)
19.64 = 26.14″ wg.

No allowance has been made for pipe deterioration or scale. The 26.14″ clean pipe backpressure is near the 27″ maximum, and is unacceptable.

Revised Calculation

The area of a 12″ muffler is 0.785 sq. ft.
Velocity = 2860 cfm (70°F) / 0.785 = 3643 fpm
Using the muffler pressure drop chart of Figure 36.2, the muffler pressure drop is 3″ wg.
Piping pressure drop using 12″ pipe:
P (psi) = 239 ft x 0.03 x (7150) / 5184 x (12) = 0.284 psi.
0.284 psi / 0.036 psi-in = 7.89 in wg.

Total engine backpressure = (muffler) 3″ + (piping) 7.89″ = 10.89″ wg. This is acceptable. Backpressure using 10″ muffler and 12″ piping = (muffler) 6.5″ + (piping) 7.89″ = 14.39″ wg, which is also acceptable.

Figure 34.3

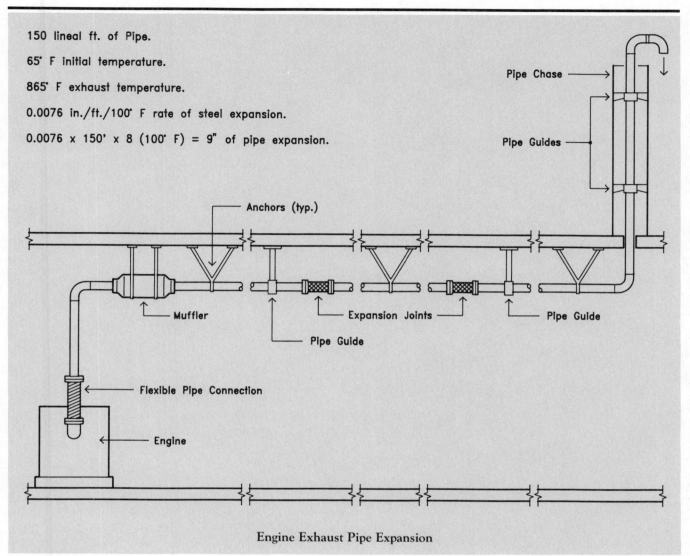

150 lineal ft. of Pipe.

65° F initial temperature.

865° F exhaust temperature.

0.0076 in./ft./100° F rate of steel expansion.

0.0076 x 150' x 8 (100° F) = 9" of pipe expansion.

Pipe Chase

Pipe Guides

Anchors (typ.)

Muffler

Pipe Guide

Expansion Joints

Pipe Guide

Flexible Pipe Connection

Engine

Engine Exhaust Pipe Expansion

Figure 34.4

Chapter Thirty-Five:

COOLING LOADS

Most HVAC systems designed since the late 1970s have included energy conservation measures in their calculations of cooling loads. Lighting loads were considered less than 2 watts per square foot, and office equipment loads were 1 watt per square foot. Outside air ventilation was based on 5 cfm per person, and occupancy was based on 1 person per 100 sf. Internal loads are now often over 3 watts/sf, with many at 5 or 6 watts per sf. Ventilation requirements are now 15-20 cfm per person. Modular work cubicles are increasing the occupancy to 1 person per 70 sf.

Most of the existing HVAC systems cannot effectively handle the increased cooling loads. Some means of adding additional cooling capacity must be provided in order to maintain comfort. Additional space cooling can be provided by:

- Increasing the supply air volume
- Lowering the supply air temperature
- Adding additional self-contained air-conditioning units

Figure 35.1 shows an example of interior versus perimeter loads.

The size of the "cloud" of conditioned air that is supplied to a space is determined by the temperature of the air supply. The psychrometric chart in Figure 35.2 describes that the air condition leaving the cooling coil at point "C" is determined by the sensible heat percentage line "R-C."

The cloud of air at "C" is warmed by the sensible heat in the space to point "Rs," and by the latent heat in the space from "Rs" to "R." To maintain room condition "R", supply air must enter the room at some point along the sloping line "C-R." The ratio of room sensible heat and latent heat that warms the supply air cloud is the same. The size of the cold air cloud entering the room depends on how close the supply air condition (on sloping line "C-R") is to room condition "R." If the supply air temperature is close to room air temperature "R," the cloud must be large, because each cubic foot of the cloud can only be warmed a small amount (small temperature change).

The relationship of the supply air cloud (cfm), the room temperature, and the cloud temperature (td) to the sensible heat in the room (btuh) is:

$$cfm = \frac{sensible\ btuh}{1.08 \times td}$$

In Figure 35.2, the 172,800 btuh sensible heat load is produced in a space to be maintained at 75°F dry bulb (db) and 50% relative humidity (rh). The sensible heat percentage line "R-C" intersects the 90% "rh" line at 57°F. This is the coil leaving condition. Fan heat warms the supply air to 59°F. This is the temperature of the air entering the room. The td is:

$$75 - 59 = 16$$

The supply air volume is:

$$\frac{172,800}{1.08 \times 16} = 10,000 \text{ cfm}$$

Let us examine the options for handling an increase of 12,000 more watts of power and lights, and 20 additional people to the space sensible heat load of 172,800 btuh. The sensible heat load is increased to 213,600 btuh.

Increasing the Supply Air Volume

$$\text{cfm} = \frac{213,600 \text{ btuh}}{1.08 \times 16} = 12,361 \text{ cfm}$$

The 2,361 cfm increase (23.6%) is beyond the existing fan's capacity. The cfm increase raises the static pressure from 3.5 to 5.34. The fan rpm would have to be increased from 1,025 to 1,250, and the motor hp would increase from 10 to 20 hp.

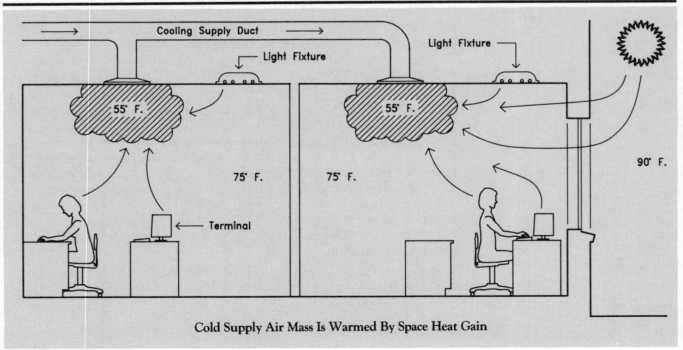

Cold Supply Air Mass Is Warmed By Space Heat Gain

Figure 35.1

The increase in fan hp increases the fan heat added to the supply air by almost 2°F. Raising the supply air temperature means that more cfm must be delivered to the space. The increase in cfm means an increase in fan hp. The increase in fan hp raises the supply air temperature. Because the supply air is warmer, more supply air is needed to cool the room. The cycle never ends.

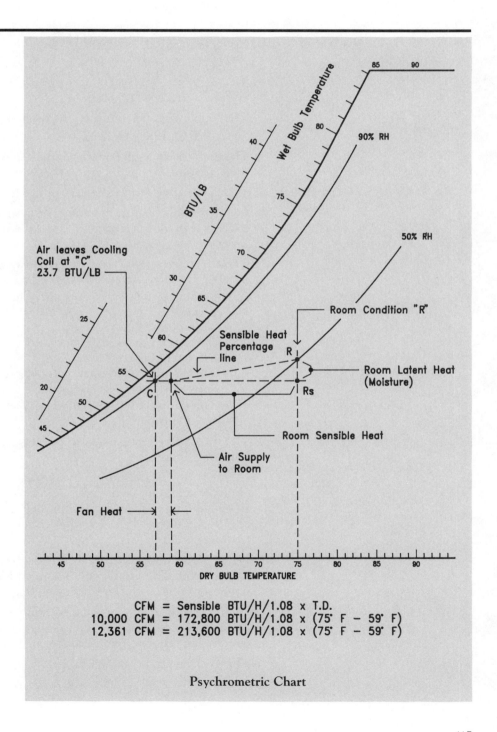

CFM = Sensible BTU/H/1.08 x T.D.
10,000 CFM = 172,800 BTU/H/1.08 x (75° F − 59° F)
12,361 CFM = 213,600 BTU/H/1.08 x (75° F − 59° F)

Psychrometric Chart

Figure 35.2

The 1.3″ increase in duct static pressure could produce excess duct air leakage. If the noise generated by the increased duct air velocity and the drafts created by the extra air supplied from the air outlets can be tolerated, adding a booster supply fan to the air distribution system will overcome the supply fan's inadequacy.

Lowering the Supply Air Temperature

Lowering the supply air temperature is the simplest method of overcoming additional space cooling loads. In Figure 35.2, the original supply air to the room is 59°F, with a coil leaving air temperature of 57°F. Figure 35.3 shows that if the original 10,000 cfm air volume is to be retained, the air must enter the room at 55.2°F in order to satisfy the new internal cooling load.

$$td = \frac{\text{Internal Sensible Heat}}{1.08 \times \text{cfm}}$$

$$td = \frac{213,600}{1.08 \times 10,000} = 19.8°F$$

The supply air temperature to the room is:

$$75°F - 19.8°F = 55.2°F$$

The coil leaving temperature will be the room entering temperature of:

$$55.2°F - \text{the fan heat of } 2°F = 53.2°F$$

The cloud of 55.2°F supply air will be heated to the 75°F room temperature by the sensible heat in the room. Because the supply air is starting at a lower coil leaving temperature, the final room air condition "R" will be at a lower relative humidity (42%).

The advantages of lowering supply air temperature versus increasing the supply air volume are as follows:

- Retaining the same supply air volume means that no fan changes are necessary.
- No booster fan is required.
- There is no increase in fan motor hp.
- Fan operating costs are lower.
- There are no additional drafts, ductwork, insulation, or air diffusers due to the increased air volume.
- There is no interruption of HVAC to the occupied space during system upgrading.

The increased cooling load can possibly be managed by investigating the following options:

Figure 35.4 is an illustration of a system using the existing cooling coil, chilled water pump and piping, and lowering the chilled water temperature to 40°F during peak load conditions. This figure also shows the installation of a new chiller and chilled water circulating pump in parallel with the existing chiller.

Figure 35.5 shows how a new chilled water coil is installed in series with the existing coil. The installation of a new chiller and chilled water pump are done in parallel with the existing chiller.

Figure 35.6 shows the installation of a new DX precooling coil in front of the existing cooling coil. The installation includes a new air-cooled condensing unit and refrigerant piping to serve the new coil.

The existing total cooling load is:

10,000 cfm x 4.5 x (29.4 btu/# − 23.7 btu/#) = 256,500 btuh (21.4 tons)

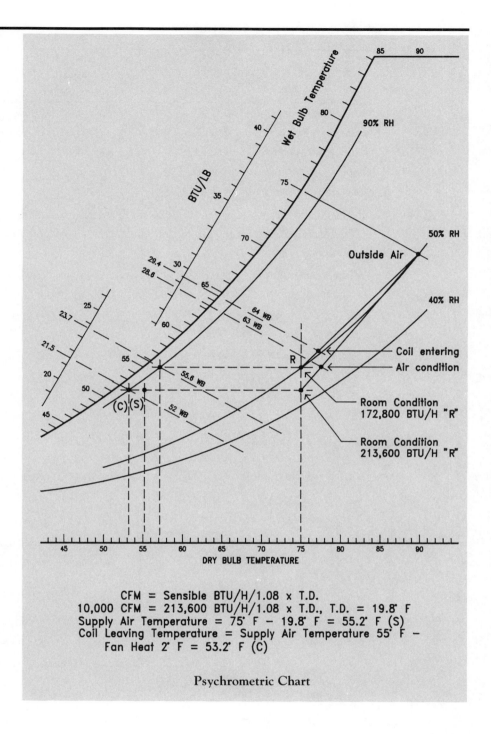

CFM = Sensible BTU/H/1.08 x T.D.
10,000 CFM = 213,600 BTU/H/1.08 x T.D., T.D. = 19.8° F
Supply Air Temperature = 75° F – 19.8° F = 55.2° F (S)
Coil Leaving Temperature = Supply Air Temperature 55° F –
Fan Heat 2° F = 53.2° F (C)

Psychrometric Chart

Figure 35.3

419

Cooling Coil Selection		
Total Load btuh	Original Load 256,500 btuh	New Load 319,500 btuh
10,000 cfm, 20.4 sq. ft. face area		
Entering air wet bulb temperature (EWB)	64°F	63°F
Leaving air wet bulb temperature (LWB)	55.6°F	52°F
Entering water temperature (EWT)	45°F	40°F
Leaving water temperature (LWT)	57°F	55°F
Water flow (gpm)	42	42
Water velocity (fps) = gpm x 1.71 / no. of circuits		

Existing Coil Selection
From manufacturer, 6 row, 8 fins/in. is 14 circuits
fps = 42 gpm x 1.71 / 14 = 5.13 fps, velocity should be 3-6 fps

$$\text{Coil capacity index (CPI)} = \frac{\text{btuh/sq ft coil area}}{.5\ (\text{EWB} + \text{LWB}) - \text{EWT}} = \frac{256,500/20.4}{.5\ (64 + 55.6) - 45} = 850$$

Coil capacity index (CPI) of 6 row, 8 fins/in, half circuit is 900

Capacity of Existing Coil With Additional Load

$$900\ (\text{CPI}) = \frac{319,500/20.4}{.5\ (63 + 52) - \text{EWT}} = \text{EWT} = 40°F$$

gpm = btuh / 500 x td, 42 = 319,500 / 500 x (LWT– 40), LWT = 55°F

Existing coil will handle 319,500 btuh with 42 gpm of 40°F water.

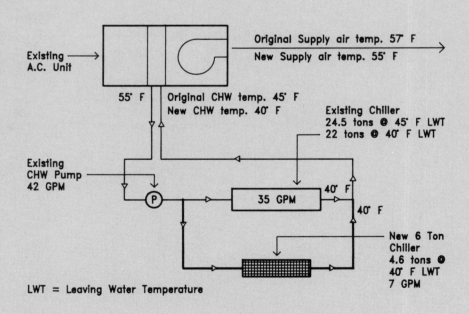

Lowering Chilled Water Temperature To Simplify Handling Increased Cooling Load

Figure 35.4

The new total cooling load is:

10,000 cfm x 4.5 x (28.6 btu/# − 21.5 btu/#) = 319,500 btuh (26.6 tons)

Adding Additional Self-Contained Air-Conditioning Units (SCAC)

If the tenant cannot exercise control over the central system to lower the air supply temperature or increase the supply air volume, he may have to provide the additional cooling capacity within his rented space. The additional cooling can be provided by:

- Adding a chilled water or DX cooling coil to each zone's duct system.
- Adding local SCAC units and air distribution systems at various locations within the conditioned space.

Figure 35.7 shows examples of adding cooling capacity with chilled water or DX booster coils.

Figure 35.8 shows how self-contained air-conditioning units can accommodate the additional cooling load.

Providing for Winter Cooling of Southern Solar Loads

Buildings with large areas of glass facing south usually experience the maximum cooling load for the south zone in January, February, March, September, October, November, and December. Figure 35.9 shows that the maximum solar heat gain for southern exposures in the 40° latitude are

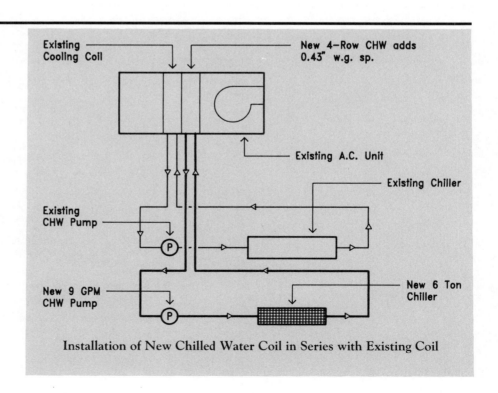

Installation of New Chilled Water Coil in Series with Existing Coil

Figure 35.5

254 btuh/sf, 241, 206, 200, 234, 250, and 253 btuh/sf, respectively. The maximum solar heat gain in June is 95 btuh/sf, 109 btuh/sf in July, and 149 btuh/sf in August.

Figure 35.10 shows that 2,300 cfm of 59°F supply air is needed to offset the exterior wall and window heat gain in July.

To offset the solar heat gain through the south glass in January, 3,400 cfm of 59°F air must be delivered to the space. Although 38,500 btuh is lost through the glass and 2,730 btuh leaks through the exterior wall these are replaced by the radiation under the window. Heat should be introduced under the windows in a cold climate so that the body will not radiate heat to a cold wall or glass surface. Should the radiator be inactive when it is 0°F, outside the glass surface temperature would be 23°F. With 20% relative humidity in the room, water will condense on the inside surface of the glass, and freeze when contacted by the 23°F glass surface temperature. Air systems must have a means of producing 59°F supply air for the south zone during the winter months.

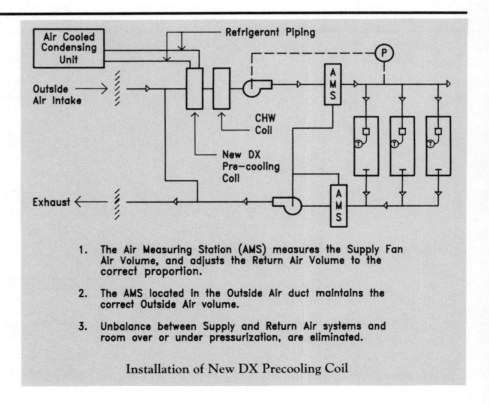

1. The Air Measuring Station (AMS) measures the Supply Fan Air Volume, and adjusts the Return Air Volume to the correct proportion.

2. The AMS located in the Outside Air duct maintains the correct Outside Air volume.

3. Unbalance between Supply and Return Air systems and room over or under pressurization, are eliminated.

Installation of New DX Precooling Coil

Figure 35.6

Methods of Producing 59°F Supply Air in Winter

- Economizer (100% outside air capability)
- Winterized cooling tower for water-cooled systems
- Air-cooled refrigeration units
- Use of cooling tower condenser water in a chilled water coil
- Cooling tower condenser water with a heat exchanger for chilled water
- Glycol closed circuit coolers with a heat exchanger for chilled water
- Free cooling with centrifugal chillers

Economizer

Using 55°F outside air to cool the south zone is the easiest method of providing winter cooling. This is most effective when the south zone is served by a separate air handling unit. With a separate air handling unit for the south zone, the south zone can be cooled while the other zones are being heated. If a single air handler serves the entire building, those zones that will be overcooled with 59°F air must have reheat available. Reheat is not energy efficient, but in this application, essential to produce space comfort.

Winterized Cooling Tower for Water-Cooled Systems

If only minimum outside air is available for winter cooling, the refrigeration systems may have to be operated to cool the south zone.

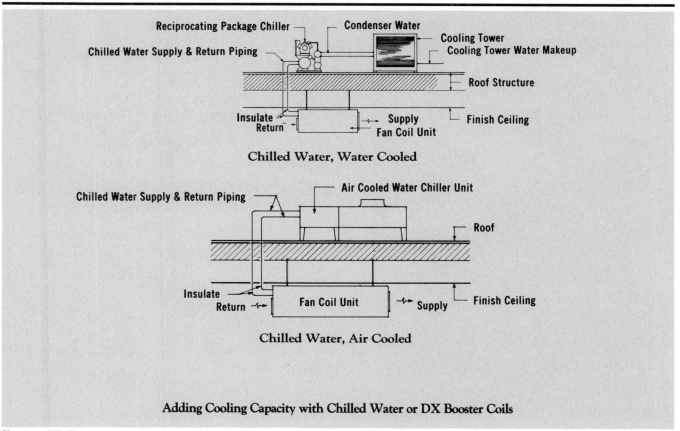

Chilled Water, Water Cooled

Chilled Water, Air Cooled

Adding Cooling Capacity with Chilled Water or DX Booster Coils

Figure 35.7

423

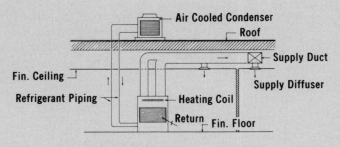

Self-contained, Air Cooled

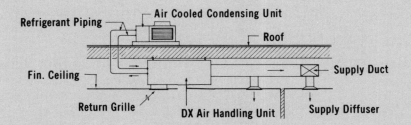

Split System, Air Cooled

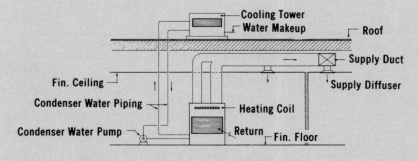

Self-contained, Water Cooled

Self-Contained Air-Conditioning Units

Figure 35.8

Cooling towers must be protected from freezing. This is accomplished by providing electric, steam, hot water, or glycol heating coils in the tower basin to prevent the water from freezing. The condenser water, makeup water, drain, and overflow pipes should be electrically heat traced and insulated to avoid freezing of the tower water. Winter tower operation can also be provided by locating the cooling tower basin inside the heated building. If the south zone is not served by a separate air handler, reheat must be available for those zones that do not require cooling.

Air-Cooled Refrigeration Units

One of the major advantages of air-cooled refrigeration systems is that they are available for operation year-round. No water is involved in the condensing process. This eliminates the necessity for winterization. Mechanical cooling is ready for use any time that it is needed.

Use of Cooling Tower Water in a Chilled Water Coil

Cooling towers are capable of cooling water to within 7°F of the outdoor wet bulb temperature. If the heat rejection load on the cooling tower is low, as would be the case for winter cooling of the south zone, the cooling tower is very efficient and 48°F water can be produced with a 39°F wet bulb temperature. There are 3,000 hours in the northern states when the wet bulb temperature is below 39°F. Figure 35.11 shows that the condenser water from the cooling tower bypasses the chiller and is circulated directly to the chilled water cooling coil.

Dirty condenser water from the cooling tower must never be permitted to enter the chiller or the cooling coil tubes. Condenser must be efficiently filtered before entering the chilled water coils.

	40°N Lat South
Jan	254
Feb	241
Mar	206
Apr	154
May	113
Jun	95
Jul	109
Aug	149
Sep	200
Oct	234
Nov	250
Dec	253

Maximum Solar Heat Gain Factor btuh/ft²
for Sunlit Glass North Latitudes

Figure 35.9

If the system is cooled by a central DX coil, and even if it is served by a central chilled water coil, chilled water coils can be installed in the south zone air distribution system. This will permit cooling tower cold water to be circulated to the south zone cooling coils without cooling the rest of the building.

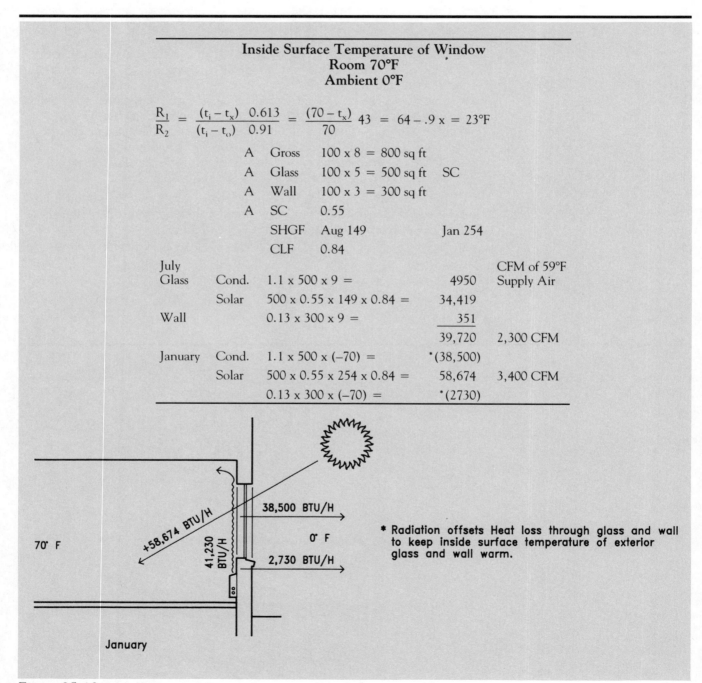

Inside Surface Temperature of Window
Room 70°F
Ambient 0°F

$$\frac{R_1}{R_2} = \frac{(t_i - t_x)}{(t_i - t_o)} \frac{0.613}{0.91} = \frac{(70 - t_x)}{70} 43 = 64 - .9 \ x = 23°F$$

A	Gross	100 x 8 = 800 sq ft		
A	Glass	100 x 5 = 500 sq ft	SC	
A	Wall	100 x 3 = 300 sq ft		
A	SC	0.55		
	SHGF	Aug 149	Jan 254	
	CLF	0.84		

July				CFM of 59°F
Glass	Cond.	1.1 x 500 x 9 =	4950	Supply Air
	Solar	500 x 0.55 x 149 x 0.84 =	34,419	
Wall		0.13 x 300 x 9 =	351	
			39,720	2,300 CFM
January	Cond.	1.1 x 500 x (–70) =	*(38,500)	
	Solar	500 x 0.55 x 254 x 0.84 =	58,674	3,400 CFM
		0.13 x 300 x (–70) =	*(2730)	

38,500 BTU/H

+58,674 BTU/H

41,230 BTU/H

0° F

2,730 BTU/H

70° F

January

* Radiation offsets Heat loss through glass and wall to keep inside surface temperature of exterior glass and wall warm.

Figure 35.10

426

Cooling Tower Condensed Water with a Heat Exchanger for Chilled Water

One way to avoid contaminating the chilled water system with dirty condenser water is by separating the condenser water and chilled water circuits via a heat exchanger. Although a 4-5°F loss of chilled water temperature results from the heat exchange process, the separation of the water loops justifies that penalty. Figure 35.12 shows the indirect "free cooling" system. All other aspects of cooling tower chilled water apply.

Glycol Closed Circuit Coolers with a Heat Exchanger for Chilled Water

Glycol closed circuit coolers cool the circulating warm glycol by passing cooler outside air over the glycol coil. Figure 35.13 shows how the cooled glycol is circulated through the refrigeration unit's water-cooled condenser to pick up refrigerant heat and then to the closed circuit cooler for

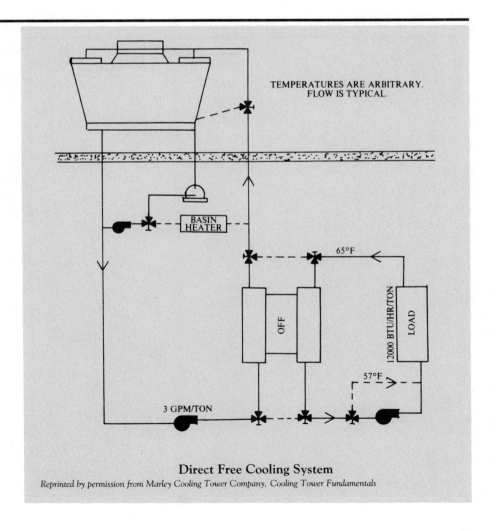

Direct Free Cooling System

Reprinted by permission from Marley Cooling Tower Company, Cooling Tower Fundamentals

Figure 35.11

another cycle. During the fall and winter seasons, the cooled glycol can be circulated directly to the chilled water coil or to a heat exchanger to cool the chilled water.

The cooled glycol can also be circulated to chilled water coils installed in the south supply air ducts to provide winter cooling without adding unnecessary cooling to the other building areas.

Circulating the glycol directly to the cooling coil will dilute the glycol by mixing it with the water that is in the cooling coil. This type of "free cooling" requires monitoring the glycol solution during the winter months to assure freeze protection is intact.

Free Cooling with Centrifugal Chillers
Centrifugal chillers can provide cooling without using the compressor. When the condenser water supply is below the desired chilled water temperature, the chiller can act as a thermal siphon. The refrigerant is condensed by the low temperature cooling tower water. The liquid

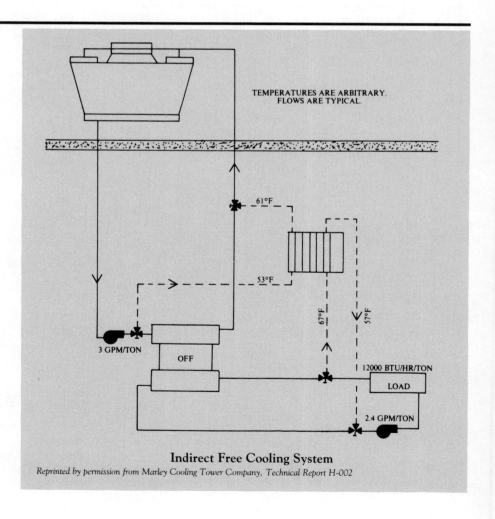

Indirect Free Cooling System
Reprinted by permission from Marley Cooling Tower Company, Technical Report H-002

Figure 35.12

428

refrigerant either flows by gravity or is pumped into the evaporator. The higher temperature chilled water return causes the refrigerant to evaporate, and the pressure difference between the evaporator and the condenser moves the refrigerant vapor back to the condenser. Between 10% and 30% of the chiller's capacity is available for free cooling. The design of the centrifugal chiller determines its use for free cooling. See Figure 35.14

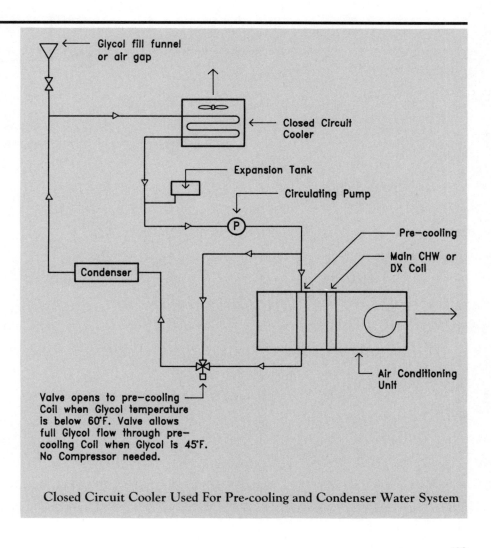

Glycol fill funnel or air gap

Closed Circuit Cooler

Expansion Tank

Circulating Pump

Pre-cooling

Main CHW or DX Coil

Condenser

Air Conditioning Unit

Valve opens to pre-cooling Coil when Glycol temperature is below 60°F. Valve allows full Glycol flow through pre-cooling Coil when Glycol is 45°F. No Compressor needed.

Closed Circuit Cooler Used For Pre-cooling and Condenser Water System

Figure 35.13

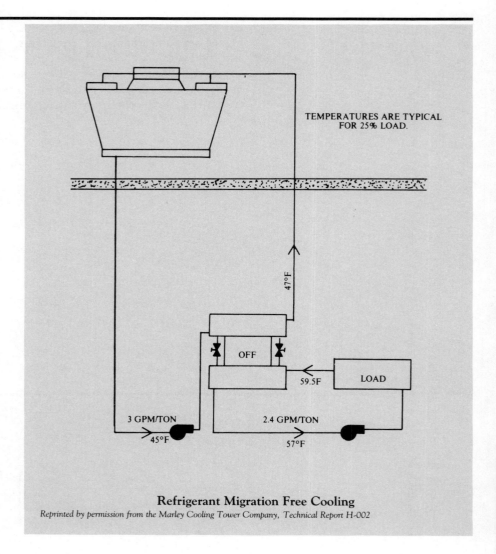

TEMPERATURES ARE TYPICAL
FOR 25% LOAD.

47°F

OFF

59.5F

LOAD

3 GPM/TON

45°F

2.4 GPM/TON

57°F

Refrigerant Migration Free Cooling
Reprinted by permission from the Marley Cooling Tower Company, Technical Report H-002

Figure 35.14

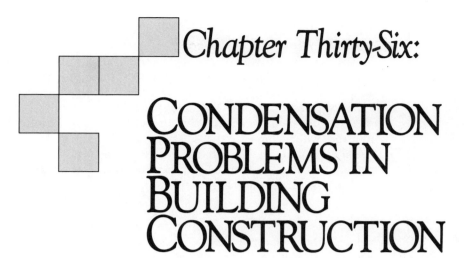

Chapter Thirty-Six:

CONDENSATION PROBLEMS IN BUILDING CONSTRUCTION

Water vapor is always present in the air and is present in most building materials. Moisture becomes a problem when it becomes visible or degrades building material. Condensation, the change from a gas to a liquid or solid, occurs when the air is cooled below its dew point temperature. This is the result of the vapor moving to a region of lower temperature or to a reduction of surface temperature. Figure 36.1 shows how room air at 72°F and 40% relative humidity at point A contains 0.0066 pounds of moisture per pound of air.

In this figure, the 70°F room air is cooled by a cold surface to point B. The air at point B has been cooled to 46°F and is saturated. That means that the 46°F air cannot contain any more than 0.0066 lbs. of moisture per lb. of air. When the air is further cooled, it follows the saturation curve down toward point C, where the air can hold only 0.006 pounds of moisture per pound of air. The 0.006 pounds of moisture per pound of air is released from the air.

Condensation in Cold Climates

Winter condensation usually takes place on window glass and frames where the thermal resistance is not great enough to keep the inside surface temperature above the dew point temperature. Figure 36.2 demonstrates that the inside surface temperature of single glass with an indoor room temperature of 70°F and an outdoor temperature of 5°F, is 26°F.

The dew point temperature at 70°F and 30% relative humidity is 35.6°F with 0.0047 pounds of moisture per pound of air. A pound of air at 26°F can hold only 0.00285 pounds of moisture, so water will drop out of the air onto the glass. Since the glass temperature is 26°F, the water on the glass will become ice.

Figure 36.3 shows that the inside temperature of double glass is 46°F, which is above the 35.6°F room dew point temperature. Therefore, condensation does not take place.

Heat should be supplied under windows to prevent the discomfort caused by the loss of body heat to the cold window surface. Condensation and ice formation on single glazed windows can *only* be avoided by delivering heat to the glass inside surface.

It is important that a good airtight vapor barrier be applied to the warm room side of the exterior wall to prevent the moisture from the room at

0.205" of mercury vapor pressure from passing through the wall to the outdoor vapor pressure of only 0.03" of mercury.

When the vapor barrier is attached to the insulation, the vapor barrier should be on the warm room side. Staples, piping, electrical boxes and wiring, and framing penetrate the vapor barrier, thereby permitting water vapor to pass through the wall. Depending on the quality and integrity of the insulation, dew point temperature may be reached inside the wall, and moisture will accumulate there. Masonry walls can crack from freezing, and metal building materials will deteriorate. All air leaks at the exterior walls should be sealed. A continuous vapor barrier and as much insulation in the wall as possible will limit the condensation.

Condensation Under Roofs

Figure 36.4 shows how condensation develops on the underside of roofs in cold weather when the air in contact with the underside of the roof is cooled below its dew point temperature.

The surface temperature of the uninsulated roof in Figure 36.4 is 50.4°F, which is below the 55°F dew point temperature of the air under the roof. The staff of this facility will not appreciate the interior precipitation.

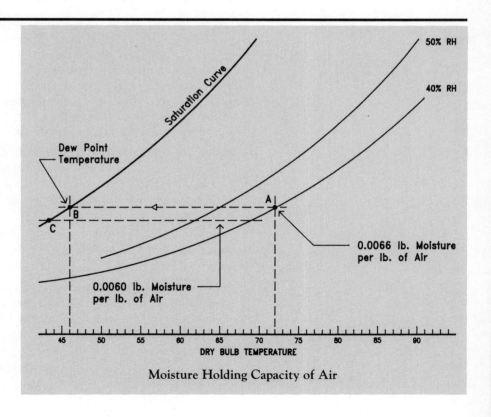

Figure 36.1

Moisture Holding Capacity of Air

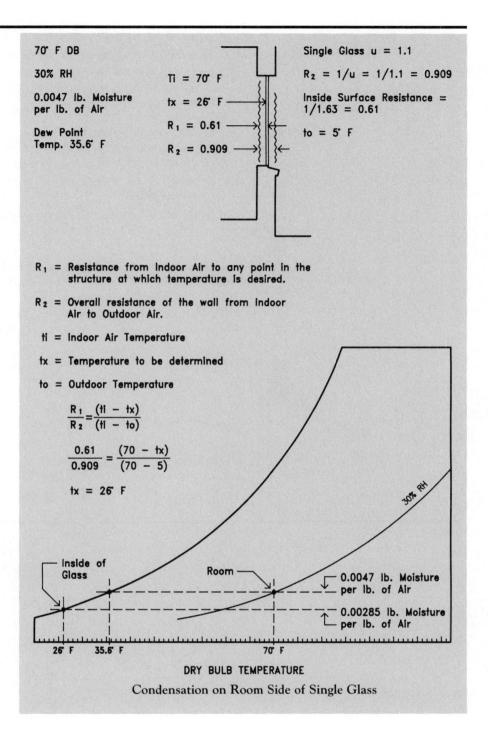

70° F DB

30% RH

0.0047 lb. Moisture per lb. of Air

Dew Point Temp. 35.6° F

$T_i = 70°$ F

$t_x = 26°$ F

$R_1 = 0.61$

$R_2 = 0.909$

Single Glass $u = 1.1$

$R_2 = 1/u = 1/1.1 = 0.909$

Inside Surface Resistance = $1/1.63 = 0.61$

$t_o = 5°$ F

R_1 = Resistance from Indoor Air to any point in the structure at which temperature is desired.

R_2 = Overall resistance of the wall from Indoor Air to Outdoor Air.

t_i = Indoor Air Temperature

t_x = Temperature to be determined

t_o = Outdoor Temperature

$$\frac{R_1}{R_2} = \frac{(t_i - t_x)}{(t_i - t_o)}$$

$$\frac{0.61}{0.909} = \frac{(70 - t_x)}{(70 - 5)}$$

$$t_x = 26° \text{ F}$$

Inside of Glass

Room

30% RH

0.0047 lb. Moisture per lb. of Air

0.00285 lb. Moisture per lb. of Air

26° F   35.6° F   70° F

DRY BULB TEMPERATURE

Condensation on Room Side of Single Glass

Figure 36.2

433

The problem can be removed by:

- Spraying two inches of polyurethane insulation on the roof. The polyurethane will not only raise the inside surface temperature to 67.2°F to prevent condensation, but will also provide a waterproof roof surface.
- Blow supply air into the air space between the roof and the suspended ceiling to produce a higher roof underside surface temperature.
- Treat the room ceilings and exterior walls with vapor retarding paint or vinyl wall covering. In this case, an ounce of prevention can keep a pound of moisture from reaching a cool surface, thereby preventing a host of problems.

Facilities that enjoy winter indoor humidities of 40-50% are prone to greater moisture damage. Insulation added to the roof will produce a higher roof underside surface temperature and prevent condensation from forming.

The roof membrane resists vapor escape. When insulation is placed between the warm interior and the roof membrane, the insulation lowers the temperature under the membrane. Temperatures below dew point may persist for some time, and moisture will accumulate in the insulation, reducing its "R" value, thereby freezing and damaging the building

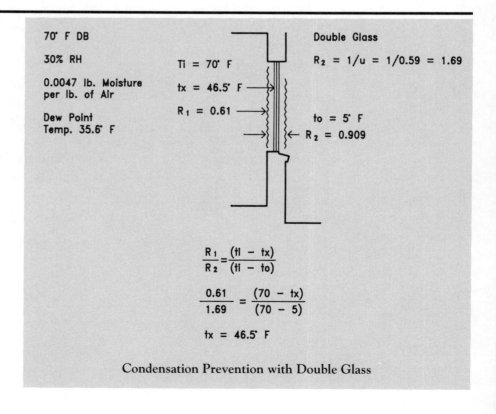

70° F DB

30% RH

0.0047 lb. Moisture per lb. of Air

Dew Point Temp. 35.6° F

$T_i = 70°$ F

$t_x = 46.5°$ F

$R_1 = 0.61$

Double Glass

$R_2 = 1/u = 1/0.59 = 1.69$

$t_o = 5°$ F

$R_2 = 0.909$

$$\frac{R_1}{R_2} = \frac{(t_i - t_x)}{(t_i - t_o)}$$

$$\frac{0.61}{1.69} = \frac{(70 - t_x)}{(70 - 5)}$$

$$t_x = 46.5° \text{ F}$$

Condensation Prevention with Double Glass

Figure 36.3

434

materials. An application of insulated roof material will keep the roof membrane temperature above dew point and prevent condensation.

Summer Moisture Problems

During summer, moist room air is trapped under the roof membrane. The hot sun expands the contained water vapor, creating a bubble in the roof membrane. Figure 36.5 shows the resulting bubble and the crack that develops–either from weather conditions or when it is stepped on.

Wall Destruction in Warm, Humid Climates

Air-conditioned buildings often experience mold and mildew growth. Mildew is produced by molds which are simple plants of the fungi group. Molds that generate mildew need liquid moisture and warm temperatures to grow. They usually develop in the still air of muggy summer weather. Air-conditioning in humid climates reduces the vapor pressure inside the building, and moisture flows from the higher exterior vapor pressure into the lower interior vapor pressure environment. In humid climates, vapor barriers should be applied to the warm exterior side of the exterior wall.

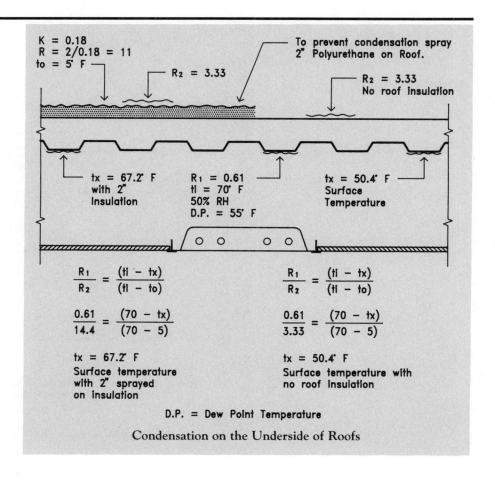

$K = 0.18$
$R = 2/0.18 = 11$
$to = 5° F$

$R_2 = 3.33$

To prevent condensation spray 2" Polyurethane on Roof.

$R_2 = 3.33$
No roof insulation

$tx = 67.2° F$ with 2" Insulation

$R_1 = 0.61$
$ti = 70° F$
50% RH
$D.P. = 55° F$

$tx = 50.4° F$ Surface Temperature

$$\frac{R_1}{R_2} = \frac{(ti - tx)}{(ti - to)}$$

$$\frac{0.61}{14.4} = \frac{(70 - tx)}{(70 - 5)}$$

$tx = 67.2° F$
Surface temperature with 2" sprayed on insulation

$$\frac{R_1}{R_2} = \frac{(ti - tx)}{(ti - to)}$$

$$\frac{0.61}{3.33} = \frac{(70 - tx)}{(70 - 5)}$$

$tx = 50.4° F$
Surface temperature with no roof insulation

D.P. = Dew Point Temperature

Condensation on the Underside of Roofs

Figure 36.4

Gypsum board can be destroyed because the vinyl wall covering trapped the moisture that poured through the exterior wall (which has no vapor barrier). Figure 36.6 describes how the trapped vapor at 0.018 pounds of moisture per pound of air is condensed by being cooled by the 72°F indoor air.

The 72°F indoor air can only contain 0.017 pounds of moisture per pound of air. The trapped air drops its water bomb at the rate of 0.001 pounds of water per pound of air, saturating the gypsum board and rusting the metal studs, until finally, the saturated collapsed gypsum board is supported only by the vinyl wall covering.

Correcting this problem involves adding vapor barrier solutions to the building exterior to keep the moisture outside the building. Relief can also be brought by removing the vinyl wall covering and allowing the water that enters the structure to be removed by the air conditioning.

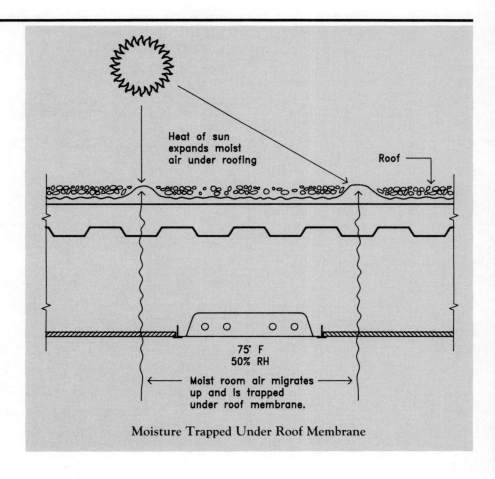

Moisture Trapped Under Roof Membrane

Figure 36.5

Air-Conditioning System Operation

Reheat is usually necessary to limit the humidity level in the building. It is best not to reset supply air up unless the humidity is controlled. When the supply air temperature is raised, dehumidification at the cooling coil is reduced, and the room humidity is increased. Humidity control will activate the cooling coil, and space overcooling will be controlled by the reheat coils.

Space Pressurization

If the inside air pressure is greater than the outdoor air pressure, the amount of outside air infiltration will be eliminated or reduced. This can be accomplished by returning less air from the space than is supplied to

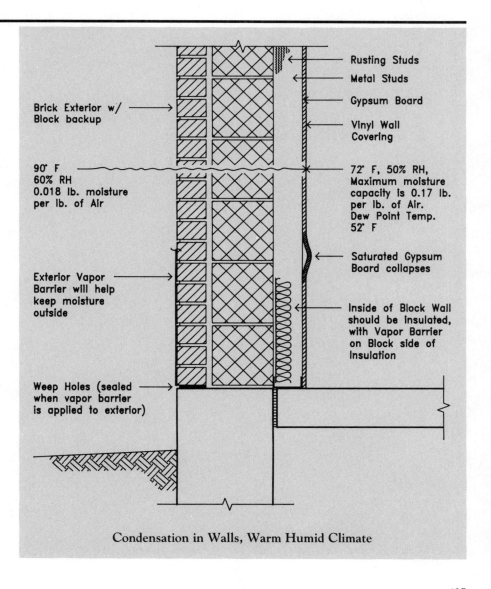

Condensation in Walls, Warm Humid Climate

Figure 36.6

the space. Outside air must be taken into the air-conditioning unit to replace the air that is exfiltrating the building.

Outside air taken into the air-conditioning unit is dehumidified by the cooling coil before it is supplied to the conditioned space. Infiltrated outside air is not conditioned, and therefore adds to the building humidity.

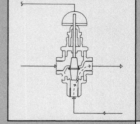

Part Three:
ELECTRICAL
COORDINATION

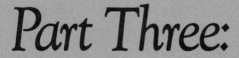

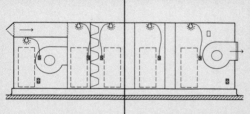

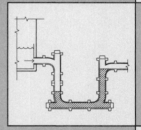

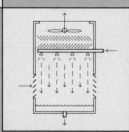

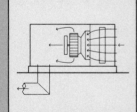

Part Three, "Electrical Coordination,"
covers electrical requirements commonly
associated with HVAC equipment. This
information is important not only to the
HVAC contractor (who is responsible for
the complete system), but also to the
electrical contractor (who must perform the
actual electrical installation). These items are
frequently overlooked at the design stage,
thereby causing conflicts, delays, and claims.

Chapter Thirty-Seven:

ELECTRICAL COORDINATION

HVAC designers often forget to advise the electrical designer that mechanical items need electrical power to function. The list below includes electrical items that are often required, but not specified. Following the list is a description of each item – both its function and basic characteristics.

- Temperature control panels
- Temperature control air compressors
- Temperature control air dryers
- Water treatment circulating pumps
- Unit and cabinet heater control wiring
- Booster circulating (freeze protection) pumps
- Domestic hot water circulating pumps
- VAV boxes
- Fan-powered air terminal units
- High efficiency filtration units
- Electric heat tracing on piping
- Electric reheat coils
- Air transfer fans
- Service lights and power in mechanical equipment rooms and air handling units
- Water flow switches
- Airflow switches
- Fan run pilot lights from airflow switches (for proof of flow)
- Interlock and alarm wiring

Temperature Control Panels

Electric, electronic, pneumatic, and direct digital control system temperature control panels require electrical power to function. Temperature control panels need 120 volt power for transformers to deliver 24 volt control circuits, as well as 120 volt power for electric to pneumatic (e-p) switches.

In all of the above systems, the 120 volt power supply should be served by a fused disconnect switch. In this way, a $.75 fuse can save a $400 control. DDC control system controls should be served by a dedicated 120 volt circuit that is not providing power to any other device. Electrical interference from fluorescent lights or copy machines will raise havoc with DDC systems.

Many temperature control panels are equipped with fluorescent lights to facilitate servicing the panel. When DDC is involved, the control power should be on a separate circuit from the control panel light power. Figure 37.1 shows the electrical requirements of temperature control panels.

Temperature Control Air Compressors

Pneumatic control systems operate with compressed air. Very often a pneumatic control system is specified, but the electrical requirements of the air compressor and the associated accessories are not provided. The air compressor hp must be determined by the HVAC designer who must give the compressor power data to the electrical designer for inclusion in the contract documents.

Temperature Control Air Dryer

Compressed air for pneumatic control systems must be free of moisture. Water or oil in a pneumatic control system prevents control air from surviving the passage through the control devices. Refrigerated dryers cool the compressed air, leaving the temperature control air compressor below the control air dew point temperature, removing the contaminating moisture from the control air. The smallest air-cooled condensing unit delivers 10 scfm of control air and uses a 1/6 hp motor. Most control systems use 1/6-1/2 hp units. Even if the air compressors are identified for the electrical designer, the small refrigerated dryer is often overlooked. Figure 37.2 shows a typical pneumatic temperature control air supply layout, with the electrical provisions.

Water Treatment Circulating Pumps

Chilled water, condenser water, hot water heating, and steam heating systems require water treatment systems. The chemicals that control the water quality for these systems are often pumped into the respective piping systems. Those injection pumps need electrical power to drive the pumps. The pumps are usually small and 120 volt power is normally adequate. Power provisions for chemical treatment systems are shown in Figure 37.3.

Unit and Cabinet Heater Wiring

One of the major conflicts in assigning responsibility for construction installations involves the question of what is control wiring, and what is power wiring? The unit heater motor must have power wiring connected to it to enable it to run. The thermostat that turns the heater fan on and off, and the aquastat that prevents the heater fan from operating unless there is heat flowing through the heater coil must interrupt the power supply. If the electrical contractor does power wiring, and the HVAC contractor does control wiring, who wires the line voltage thermostat and aquastat? It is not desirable to have two different contractors running separate conduits for wiring the same piece of equipment, when a single conduit can accommodate all of the wiring. It is better to have the electrical contractor perform the power and control wiring of all equipment where the power supply to the equipment is activated by control devices. This work must be identified on the electrical construction documents. Figure 37.4 shows an example of cabinet heater wiring.

Booster Circulating (Freeze Protection Pumps)

In-line hot water circulating pumps are often installed in the preheat coil piping circuit to induce water circulation through the coil, when coil freezing is possible. Outdoor thermostats activate the circulating pumps when the outdoor temperature is below 30°F. Unfortunately, the circulating pumps are often installed before it is realized that there is

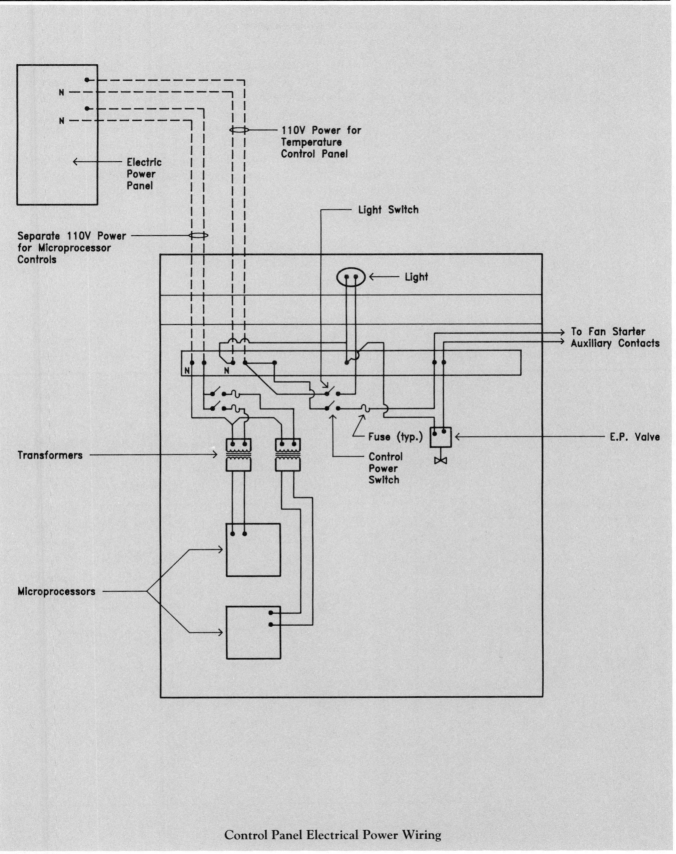

Control Panel Electrical Power Wiring

Figure 37.1

no power to run the pumps. Figure 37.5 shows a freeze protection pumping system with the associated electrical work.

Domestic Hot Water Circulating Pumps

Domestic hot water circulating pumps occasionally suffer the fate of their freeze protection cousins. Domestic hot water circulating pumps are usually located near the hot water generating unit, and are controlled by an aquastat in the hot water return pipe. The pumps are usually small and could be another case where even if the wiring is addressed, it may be subjected to the power and control debate. If the aquastat breaks the power supply to the pump in order to cycle the pump, is it power or control wiring? Figure 37.6 shows the electrical needs of a domestic hot water circulating system.

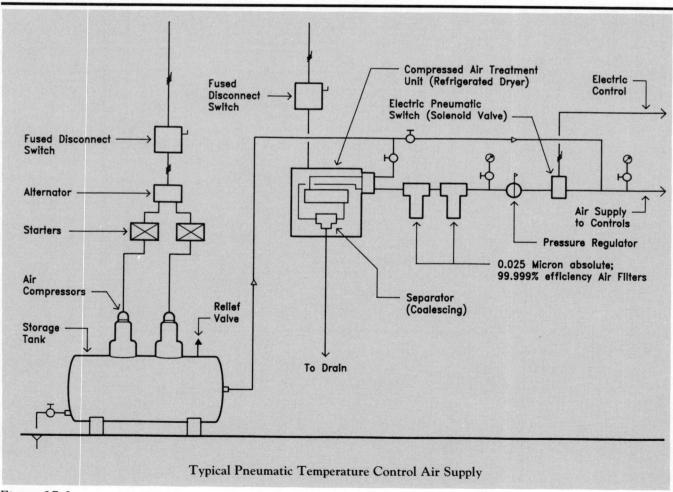

Fused Disconnect Switch

Fused Disconnect Switch

Alternator

Starters

Air Compressors

Storage Tank

Relief Valve

To Drain

Compressed Air Treatment Unit (Refrigerated Dryer)

Electric Pneumatic Switch (Solenoid Valve)

Electric Control

Air Supply to Controls

Pressure Regulator

0.025 Micron absolute; 99.999% efficiency Air Filters

Separator (Coalescing)

Typical Pneumatic Temperature Control Air Supply

Figure 37.2

VAV Boxes

Each VAV control box should be provided with 110 volt power so that the transformer at each VAV controller isolates each of the units from the other units. Electric and electronic controlled air terminal units must be supplied with 120 volt or 24 volt power for the unit control. If the HVAC contractor is to install the temperature control wiring, 120 volt power must first be made available for VAV wiring. DDC shielded wiring must not be run in the same conduit as the 120 volt power supply wiring.

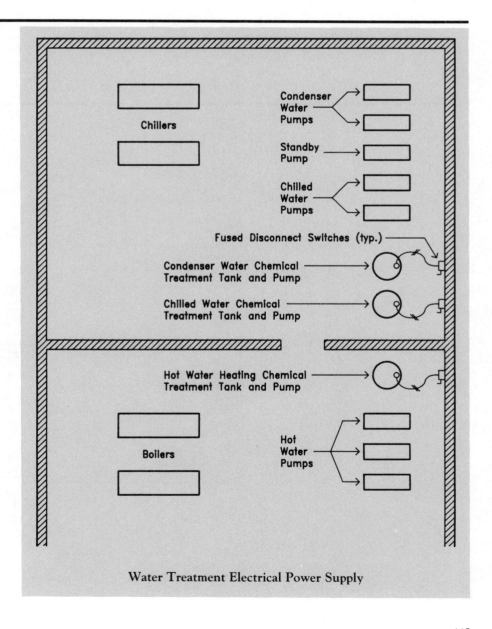

Water Treatment Electrical Power Supply

Figure 37.3

Fan-powered boxes must be provided with 110 volt power. DDC control power must not be taken from the power supply at the VAV fan, but should be from a completely separate supply.

High Efficiency Filtration Units

Specialty HVAC items such as high efficiency filtration units must have 120 volt power to operate. High efficiency filtration systems are installed within the air handling unit between the cooling coil and the fan. In addition to the power supply required to energize the filtration unit, interlock wiring is required to ensure interruption of the power supply to the filtration unit whenever the supply fan is off or the access door to the air handling unit coil or fan section is opened. This is yet another source of possible confusion in determining responsibility for power supply/control wiring. The electrical contractor is usually assigned the responsibility of

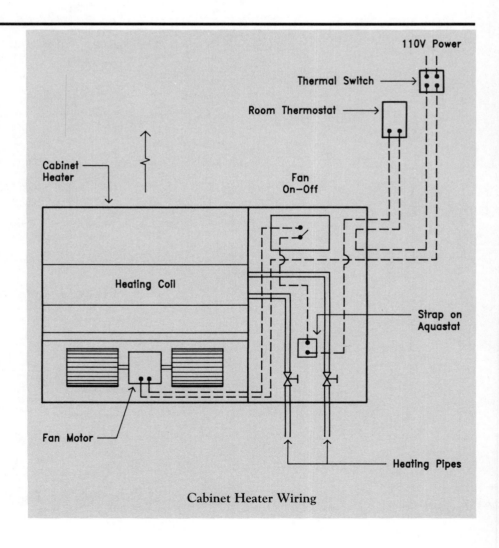

Cabinet Heater Wiring

Figure 37.4

bringing 120 volt power to the filtration unit. The extent of control and interlock wiring should be identified in the construction documents and performed by the electrical contractor. Figure 37.7 identifies the extent of the electrical work that should be presented to the contractors.

Electric Heat Tracing for Piping

Piping systems that may be subject to freezing conditions are very often protected by electric heat tracing systems. These are electric heating elements that are secured to the pipe exterior and controlled by a thermostat attached to the pipe. Pipe insulation covers the heat tracing elements and the piping to preserve the generated heat. A similar application involves water pipes that are heat traced to maintain water

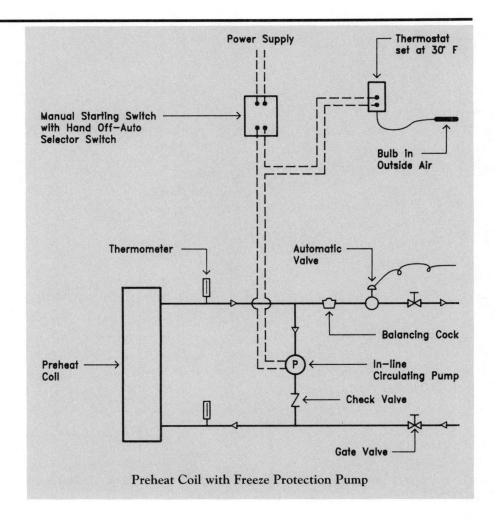

Preheat Coil with Freeze Protection Pump

Figure 37.5

temperature in lieu of a hot water recirculating pipe and pump. Electrical power supply, as well as control wiring, must be provided for these systems.

Applications for heat tracing include winterized cooling tower condenser water, makeup water, and drain piping that are exposed to outdoor conditions. If the cooling tower is drained during extreme winter conditions, a float switch in the condenser water and makeup water piping should de-energize the heat tracing electric power so that energy is not wasted protecting an empty basin or piping. Figure 37.8 is a schematic illustration of a condenser water heat tracing system.

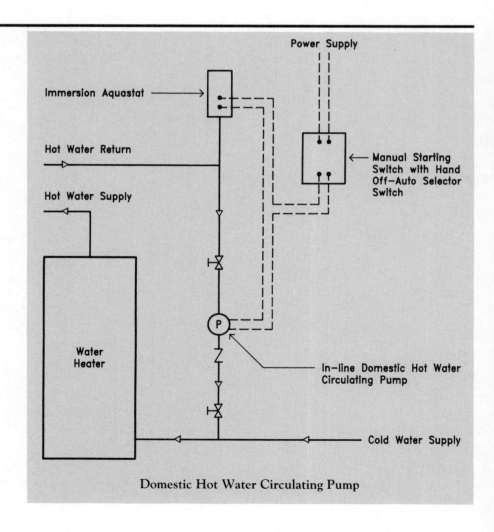

Domestic Hot Water Circulating Pump

Figure 37.6

Electric Reheat Coils

Electric reheat coils are often located in the branch duct system. Since they are not major components of HVAC equipment, it is possible to overlook the electrical coordination required to provide for power supply to the reheat coils.

Air Transfer Fans

Air transfer fans are used to overcome the ductwork and acoustic treatment resistance of the duct system used to transfer air from an enclosed room to the ceiling return plenum. The rooms are enclosed for acoustical reasons, either for privacy, or to prevent equipment-generated noise from an enclosed room from being transferred to the ceiling plenum and the adjacent rooms. Enclosed rooms that may utilize air transfer ducts and fans include conference rooms, executive offices, doctors' examining rooms, copy rooms, office machinery rooms (such as automatic mail handling), air compressor rooms, emergency generator equipment rooms, and boiler rooms.

The air transfer fans should be interlocked with the associated supply fan, so that the transfer fan operates only when the supply fan operates. Responsibility for installing the interlock wiring must be identified.

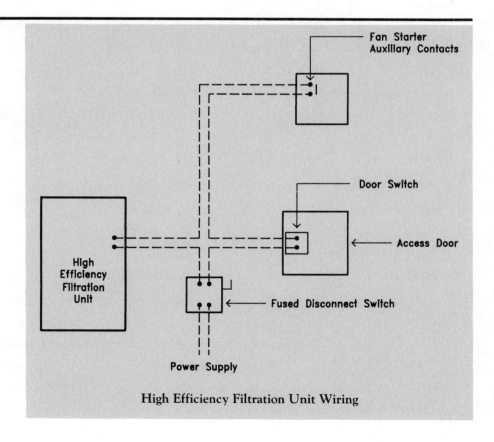

High Efficiency Filtration Unit Wiring

Figure 37.7

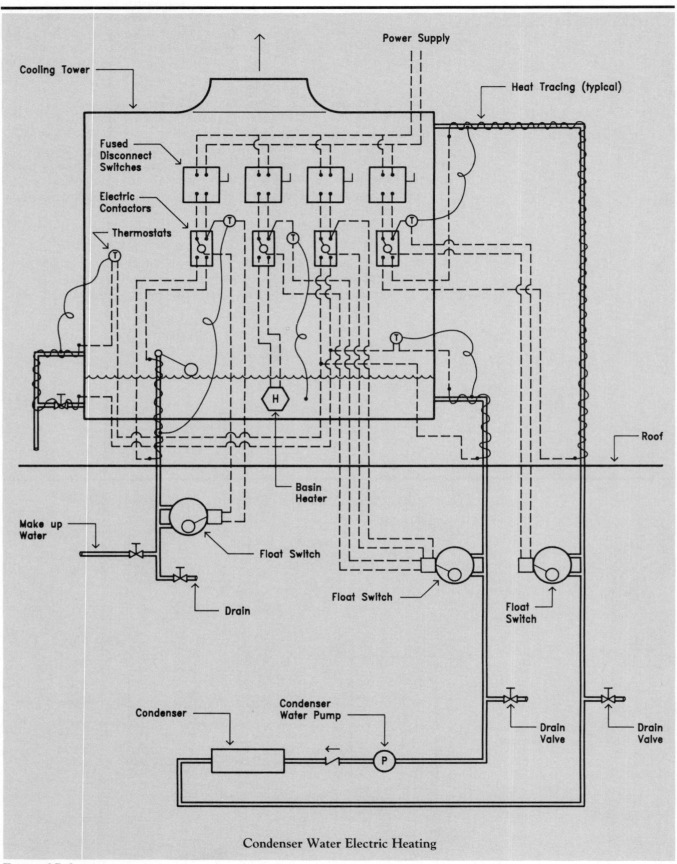

Cooling Tower

Power Supply

Heat Tracing (typical)

Fused
Disconnect
Switches

Electric
Contactors

Thermostats

Roof

Basin
Heater

Make up
Water

Float Switch

Float Switch

Float
Switch

Drain

Drain
Valve

Drain
Valve

Condenser

Condenser
Water Pump

Condenser Water Electric Heating

Figure 37.8

Figure 37.9 illustrates air transfer fan applications. Air transfer fans fall into the minor HVAC equipment category, and are also usually last minute additions to the design. These factors contribute to the possibility that the electrical designer may not know to include the air transfer fans in his construction documents.

Service Lights and Power

Service lights and power in mechanical equipment rooms and air handling units are an electrical item, responsibility for which is often not clearly specified. Maintenance of filter, fans, automatic valves and dampers, insulation and casing can be greatly enhanced by service lights installed in the access chambers for those items. Having 120 volt receptacles available near the units for electric tools and extension lights will be very helpful to service personnel.

Figure 37.10 shows some examples of desirable locations for service lights and power. This information should be given to the electrical designer to be included in the construction documents.

Chiller Oil Pump, Oil Heater, Control Panel, and Purge Pump

Centrifugal chillers must have separate power supplies for oil pumps, oil heaters, the chiller control panel, and for the purge pump if there is one. The power supply required for these items is in addition to the main power to operate the chiller. If the HVAC designers are not aware of the additional separate power needs of the chiller, the electrical designer will not provide them.

Oil Pump

The oil pump provides lubrication under pressure to the chiller's bearings and gears. It operates both before the chiller motor is energized and during machine coast-down. A separate three-phase power supply to the oil pump is necessary to ensure operation independent of the chiller. If the oil pump starter is not factory-assembled on the machine, provisions must be made in the construction documents for that work. A fused disconnect for the oil pump is required.

Oil Heater

The oil heater must be energized at all times by a dedicated 120 volt power source. Proper oil temperature must be maintained, especially when the chiller is not operating, to ensure effective oil circulation. A fused disconnect switch must be installed per code requirements.

Purge Pump

Chillers that use refrigerants R-11, R-113, or R-114 require purge units because the evaporator pressure is below atmospheric. The purge unit prevents the accumulation of air and moisture in the refrigerant side of the chiller to ensure internal cleanliness. The purge unit operates when the chiller is off and must have a separate source of power.

Chiller and Condenser Water Flow Switches

Flow switches are installed in the condenser water and chilled water piping at the chiller. The switches prevent the chiller from operating unless the correct volume of chilled and condenser water is circulating through the chiller unit. The responsibility for this wiring must be addressed. It is considered control wiring, and should be performed by the mechanical contractor. Figure 37.11 describes the extent of chiller-associated wiring.

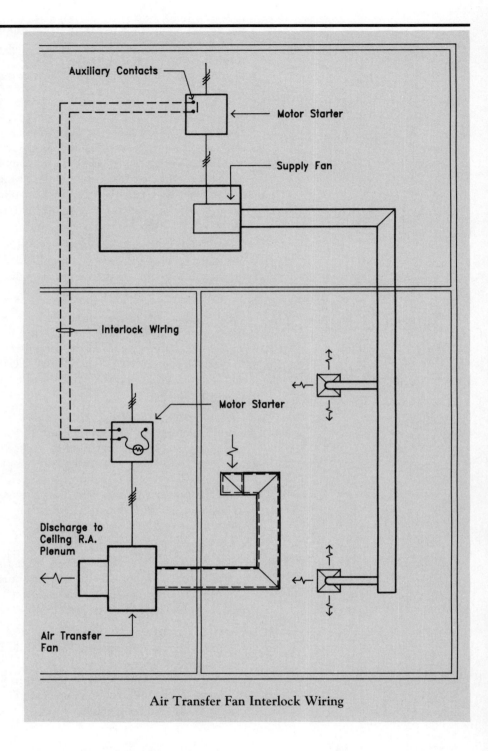

Air Transfer Fan Interlock Wiring

Figure 37.9

Airflow Switches

Airflow switches are installed in air distribution systems to establish proof of airflow for control purposes. The major applications for airflow switches are:

- Electric heating coil protection
- Proof of fan operation
- Dirty filter indication

Electric Heating Coil Protection

Airflow switches are installed at the electric heating coil to protect electric heaters from operating without air movement over the heating element. Should the electric heating coil be energized without airflow over the heating elements, the generated heat will not be removed. Overheating of the elements will activate the high temperature switch at the heater to disengage the heater control circuit.

Proof of Fan Operation

It is desirable that operating personnel know if a fan is running. Fan operation is usually indicated by fan running lights. The lights can be energized in several ways:

- *Pilot lights activated by the fan starter auxiliary contacts.* This is the least expensive and most popular method of providing an indication of fan operation. When the pilot light is activated, however, it indicates only that the *starter* is energized. The fan itself may not be operating because fan belts may be off or the motor disconnect switch or overload protector may be disengaged.

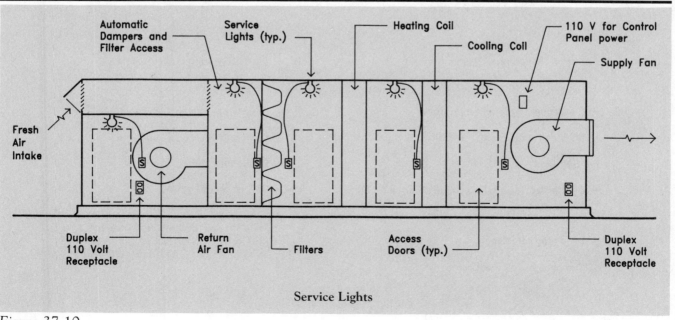

Service Lights

Figure 37.10

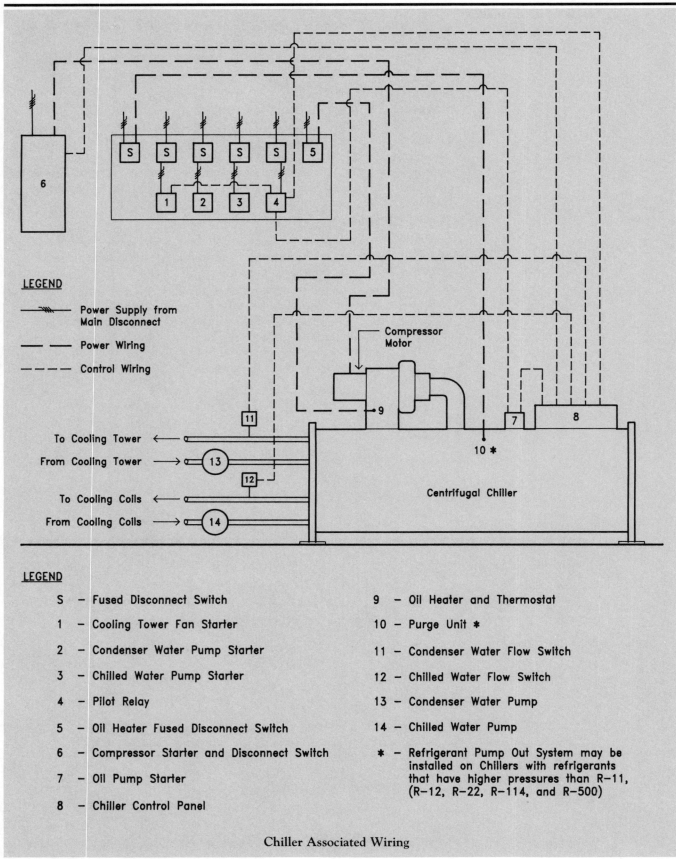

LEGEND

⟶Ɱ⟶ — Power Supply from Main Disconnect

⟶ ⟶ — Power Wiring

----- — Control Wiring

Compressor Motor

To Cooling Tower ⟵

From Cooling Tower ⟶

To Cooling Coils ⟵

From Cooling Coils ⟶

Centrifugal Chiller

10 *

LEGEND

S — Fused Disconnect Switch

1 — Cooling Tower Fan Starter

2 — Condenser Water Pump Starter

3 — Chilled Water Pump Starter

4 — Pilot Relay

5 — Oil Heater Fused Disconnect Switch

6 — Compressor Starter and Disconnect Switch

7 — Oil Pump Starter

8 — Chiller Control Panel

9 — Oil Heater and Thermostat

10 — Purge Unit *

11 — Condenser Water Flow Switch

12 — Chilled Water Flow Switch

13 — Condenser Water Pump

14 — Chilled Water Pump

* — Refrigerant Pump Out System may be installed on Chillers with refrigerants that have higher pressures than R–11, (R–12, R–22, R–114, and R–500)

Chiller Associated Wiring

Figure 37.11

454

- *Airflow Switch.* The most effective proof of fan operation is produced by the airflow switch. The switch can be in the form of a "sail switch," in which the air pressure created by the rotating fan moves a metal plate in the air stream. The plate movement engages a switch which activates a signal light.
- *Air Pressure Differential Switch.* This is another device used to prove airflow. A tube is connected between the pressure differential switch and the fan inlet, and another tube is connected to the fan discharge and the switch. When the fan differential pressure is at the set point, the switch is closed, energizing the signal light.

Fan airflow and pressure differential switches are the most effective means of realizing airflow indication. When the signal light is on, one can be reasonably certain that the fan is operating.

Dirty Filter Indication

A system to indicate that filters are dirty and that it is time to replace them is a very helpful maintenance tool. As the filters accumulate dirt, the filter resistance to airflow increases. When the filter air pressure drop reaches the "time to change" set point (usually 0.80″ wg), an electric circuit is made and a signal light is energized. The wiring for the above work must be assigned to the electrical contractor. The contractor must also know where the control devices and the indicating lights are to be located in order to establish his scope of work. Figure 37.12 presents airflow switch applications and the electrical requirements.

Interlock and Alarm Wiring

Interlock wiring determines which piece of mechanical equipment will be enabled when a particular HVAC unit operates. A device is "enabled" when it is told it is given a signal that it can control. A chiller may be interlocked with a condenser water pump. The condenser water pump starter auxilliary contacts close and "enable," sending a signal to the chiller that it can operate. The interlock allows the chilled water temperature control to activate the chiller.

Alarm wiring activates a visual and/or audible alarm when a control device set point is reached. This phase of work can be quite extensive, especially when a central control and alarm panel is involved. The scope of interlock wiring must be clearly identified in terms of both its extent and who is responsible for the work. Figure 37.13 presents examples of interlock and alarm wiring.

Examples include:
- The return air fan and the exhaust fans are to operate when the supply fan runs.
- When the chiller control is activated, the chilled water pump, the condenser water pump, and the cooling tower fan shall start in sequence.
- The control air electric-pneumatic switch shall be activated when the supply fan operates.
- The alarm bell at the central control panel shall be energized whenever:
  — High or low space temperatures are sensed.
  — Airflow is not established at an enabled fan.
  — Water flow is not established at an enabled circulating pump.
  — High or low pressure switches in boilers or air handling units have made contact.
  — Freezestat or firestat thermostats have opened their contacts.

— Any boiler/burner alarm, such as low water, high pressure, flame failure, high water temperature, low or high fuel tank level, or high or low expansion tank water level, are activated.

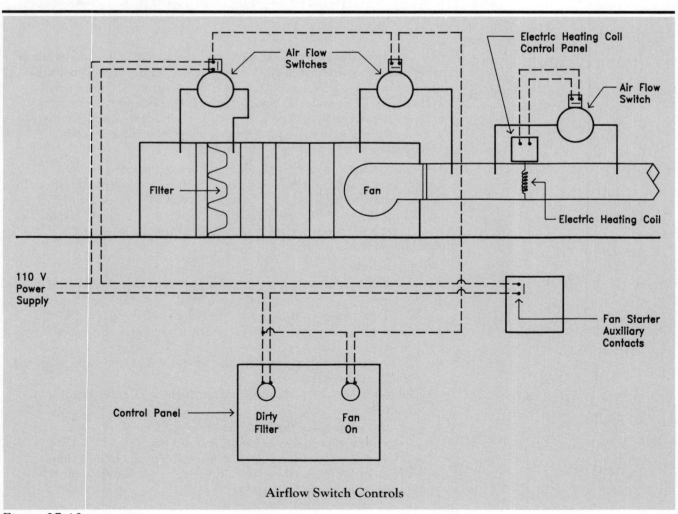

Airflow Switch Controls

Figure 37.12

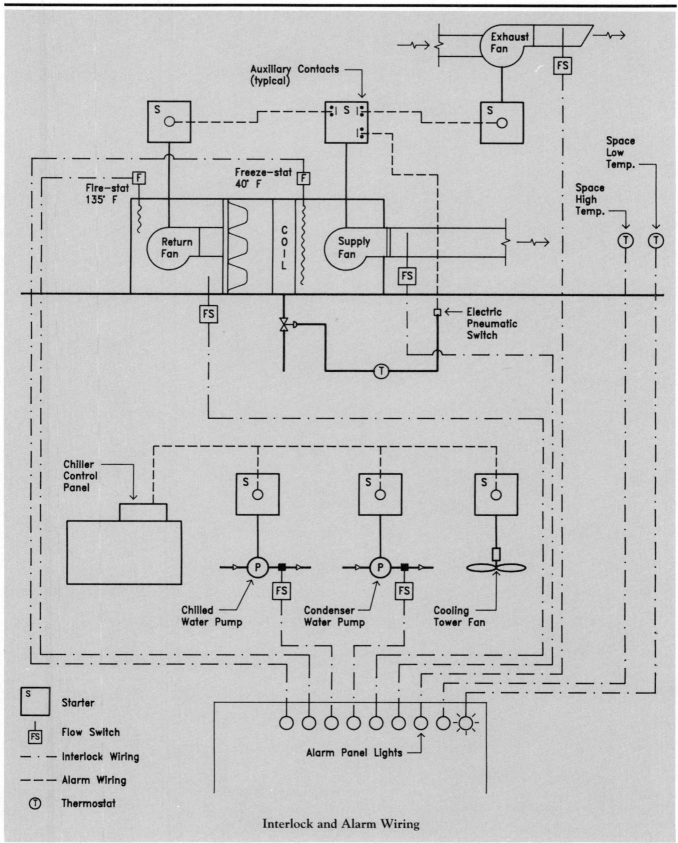

Figure 37.13

Interlock and Alarm Wiring

457

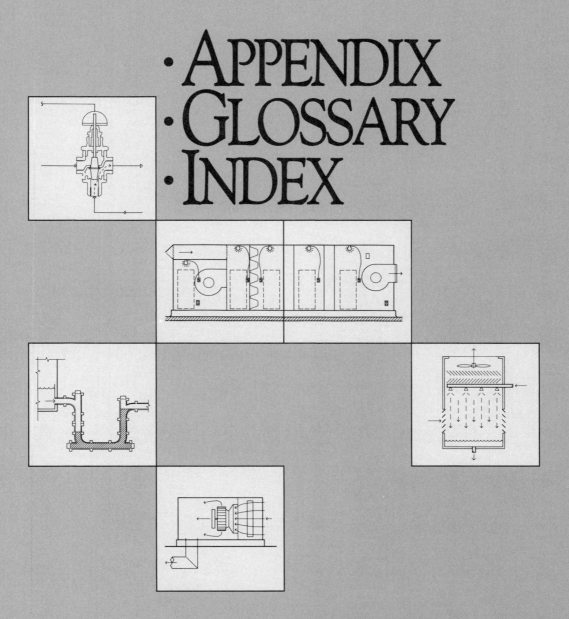

· Appendix
· Glossary
· Index

>	Greater than
<	Less than
#	Pound; Number
@	At
~	Approximately
%	Percent
A.C.	Alternating current
Adj.	Adjustable
amb	Ambient
Amp	Ampere
Approx.	Approximately
ASHRAE	American Society of Heating, Refrigeration & Air Conditioning Engineers
atm	Atmosphere
Avg.	Average
bhp	Brake horsepower
bp	Boiling point
btu	British thermal unit
btuh	Btu per hour
C_v	Coefficient, valve flow
cf, C.F.	Cubic feet
cfm	Cubic feet per minute
CHW	Chilled water
coef	Coefficient
Compr.	Compressor
COP	Coefficient of performance
cprsr	Compressor
CRT	Cathode-ray tube
Cu	Cubic
CWP	Chilled water pump
dB	Decibel
dbt	Dry bulb temperature
dens	Density
DHW	Domestic hot water
Diam.	Diameter
Dis., Disch.	Discharge
dpt	Dewpoint temperature
Ea.	Each
EAT	Entering air temperature
Elec.	Electrician, Electrical
ENT	Entering water temperature
Est.	Estimated
ET*	Effective temperature
evap	Evaporate
EWT	Entering water temperature
exp	Expansion
F	Fahrenheit
f/i	fins per inch
Fig.	Figure
Flr.	Floor
fpm	Feet per minute
fps	Feet per second
FPT	Female pipe thread
Ft.	Foot, Feet
ft. lb.	Foot pound
Ga., ga	Gauge
Gal.	Gallon
gpd	Gallons per day
gph	Gallons per hour
gpm	Gallons per minute
Hdr.	Header
HEPA	High Efficiency Particulate Air Filter
HG	Heat gain
Hg	Mercury
HL	Heat loss
hp	Horsepower
hps	High pressure steam
hr	Hour
Hrs./Day	Hours per day
HT	Heat
Htg.	Heating
Htr	Heater
Htrs.	Heaters
HVAC	Heating, Ventilation, Air Conditioning
HW	Hot water
ID, I.D.	Inside diameter
I.D.	Inside dimension
In.	Inch
Incl.	Included; Including
Insul.	Insulation
I.P.	Iron pipe
IPS	International Pipe Size
I.P.S.	Iron pipe size
I.P.T.	Iron pipe thread
K	Thousand
kw	Kilowatt
kwh	Kilowatt hour
LAT	Leaving air temperature
lb., #	Pound
lb./hr.	Pounds per hour
lf, L.F.	Linear foot
Lg.	Long, Length
LH	Latent heat
LHG	Heat gain, latent
L.P.	Low pressure
LPS	Low pressure steam
lthw	Low temperature hot water
LWT	Leaving water temperature
M	Thousand
Mag. Str.	Magnetic Starter
Max.	Maximum
Mbh	Thousand btu's per hour
Med.	Medium
Misc.	Miscellaneous
Mo.	Month
MPT	Male pipe thread
Mtd.	Mounted
Mult.	Multiple
NA	Not applicable
NC	Normally closed
NC	Noise criteria
NO	Normally open
No.	Number
O & P	Overhead and profit
oa	Outside air
OD, O.D.	Outside diameter
O.D.	Outside dimension
Oper.	Operator
Opng.	Opening
O.S. & Y.	Outside screw and yoke
p.	Page
P.E.	Professional engineer
ph	Phase
Pkg.	Package
p.m.	Post Meridian
Prefab.	Prefabricated
psf	Pounds per square foot
psi	Pounds per square inch
psig	Pounds per square inch gauge
PVC	Polyvinyl Chloride
Pwr.	Power
Q	Quantity heat flow
Q.C.	Quick coupling
Quan., Qty.	Quantity
R	Radius
R12, R22	Refrigerant (12, 22, etc.)
Reg.	Regular
Req'd.	Required
rh	Relative humidity
Rnd.	Round
rpm	Revolutions per minute
rps	Revolutions per second
S., suct	Suction
SA	Supply air
sat	Saturation
scfm	Standard conditions, cfm
SEF	System Effect Factor
sf, S.F.	Square foot
sg	Specific gravity
SH	Sensible heat
SHG	Sensible heat gain
sp, s.p.	Static pressure
spdt	Single pole, double throw
spst	Single pole, single throw
Std.	Standard
T STAT	Thermostat
td	Temperature difference
TE	Temperature entering
temp	Temperature
Ti	Inside Air Temperature
TL	Temperature leaving
To	Outside Air Temperature
tons	Tons of refrigeration
Tot.	Total
Transf.	Transformer
V	Volt
VA	Volt ampere
vac	Vacuum
var	Variable
VAV	Variable air volume
vel	Velocity
Vent.	Ventilation
vert	Vertical
Vol.	Volume
W	Watt
WB	Wet bulb
WBT	Wet bulb temperature
wg	Water gauge
WT, Wt.	Weight
Y	Wye
yd	Yard
yr	Year

HVAC

Valves, Fittings & Specialties

Gate	Pipe Pitch Up or Down (Up/Dn)
Globe	Expansion Joint
Check	Expansion Loop
Butterfly	Flexible Connection
Solenoid	Thermostat (T)
Lock Shield	Thermostatic Trap
2-Way Automatic Control	Float and Thermostatic Trap (F&T)
3-Way Automatic Control	Thermometer
Gas Cock	Pressure Gauge
Plug Cock	Flow Switch (FS)
Flanged Joint	Pressure Switch (P)
Union	Pressure Reducing Valve
Cap	Humidistat (H)
Strainer	Aquastat (A)
Concentric Reducer	Air Vent
Eccentric Reducer	Meter (M)
Pipe Guide	Elbow
Pipe Anchor	Tee
Elbow Looking Up	
Elbow Looking Down	
Flow Direction	

HVAC

Hot Water Heating Supply	—— HWS ——
Hot Water Heating Return	—— HWR ——
Chilled Water Supply	—— CHWS ——
Chilled Water Return	—— CHWR ——
Drain Line	—— D ——
City Water	—— CW ——
Fuel Oil Supply	—— FOS ——
Fuel Oil Return	—— FOR ——
Fuel Oil Vent	—— FOV ——

HVAC

—— FOG ——	Fuel Oil Gauge Line
—o— PD —o—	Pump Discharge
– – – – – –	Low Pressure Condensate Return
—— LPS ——	Low Pressure Steam
—— MPS ——	Medium Pressure Steam
—— HPS ——	High Pressure Steam
—— BD ——	Boiler Blow-Down

HVAC Ductwork Symbols

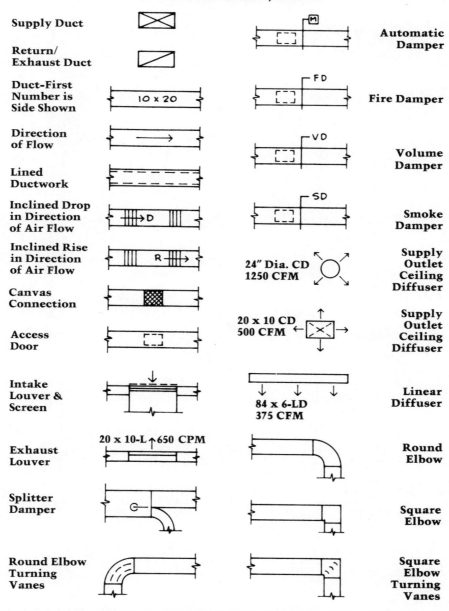

Supply Duct	Automatic Damper
Return/Exhaust Duct	Fire Damper
Duct-First Number is Side Shown (10 x 20)	Volume Damper
Direction of Flow	Smoke Damper
Lined Ductwork	
Inclined Drop in Direction of Air Flow (D)	
Inclined Rise in Direction of Air Flow (R)	24" Dia. CD 1250 CFM — Supply Outlet Ceiling Diffuser
Canvas Connection	20 x 10 CD 500 CFM — Supply Outlet Ceiling Diffuser
Access Door	
Intake Louver & Screen	84 x 6-LD 375 CFM — Linear Diffuser
Exhaust Louver (20 x 10-L 650 CPM)	Round Elbow
Splitter Damper	Square Elbow
Round Elbow Turning Vanes	Square Elbow Turning Vanes

Double Duct Air System

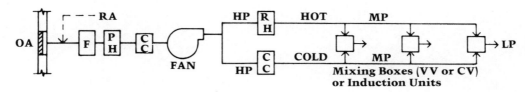

OA = Outside Air	CC = Cooling Coil	LP = Low Pressure Duct
RA = Return Air	RH = Reheat Coil	VV = Variable Volume
F = Filter	HP = High Pressure Duct	CV = Constant Volume
PH = Preheat Coil	MP = Medium Pressure Duct	

Glossary

Acoustic lining
Insulating material secured to the inside of ducts to attenuate sound and provide thermal insulation.

Actuator
A controlled motor that can effect a change in the controlled variable (temperature, pressure) by operating a control element such as a valve or damper.

Air density
The weight per unit volume of air, in pounds per cubic foot.

Airfoil fan
A backward-curved fan with blades of airfoil design.

Ambient temperature
The temperature of the outdoor air.

Anti-floatation pads
Concrete pads secured to underground tanks to add sufficient weight to the tank to overcome buoyancy.

Aspect ratio
The ratio of the width of a duct to its height.

Axial fans
Fans that produce pressure from a velocity passing through the impeller, with no pressure being produced by centrifugal force.

Backward-curved fan
Centrifugal fans with blades that curve backward away from the direction of the rotation.

Balancing damper
A plate or adjustable vane installed in a duct branch to regulate the flow of air in the duct.

Blank-off
A blank plate sealing off a sector of a diffuser to prevent airflow in the direction of the blank-off plate.

Boiler economizer
The last pass of boiler tubes or a heat exchanger located in the flue pipe that extracts some of the heat from the flue gases before they are vented to the atmosphere.

Booster circulating pump
A pump installed in series with another pump to raise the system pump head.

Cabinet units
Small air handling units that house an air filter, heating coil, and a centrifugal blower.

Central systems
Systems composed of prime movers that convert energy (in the form of heating or cooling) from fuel or electricity. They are located in a single area, to serve distribution systems that deliver the heating and cooling to the conditioned space.

Centrifugal chiller
A gas compressor in which the compression is obtained by means of centrifugal force, the force away from the center of a rapidly rotating impeller.

Chillers

– compression
Water is cooled by liquid refrigerant that vaporizes at a low pressure and is driven into a compressor. The compressor increases the gas pressure so that it may be condensed in the condenser.

– absorption
Heat-operated refrigeration chillers that use an absorbent (lithium bromide) as a secondary fluid to absorb the primary fluid (water) which is a gaseous refrigerant in the evaporator. The evaporative process absorbs heat, cooling the refrigerant

(water) which cools the chilled water circulating through the heat exchanger.

Closed loop control
Measures the actual changes in the controlled variable (temperature, pressure) and actuates the control device (valve, damper) to bring about a change.

Compressor
The pump in a mechanical refrigeration system that compresses the refrigerant vapor into a smaller volume, thereby increasing the refrigerant pressure and the boiling temperature. The compressor is the separation between the refrigerant system's high and low side.

Condenser
The heat exchanger in a refrigeration system that removes heat from the hot high pressure refrigerant gas and transforms it into a liquid.

Constant volume reheat
The volume of the supply air is unchanged, while the supply air temperature is raised as the local zone cooling load decreases.

Cooling plant
The machinery that produces chilled water or cool refrigerant gas (chiller or compressor), the condenser, cooling tower and condenser water pumps for water-cooled plants, air-cooled condensers for air-cooled systems, and chilled water pumps and expansion tanks for chilled water systems.

Cooling tower
An outdoor structure frequently placed on roofs or on the ground, over which warm water is circulated to cool it by evaporation and exposure to outdoor air.

Crystallization
The precipitation of salt crystals from the absorbent in an absorption chiller; a slush-like mixture that plugs fluid passages within the chiller and renders it inoperable.

Cylinder unloaders
Automatic devices used to hold open the reciprocating compressor valves of a number of cylinders in order to reduce compressor pumping capacity.

DDC (direct digital control)
A system that receives electronic signals from sensors, converts those signals to numbers, and performs mathematical operations on the numbers inside a computer. The computer output (in the form of a number) is converted to a voltage or pneumatic signal to operate the actuator.

De-superheaters
Heat exchangers that remove only some of the sensible heat from the hot high pressure refrigerant gas.

Diffuser
A circular, square or rectangular air distribution outlet, generally located in the ceiling and comprised of deflecting members to discharge supply air in various directions.

Direct acting controllers
Increase control pressure as the control variable (temperature, pressure) increases.

Direct drive
A system in which the driver and driven (motor and fan) are positively connected in line to operate at the same speed.

Direct expansion (DX)
Refrigeration systems that employ expansion valves or capillary tubes to meter liquid refrigerant into the evaporator.

Double suction riser method
The use of two pipes for the upward flow of refrigerant suction gas. During periods of low refrigeration load, the refrigerant flow of one of the risers is blocked by the accumulation of oil in the base of the riser, diverting refrigerant flow at a higher velocity to transport oil up the other riser.

Drift
The entrained unevaporated water carried from a cooling tower by the air moving through the tower.

Drift eliminators
Baffles in a cooling tower through which air passes before exiting the tower, to remove entrained water droplets from the exhaust air.

Dry basin systems
Condenser water systems in which the cooling tower basin is located indoors, remote from the cooling tower.

Duct static pressure
The pressure acting on the walls of a duct; the total pressure less the velocity pressure; the pressure existing by virtue of the air density and its degree of compression.

Economizer cycle
A system of dampers, temperature and humidity sensors, and actuators which maximizes the use of outdoor air for cooling.

Entrainment
The capture of part of the surrounding air by the air stream discharged from an air outlet.

Equal friction method
A way of sizing ductwork systems for a constant pressure loss per unit length of duct.

Equalizing grid
A series of individual adjustable blades installed at the diffuser inlet to produce uniform velocity through the diffuser neck.

Expansion tank
A device to control pressure in a hydronic system by storing excess water volume resulting from increased operating temperatures.

Fan brake horsepower
The horsepower output of a fan as measured at the pulley or belt. Bhp includes losses due to turbulence and other inefficiencies.

Fan coil unit
An air handling unit that houses an air filter, heating or cooling coil, drain pan, and centrifugal fan, and operates by moving air through an opening in the unit and across the filter and coils.

Fan pressure curve
The curve that represents the pressure developed by a fan operating at a fixed rpm at various air volumes.

Fan suction box
A specially designed 90° fitting installed at the fan inlet that reduces the inlet pressure loss created by system effect.

Fan surge
Turbulent flow caused when insufficient air enters the fan wheel to completely fill the space between the blades. Some of the air reverses its direction over a portion of the fan blade.

Fibrous glass duct
Ductwork constructed of fiberglass material.

Fill packing
The portion of a cooling tower that provides heat transfer surface.

Fire tube boiler
Boilers in which flue gas flows through the tubes and water surrounds the tubes.

Fixed plate heat exchanger
A static device that transfers sensible heat through plates separating a warm air stream from a cold air stream.

Flash
The conversion of condensate into steam.

Fog
The visibility and path of the effluent air stream exiting a cooling tower and remaining close to the ground.

Forced draft
Air movement from the fan discharge through the heat exchanger or cooling tower.

Forward-curved fan
Centrifugal fan with blades that curve forward in the direction of the wheel rotation.

Free cooling
Cooling without the use of mechanical refrigeration.

Freeze-stat
A thermostat set to stop the fans before freezing air temperature is sensed.

Glycol
Liquid with a very low freezing point that is miscible with water.

HEPA (high efficiency particulate air) filter
Very high efficiency dry filter (99.9%) made in an extended surface configuration of deep space folds of submicron glass fiber paper.

Head pressure
The operating pressure in the discharge line of a refrigeration system.

Heat exchanger
A device designed to transfer heat between two physically separated fluids.

Heat pump
A refrigeration system designed to utilize alternately or simultaneously the heat extracted at a low temperature and the heat rejected at a higher temperature.

Heat recovery
A method of extracting and using wasted heat.

Helical rotary or screw-type compressor
Refrigeration compression achieved by trapping the refrigerant gas in the space formed by the flutes of meshing screws, reducing the gas volume and compressing the gas.

Induced draft
A process in which air is drawn through the cooling tower into the fan.

Induction air terminal units
A factory assembly consisting of a cooling coil and/or heating coil that receives preconditioned air under pressure that is mixed with recirculated air by the induction process.

Interference
Cooling tower effluent that enters the intake of an adjacent tower.

Interlock wiring
Control wiring permitting secondary sequence to be enabled by a primary action.

Linear diffuser
An elongated diffuser with parallel slots with deflectors to divert airflow in various directions.

Lithium bromide
A chemical compound (salt) with the ability to absorb water and cool it by evaporation.

Logic panels
Electronic control panels that are designed to perform a specific control sequence.

Low temperature supply air
Supply air below 50° F.

Mechanical draft
The movement of air through a cooling tower by means of a fan or other mechanical device.

Metal nosings
Metal enclosures over the cut ends of acoustic lining sections in ductwork.

Mixing box
A chamber usually located upstream of the filters that collects outside air and return air.

Modulation
The tendency of a control to adjust by increments and decrements.

Multizone units
Air handling units with parallel heating and cooling air paths providing individual mixing of air distribution circuits into a single duct for each zone.

Natural draft
Refers to the movement of air through a cooling tower by the force of air density differential.

Noise criterion curve
Defines the limits which the octave band spectrum of a noise source must not exceed if a certain level of occupant acceptance is to be achieved.

Normally closed valves
Valve ports are closed to flow when external power or pressure is *not* being applied.

Normally open valves
Valve ports are open to flow when external power or pressure is *not* being applied.

Open loop control
The control system does not have a direct link between the valve of the controlled variable (temperature, pressure) and the controller.

Outgassing
The driving out or freeing of gases from fabrics and building materials.

Oxidation
The combining of oxygen with another element to form a new substance, such as in burning and rust formation.

Plenum or plug fans
Single-inlet, single-width centrifugal fans without the scroll, permitting 360 degree air delivery from the fan wheel.

Plume
The effluent mixture of heated air and water vapor discharged from a cooling tower.

Pneumatic control
A process in which compressed air is distributed through pipes to supply energy for the operation of valves, motors, and relays.

Pre-cooling coil
A cooling coil located at the air entering side of the primary cooling coil.

Pressure differential valve
A valve controlled by the supply and return main pressure difference to divert flow from the supply main to the return main.

Proportional control
The controlled device (valve or damper) is positioned proportionally in response to slight changes in the controlled variable (temperature, pressure).

Proportional plus derivative control (PI)
Proportional plus integral control, with the added feature of derivative control which varies with the derivative of the error (offset).

Proportional plus integral control (PI)
Proportional control with an added component that eliminates the offset (deviation from the throttling range) of proportional control.

Psychrometric
Relating to the measurement of atmospheric conditions, particularly regarding the moisture mixed with air.

Pulse tube boiler
A boiler using a sealed combustion system in which residual heat from the initial cycle ignites all subsequent air gas mixtures, and flue gas condensation takes place in the heat exchanger.

Pump head
The pressure differential produced by an operating pump.

Purge pump
A compressor that removes non-condensibles from a refrigeration system.

Radial-curved fans
Backward-curved centrifugal fans with flat blades.

Radiant heating systems
Systems with heating terminals that deliver heat by radiation from a hot surface.

Range
The difference between the hot water temperature entering a cooling tower and the cold water leaving a cooling tower.

Reciprocating (chiller, compressor)
Single acting compressor using pistons that are driven by a connecting rod from a crankshaft.

Refrigerant charge
The quantity of refrigerant in a refrigerant system.

Reverse acting controllers
Decrease in control pressure as the control variable (temperature, pressure) increases.

Runaround system
A heat recovery system in which coils in an exhaust duct transfer a portion of the exhaust air heat to a fluid. The fluid is then circulated to coils that give up a portion of the fluid's heat to the air in a cold air duct.

Saturation temperature
The temperature at which vapor and liquid coexist in stable equilibrium.

Scroll compressor
A rotary positive displacement compressor with a fixed and a rotating scroll, in which compression takes place by reducing the gas volume by the meshing of the scrolls.

Seating pressure
The pressure generated by the action of a spring and control air to close the automatic control valve plug against its seat.

Secondary containment
A chamber for the collection of oil that has leaked from a fuel oil tank.

Set point
The desired value of the controlled variable (temperature, pressure).

Sight glass
A glass tube sealed within a fluid system, providing a means to examine – visually – the fluid in the system.

Single-stage absorption
Absorption chillers with one generator to evaporate refrigerant (water) from the solution.

Solenoid
An electromagnetic coil that raises a valve stem when energized.

Splitter damper
A single blade damper hinged at one end, installed to divert air from a main duct into a branch duct.

Stackhead
A vertical duct that discharges exhaust air into the atmosphere at high velocity.

Standard air
Dry air at a pressure of 29.92 in. Hg. at 68°F temperature and a density of 0.0753 pounds per cubic foot.

Static pressure
The normal force per unit area at a small hole in the wall of a duct through which fluid flows.

Steam traps
Devices that discharge condensate (that forms when steam gives up some of its heat) and direct air and non-condensible gases to a point of removal.

Stoichiometric combustion
Fuel is reacted with the exact amount of oxygen required to oxidize all carbon, hydrogen, and sulfur in the fuel to carbon dioxide, water, and sulfur dioxide.

Stratification
The flow at the junction of two air streams in layers that prevent the proper mixing of the two air streams.

Subcooling
Cooling a refrigerant below its saturated condensing temperature.

Superheaters
Devices that add heat to saturated fluids.

Supply air outlets
Air terminals (such as grilles and diffusers) for the discharge of supply air.

System effect factor
A pressure loss factor which recognizes the effect of fan inlet and outlet restrictions that influence fan performance.

Terminal units
Devices located near the conditioned space that regulate the temperature and/or volume of supply air to the space.

Thermal lag
The time required to add or remove heat from a mass before it reaches the design set point temperature.

Thermal shock
The stress created by the introduction of large quantities of cold water into a hot boiler, or large quantities of hot water into a cold boiler.

Tilt controls
Operators that change the angle of a heat pipe tube to allow gravity to vary the velocity of liquid refrigerant flow down the tube.

Tower equalizing lines
Pipes installed to directly connect adjacent cooling tower basins to maintain a common basin water level.

Tube axial fans
Single width airfoil wheel fans arranged in a cylinder to discharge air radially against the inside of the cylinder.

Two-stage absorption
A refrigeration system in which the hot refrigerant (water) vapor travels to a second generator where, upon condensing, it supplies heat for further refrigerant vaporization from the absorbent of intermediate concentration that flows from the first generator.

Unit ventilators
Fan coil units with provisions for introducing large percentages of outside air into the cabinet.

V bank filter box
An air handling unit section in which the filters are arranged in a "V" configuration in order to provide a greater filter surface.

Vaneaxial fan
A fan that consists of a disk-type wheel within a cylinder and a set of air guide vanes placed on one side of the wheel. The fan is either belt-driven or connected directly to the motor.

Vaned elbow
A square 90° duct elbow with turning vanes to uniformly direct the airflow.

Variable air volume (VAV)
An air distribution system capable of automatically delivering a reduced volume of constant temperature cool air to satisfy the reduced cooling load of individual zones.

Variable inlet vanes (VIV)
Vaned dampers installed at the inlet to a fan to produce a spin to the air entering the fan and reduce the fan performance.

Volume dampers
Devices installed in duct system circuits to add resistance to the circuit for air balancing.

Waste heat boiler
A boiler that is fired by the hot exhaust gases of incinerators, engines, or industrial processes.

Water tube boiler
A boiler in which water circulates through the tubes, and combustion gases surround the tubes.

Wet bulb temperature
The reading on a thermometer, the bulb of which is enclosed in a layer of wet fabric; the ambient temperature of an object cooled by evaporation.

INDEX